PERTURBATION ANALYSIS OF DISCRETE EVENT DYNAMIC SYSTEMS

THE KLUWER INTERNATIONAL SERIES IN ENGINEERING AND COMPUTER SCIENCE

DISCRETE EVENT DYNAMIC SYSTEMS
Consulting Editor

Yu-Chi Ho
Harvard University

Gradient Estimation Via Perturbation Analysis, *P. Glasserman*
ISBN: 0-7923-9095-4

PERTURBATION ANALYSIS OF DISCRETE EVENT DYNAMIC SYSTEMS

by

Yu-Chi Ho
Harvard University

and

Xi-Ren Cao
Digital Equipment Corporation

KLUWER ACADEMIC PUBLISHERS
Boston/Dordrecht/London

Distributors for North America:
Kluwer Academic Publishers
101 Philip Drive
Assinippi Park
Norwell, Massachusetts 02061 USA

Distributors for all other countries:
Kluwer Academic Publishers Group
Distribution Centre
Post Office Box 322
3300 AH Dordrecht, THE NETHERLANDS

Library of Congress Cataloging-in-Publication Data

Ho, Yu-Chi, 1934–
    Perturbation analysis of discrete event dynamic systems / by Yu
-Chi Ho and Xi-Ren Cao.
        p. cm. — (The Kluwer international series in engineering and
computer science; SECS 145. Discrete event dynamic systems)
    Includes index.
    ISBN 0–7923–9174–8
    1. System analysis. 2. Discrete-time systems. 3. Perturbation
(Mathematics) 4. Control Theory. I. Cao, Xi-Ren. II. Title.
III. Series: Kluwer international series in engineering and computer
science; SECS 145. IV. Series: Kluwer international series in
engineering and computer science. Discrete event dynamic systems.
T57.6.H6   1991
620 '.001 '1—dc20   91-18008
                                                                          CIP

Printed on acid-free paper.

Printed in the United States of America

to Sophia
and
to the memory of mother
Wen-Ying Guo Cao

TABLE OF CONTENTS

Foreword to the Kluwer DEDS Series

Consider the daily operation of a large international airport. On the air side, there are a multitude of aircraft, jumbo or small, international or domestic, private or commercial, including helicopters using airspace, sectors, runways, and gates. On the land side, there are all kinds of vehicles, buses or cars, rental , private or service, taxis or limos, crowding the access ways and parking spaces. Inside the airport building, passengers and luggage require services from ticket counters, security gates, and transport mechanisms. All of these "jobs" requiring different kinds of "services" from a multitude of "resources" interact with each other continuously and dynamically at every instant. Anyone who has experience with such a system on a hot Sunday afternoon in July at the Kennedy airport in NYC or a stormy weekend in January at O'Hare Airport in Chicago can easily visualize the complexity and challenge in the management of such a system. Slightly less familiar but equally real is the example of a large automated manufacturing plant whether it is involved in the making of microchips or jet engines. Here again, the system evolves dynamically in time depending on a host of events such as the arrival of a batch of parts to be machined, the breakdown of a piece of equipment, the outcome of a test, etc. Finally, a more abstract but no less important example is the functioning of many worldwide telecommunication / computer networks. The fact that we can place a phone call to almost anyone in the world within range of a telephone while flying 30,000 feet above in a plane simultaneously with millions of other calls is a technological feat worth appreciating. Other examples of such Discrete Event

Foreword

Dynamic Systems (DEDS) are almost endless: military C3I / logistic systems, the emergency ward of a metropolitan hospital, back offices of large insurance and brokerage firms, service and spare part operations of multinational firms . . . . the point is the pervasive nature of such systems in the daily life of human beings. Yet DEDS is a relatively new phenomenon in dynamic systems studies. From the days of Galileo to Newton to quantum mechanics and cosmology of the present, dynamic systems in nature are primarily differential equations based and time driven. A large literature and endless success stories have been built up on such Continuous Variable Dynamic Systems (CVDS). It is, however, equally clear that DEDS are fundamentally different from CVDS. They are event driven, asynchronous, mostly man-made and only became significant during the past generation. Increasingly, however, it can be argued that in the modern world our lives are being impacted by and dependent upon the efficient operations of such DEDS. Yet compared to the successful paradigm of differential equations for CVDS the mathematical modelling of DEDS is in its infancy. Nor are there as many successful and established techniques for their analysis and synthesis.

The purpose of this series is to promote the study and understanding of the modelling, analysis, control, and management of DEDS. The idea of the series came from editing a special issue of the Proceedings of IEEE on DEDS during 1988. The kind reception by the scientific public of that special issue suggests that a more substantial publication effort devoted to DEDS may be in order. I am delighted that Kluwer has decided to undertake such an effort. As readers of the foregoing paragraph can infer, our intention is to publish both engineering and mathematically oriented books. We firmly believe in the synergistic interplay between the serendipitous application of the pure and the discovery of new knowledge by the applied.

Perturbation Analysis (PA) is a technique for the efficient performance analysis of DEDS trajectories. PA had its beginning in a real world application which is more engineering oriented and less formal. Although it was recognized at the outset as a general approach by its proponent, it wasn't until much later in the work of Cao, Zazanis, Suri, and Glasserman that the foundation of infinitesimal perturbation analysis (IPA) was established. This particular book collects in one place essentially the state of the art of perturbation analysis as of late 1990. A companion and more rigorous book on IPA by P. Glasserman has already been published in this series. There is little overlap between these two books and they should complement each other.

The succeeding volumes of this series will deal with other topics of DEDS.

Preface

During the post-sputnik years since 1957, we have witnessed a tremendous growth in the subject of control systems, both in academia and in industry. As the subject is generally understood, it is concerned with the analysis, design, optimization, and control of dynamical systems, ranging from concrete and physical systems, such as aerospace vehicles, chemical processes, engines, and gas turbines to the more abstract systems, such as heat transfer, fluid flow, and national economies. The underlying commonality of such dynamical systems is that they can all be adequately described by a set of nonlinear ordinary or partial differential equations. We denote these systems as Continuous Variable Dynamic Systems (CVDS). The success of this discipline need not be repeated or elaborated upon. In fact, this very success has caused us almost to take such accomplishments for granted. Concurrently, however, since World War II, we have increasingly created large man-made systems which evolve in time but which cannot easily fit the differential equation paradigm. The foreword to this series gave an ample number of such examples. These man-made systems are now classified as Discrete Event Dynamic Systems (DEDS). This book is devoted to one set of DEDS analysis techniques called Perturbation Analysis (PA).

Despite the dissimilarity of the mathematical nature of DEDS and that of the differential equation paradigm of CVDS, we submit that many useful analogs and conceptual approaches can be successfully transplanted from the CVDS world. This book is written from that viewpoint and as a natural outgrowth of the control and optimization ideas from the control system theory. As such, the approach is time domain oriented and sample-path based; it emphasizes the dynamic nature of such systems. However, this does not mean we downplay the probabilistic contributions of the queueing theory and the statistical analysis of discrete event simulation. We firmly believe in the synergistic interplay of the two world-views. In fact, the existence of this book can be offered as a piece of such evidence. Furthermore, we have consistently used queueing networks as examples of DEDS in this book. This enables us to provide an independent check on the validity of PA as well as an alternative viewpoint more familiar to operations researchers.

Meanwhile, the optimal control of DEDS is less advanced than the corresponding effort in CVDS. Although there are fair amount of work on the control of queues and queueing networks, we have deliberately restricted the techniques treated in this book only to parametric analysis and optimization. Comprehensive dynamic feedback control for general DEDS awaits separate treatment and further development.

The primary readers of this book are first year graduate engineering students and undergraduate seniors planning research in this area as well as practicing engineers interested in non-routine advanced applications. As such, exercises are placed throughout the text both as reinforcement of the ideas discussed and as further extensions. A solution manual is available to instructors who request it. Appendixes A through E are meant to be an integral part of the book and instruction if the students are not familiar with the material, but the book can be read without first going through the appendixes. See the logical dependence chart at the end of this preface and plan for using the book as a semester course on performance analysis of DEDS. The book can also serve as a source of ideas for theoreticians to formalize, generalize, and prove; this is consistent with the early stage of development of the subject matter.

Since the book is written by engineers for engineers. the style is decidedly informal. Nevertheless, we realize the flavor of this book is academic. Practicing engineers will find few ready-made formulas to plug in or algorithms to code without a certain amount of thinking and work on their part. In other words, at this stage custom tailoring is necessary for applications. It is our conviction that the purpose of theories in engineering is conceptual insight. Engineers use such insight for creative applications. An analogous case can be found in the difference between the developer of a sophisticated software program such as Lotus 1-2-3® and a very advanced user of the program. The former person must practice precision and discipline; the latter must have an appreciation and understanding of the same in order to use the tool creatively. Advanced engineers need to have the language ability to read theoretical text so that they can make informed judgments on the usefulness of a theory and on valid approximations in applications. In this sense, we are believers of "rigor" in technical exposition. Given the state of the art of DEDS/PA, routine designs and application of the theory are likely to be the exception rather than the rule in the near future. We apologize to readers who are looking for instant applications.

One decision that faces all authors of a technical book is the size of the intended reader population. On the one hand, one can strive for generality, precision, and minimal redundancy. Results are elegance but often have restricted applicability. Readers are asked to pay a high one-time setup cost for notations and definitions. On the other hand, one can emphasize conceptual insight and rely more on intuition, common sense, and a certain amount of deliberate repetition and verbosity. There is no unique optimal tradeoff. We have opted for the latter. A companion book in this series by P. Glasserman treats Infinitesimal Perturbation Analysis from the viewpoint of theoretical foundation and complements our book with a minimal overlap.

The material in this book has been used as course material for a first-year graduate course in performance analysis at Harvard University by both authors and at the University of Texas at Austin where the first-named author spent a sabbatical. We are grateful for the support extended by these two fine universities. The second-named author also appreciates Digital Equipment Corporation's support of his continuing research in the topics related to this book. Although it will be obvious to the readers that this book owes a great deal to the pioneering works in PA by many people, all of whom had their academic origin at Harvard University during the 1980's, we would be remiss if we did not mention their names here. In the chronological order of their twelve Ph.D. theses, A. Eyler, Christos Cassandras, Michael Zazanis, Jim Dille, Wei-bo Gong, Shu Li, Pirooz Vakili, Steve Strickland, Paul Glasserman, Micheal Fu, Jiang-Qiang Hu, and Bin Zhang can be thought of as coauthors of this book. We particularly wish, however, to acknowledge Rajan Suri who contributed to PA in its early development and originally agreed to be a third coauthor of this book. He had to withdraw because of other commitments. Nevertheless, his contributions are indelible, and we still consider him a coauthor in spirit, if not in name.

Finally, the research for this book could not have been carried out without continuous funding by the Joint Service Electronics Program, the Office of Naval Research, the Army Research Office, the National Science Foundation, and the constant support of their visionary program officers.

Yu-Chi Ho

Xi-Ren Cao

Cambridge, Massachusetts

December 1990

Prerequisites And Appendices:

The book is self-contained beyond linear algebra, multivariate calculus, and a first course in probability and stochastic processes. The Appendices furnish additional materials for definitions, notations, and basic knowledge assumed. Appendix D together with Chapter 1 provides for additional material on DEDS.

Suggested Schedules:

A possible schedule for one semester course on **the performance analysis of DEDS** for the first year graduate or senior undergraduate students in engineering:

Chapter 1 and Appendix D	2.0	weeks
Appendix A and B	3.0	weeks
Chapter 2 and 3	1.5	weeks
Chapter 5	1.0	weeks
Chapter 6.1-2	2.0	weeks
Chapter 7.1 and 7.3	2.0	weeks
Chapter 8	1.0	week

A more application oriented one semester course can be

Chapter 1 and Appendix D	2.0	weeks
Appendix A and B	3.0	weeks
Chapter 2 and 3	1.5	weeks
Chapter 7.1 and 7.3	2.0	weeks
Implementation and software engineering issues	1.5	weeks
Applications to selected areas	2.5	weeks

LOGICAL DEPENDENCE OF THE CHAPTERS OF THE BOOK

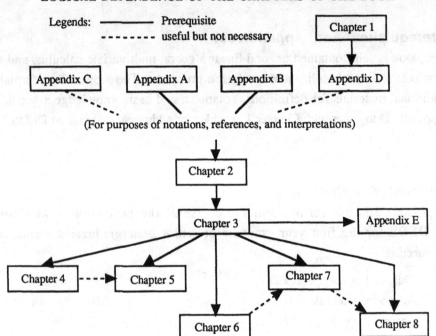

PERTURBATION ANALYSIS
OF DISCRETE EVENT
DYNAMIC SYSTEMS

CHAPTER ONE

Introduction To Discrete Event Dynamic Systems

1.1 Introduction

Picture yourself with the mythical Mr. T. S. Mits (The Scientific Man In The Street) and the task of explaining to him the phenomena and workings of (i) the Gulf Stream and (ii) a computer-controlled flexible manufacturing system (FMS). Both phenomena are real and both are not completely understood. However, for task (i), you face an easy assignment since you can draw upon a knowledge of calculus and differential equations to provide a succinct description of ocean currents and fluid dynamics. For task (ii), no such ready-made models are in existence[1]. One is essentially reduced to using an algorithmic description not very different from writing a computer program to simulate the FMS. In fact, modern technology has increasingly created dynamic systems which cannot be easily described by ordinary or partial differential equations. Examples of such systems are production or assembly lines, computer communication networks, traffic systems on both the air and land side of a large airport, military C^3I (Command-Control-Communication-Intelligence) system, etc. The evolution of these systems in time depends on the complex interactions of the timing of various discrete events, such as the arrival or the departure of a job, and the initiation or the completion of a task or a message. The "STATE" history of such dynamic systems, instead of changing continuously, is piecewise constant and changes only at discrete instants of time at which the events occur. We shall call such man-made systems Discrete

---

[1] We do not regard queueing networks or automata models of an FMS as being accessible to Mr. T. S. Mits. More will be written about these models in Chapter 2 and Appendices A and D.

Event Dynamic Systems (DEDS) as opposed to the more familiar Continuous Variable Dynamic Systems (CVDS) in the physical world which are described by differential equations. Although systems governed by difference equations are often referred to as discrete time systems, they have more in common conceptu-ally with CVDS than with DEDS despite the name similarity. (For the purpose of this book, we shall not distinguish systems modeled by differential and difference equations. They are continuous time or discrete time CVDS). To help fix ideas for DEDS, consider a Flexible Manufacturing System (FMS) with several work stations, each consisting of one or more identical machines. These stations are attended by operators and/or inspectors. Following some routing plan, parts belonging to different classes arrive at these stations via computer control. The parts queue up in the buffers according to some priority discipline until a machine and an operator are available to work on them. Stations are connected by some kind of material handling system (MHS) or automatic guide vehicle (AGV) which transports parts from station to station until they finish processing and leave the system. Typical performance criteria of interest for such an FMS are average through-put (TP), flow or wait time (WT), and work In progress (WIP). Note that with some simple changes in terminology from parts to messages, work stations to nodes, routes to virtual circuits, MHSs to communication links, AGVs to packets, fixtures to tokens, etc., the above description could be for a communication network which moves and processes information packets rather than material parts. Multi-programmed computers operating simultaneously in batch and time-shared mode, paper work flow in complex bureaucracies such as the back office of insurance or brokerage offices, and passengers, ground and aircraft traffic at any airport are other examples fitting this description. These examples illustrate the pervasive nature of such systems in the modern world and the relative lack of good analytically and dynamically oriented model for their description.

Conceptually, we can visualize such a DEDS as consisting of **jobs** (parts or packets) and **resources** (machines or transmitters, operators or nodal CPUs, AGVs, MHSs or communication links, and buffer spaces). Jobs travel from resource to resource demanding and competing for service. The dynamics of the system are determined by the complex interactions of the timings of various discrete events associated with the jobs and the resources. In this sense, DEDS are simple. There are only two objects in DEDS, jobs and resources, which interact. When they interact, a job occupies (receives service from) a resource for a random/deterministic period of time. If we let the number of jobs waiting at a resource be a state variable, $x(t)$, and furthermore consider the approximation that there are an infinite number of jobs, each infinitesimally small, then we have the differential equation

$$\frac{dx}{dt} = \begin{cases} \lambda - \mu & \text{if } x>0, \\ 0 & \text{if } x=0 \text{ and } \lambda-\mu<0, \\ \lambda - \mu & \text{if } x=0 \text{ and } \lambda-\mu>0, \end{cases} \tag{1}$$

where λ is the arrival rate of the jobs and μ is the service rate of the resource. This is a deceptively simple model of one component of a queueing system or a DEDS. The complications come because λ and μ can depend in arbitrary ways on other state variables of the DEDS. For example,

   (i)   admission into a queue

   (ii)   routing to one of two resources

   (iii)   order of service by a resource

may depend on the "state" of the system. When there is a multitude of such decisions interacting in a system, the resultant resource allocation problem can be very complex. Here we have an endless variety of queueing disciplines, priorities, service requirements, routing, scheduling, resource sharing, and general logical conditions that need to be met for interactions to take place. Any attempts to describe a real world DEDS by (1), even in cases where the approximation is appropriate, often result in making the right-hand side of (1) so complicated as to be useless. Such a description amounts to the writing of a computer program to simulate the DEDS. In fact, general purpose discrete event simulation languages provide the constructs of jobs, resources, timing of events, and logical tests (see Appendix B) while the coding in such languages produces the description of the specific systems. At this point we simply have no convenient way to capture these descriptions mathematically with the same degree of efficiency as in the case of CVDS with differential equations. Thus, as mentioned earlier, the workings of such DEDS are described through a system of "rules of operation" or "algorithms." This is essentially a brute force approach. However, it is important to emphasize that DEDS are nevertheless "dynamic systems" in the usually understood sense of the term; that is, it is a quintuple consisting of {input set, output set, state set, state transition map, and output map}[2]. But the specification of these five objects is far from succinct and pristine as is the case in mathematical system theory. Nevertheless the fact that we can implement *general purpose* discrete event simulation languages can be construed as testimony to the "dynamical" nature of these systems. Appendix D shall also give some glimpses into various formal mathematical models that have been proposed for DEDS. All of these models ultimately emphasize the dynamical nature of the system under study by concentrating on its "trajectory or behavior," i.e., its sample path. It is instructive to look at a typical example trajectory of a DEDS in terms of the operations of a simple communication system, which we have illustrated in Fig .1. This trajectory is contrasted with a typical CVDS trajectory in Fig. 2, which represents a solution of some differential equations. For DEDS, the trajectories are piecewise constant and event-driven[3]. The sequence of piecewise constant segments represents the **state sequence,** and the duration of each segment represents the **holding time** in each state.   While its duration is generally a continuous variable, the state takes value in a discrete set. Thus a sequence of two numbers (state and its holding time) basically characterizes the trajectory. On the other

---

[2] The transition and the output map for a DEDS have to be interpreted to include objects such as the time advance and the event selection mechanism, which are commonly found in discrete event simulation languages.

[3] Reader should be cautioned not to read too much into Fig. 1. Since the state space of a DEDS is simply an arbitrary collection of discrete states and no metric is defined on the state space, mathematical operations, such as integration and differentiation, on the trajectory makes no sense. Thus, the comparison of Figs. 1 and 2 should not be taken too literally.

hand, a CVDS trajectory is constantly changing, with the state taking value in R^n, and is generally driven by continuous inputs. The definition of state is also different in DEDS and CVDS. In the former, it is the "physical" state (e.g., the number of messages at each node in the communication network awaiting transmission); while in the latter, it is the "mathematical" state, which is defined to be all the information required at a given time (other than external inputs) to uniquely specify the future evolution of the system. The mathematical definition of a state includes the "physical" definition, but not vice versa.

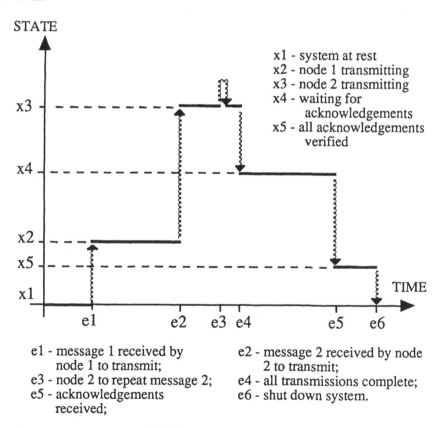

Fig. 1 An Example of a DEDS Trajectory

x1 - system at rest
x2 - node 1 transmitting
x3 - node 2 transmitting
x4 - waiting for acknowledgements
x5 - all acknowledgements verified

e1 - message 1 received by node 1 to transmit;
e2 - message 2 received by node 2 to transmit;
e3 - node 2 to repeat message 2;
e4 - all transmissions complete;
e5 - acknowledgements received;
e6 - shut down system.

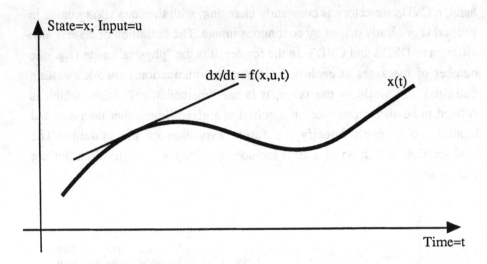

Fig. 2 An Example of a CVDS Trajectory

As a possible generalization, we can replace each piecewise constant segment of the DEDS trajectory in Fig. 1 with a trajectory generated by a differential equation with a starting and ending state. This will result in a mixed CVDS and DEDS possibly modeling a large chemical plant or a power distribution system. The continuous trajectory segments can represent the system behavior under various modes of operations such as the normal and emergency modes, etc., while the discrete event transitions capture the high level supervisory controls. This is another example of the reason for a "dynamical" emphasis in the study of DEDS in this book.

Compared with CVDS, which is the underpinning of physics, DEDS is a relatively recent phenomena. It belongs in the domain of Operations Research (OR). Although a brief examination of any OR textbook tends to give the impression that OR represents a collection of useful techniques, such as decision analysis, mathematical programming, Markov chains, and queueing theory, OR can be and was originally thought of as the *science of operations and events* of man-made systems. In fact, the father of OR, Professor P. Morse [Morse and Kimball 1951], conceived OR as a parallel development to physics, which is the *science of nature and materials*. Since it is primarily represented by man-made systems governed by "rules of operations," DEDS is in this sense properly a subject of OR. However, the developments of DEDS also received a large impetus from control system theory. In particular, we submit that the concepts of dynamics, such as time constants, time and frequency responses, controllability and observability, have played and will continue to play important roles in the development of models and tools for DEDS.

1.2 Models of DEDS

We cannot overemphasize the importance of a well-established modeling framework or paradigm such as the one found in Continuous Variable Dynamic Systems (CVDS) using differential equations. This framework permeates the entire scientific culture, facilitates cross disciplinary communications and influences our approach to problems. We submit that it is precisely the lack of a commonly accepted paradigm that underlies the relatively primitive analysis and synthesis effort we have with Discrete Events Dynamic Systems. While there has been no lack of attempts at constructing general models for DEDS, no consensus has been developed as to which one of the models has the potential to eventually serve as the analogy of the differential equation paradigm for CVDS. Among these modelling efforts we count

* Markov chain/Automaton model — To this group we also assign the Petri net and the Finite-state machine models.
* Min-max or Dioid algebraic models
* Communicating sequential process models
* Queueing network models
* Generalized Semi-Markov Process (GSMP) models — To this category we include all efforts relative to general discrete event simulation.

These can be further classified as shown in Fig. 3.

	Timed	Untimed
Logical	Temporal Logic -> Times Petri Nets ->	Finite State Machines Petri Nets
Algebraic	Min-Max Algebra	Finitely Recursice Proc Comm. Sequential Proc
Performance	Markov Chains Queueing Networks GSMP/Simulation Stochastic Petri Nets	

Stochastic => <= Nonstochastic

Fig. 3   Different Models of DEDS

The distinction between untimed and timed models of DEDS is clear. The untimed models emphasize the "state-event sequence" of a DEDS and de-emphasize or omit entirely the "holding time" specifications for every state. Thus in such models we primarily ask questions of a qualitative or logical nature (e.g., "yes" or "no," "true" or "false"). Timed models, on the other hand, incorporate "time" as an integral part of the model and are more suited to answering quantitative performance-related questions. Finally, algebraically based models, such as finitely recursive processes (or communicating sequential processes) aim to capture the description of the trajectories of DEDS in terms of a small set of algebraic operations on functions of states and events in very much the same way that CVDS are succinctly described by differential and algebraic operations on functions of state and inputs. However, one should be cautioned not to read too much into the classification. The boundaries of Fig. 3 are by no means absolute. For example, timed Petri nets often contain stochastic effects, and qualitative performance can be deduced from using untimed models.

To describe adequately the features of these models would require several chapters; that is not the purpose of this book. Instead, for interested readers, we shall choose to illustrate in Appendix D the working of these models in terms of a single and relatively simple example. However, Appendix D is not crucial for the understanding of the rest of the development in this book.

There are several reasons for the lack of a universal modelling framework for DEDS. They can be understood by an examination of the desiderata for a DEDS model of all seasons.

The discontinuous nature of Discrete Events

The physical states and events of a DEDS are inherently discrete. Furthermore, metric notions such as neighborhood and distance can not be easily defined for the state space. Although it is possible on occasion to successfully analyze a DEDS by continuous approximations via diffusion models or by dealing exclusively with rates rather than explicit constituent parts, it does not seem possible to avoid the discrete nature of the problem entirely. Any kind of universal DEDS model must acknowledge this. Also, "discrete" is to be distinguished from the commonly used adjective "digital" in CVDS where digital simply means that we do numerical analysis rather than real variable analysis.

The continuous nature of most Performance Measures

The performance measures for DEDS are mostly formulated in terms of continuous variables such as average throughput (the rate of occurrence of certain events), waiting time (duration of certain events or states), utilization (percentage of the duration of certain events or states), inventory or queue length, profit, etc. In fact, "time," a fundamental performance variable, is by definition and convention a continuous variable. There is little technological or mathematical advantage in considering a discrete time or "sample-data" model of a DEDS. In fact, many tools such as queueing network theory and perturbation analysis explicitly rely on the smoothing properties of the "expectation" or "average" operator to make analysis possible.

The importance of Probabilistic Formulation

As Murphy's law of human endeavor would have it, "Anything that can go wrong, will." Unscheduled disturbances and/or breakdowns are facts of life in DEDS operations. Almost all performance-oriented approaches to DEDS incorporate stochastic effects as integral parts of modeling. This, however, does not imply that a totally deterministic approach (e.g., Min-max algebra and Petri nets) has no place in the scheme of things. It does mean that unless useful stochastic extensions are made, such models cannot be expected to replace entirely probability-based models such as GSMP in quantitative performance studies.

The need for **Hierarchical Analysis**

Many man-made operations, such as factories, operate on yearly - quarterly - monthly - weekly - daily schedules. Military commands are divided into theater - army - battalion - platoon levels. The details required for modeling and control are different for each level. Furthermore, different aspects of DEDS analysis (e.g., planning vs. operation) often take place at different time scales leading to different levels of detail. A good hierarchical DEDS model should have the property of being able to operate at any level by using the same set of constructs without the necessity of inventing new elements for each level. This is also important for aggregation and decomposition analysis.

The presence of **Dynamics**

As the late C.S. Draper of MIT was fond of saying[4], "So damn few things in this world are static!" If we are to reach the same level of development for the dynamic control of DEDS as we have with CVDS, then it seems obvious that we must adopt models that can capture the dynamics and transient behaviors of DEDS. We submit that by ignoring the dynamical nature of DEDS and concentrating on the output analysis of the system as a black box, extant discrete event simulation literature gives up a considerable amount of valuable information inherent in the system output. Single run gradient estimation, such as likeli-

---

[4] As quoted in the video tape "Thinking about the Future" MIT Development Office 1988.

hood ratio methods (see Chapter 7) and perturbation analysis (Chapters 2 - 8), have demonstrated this. See also [Glynn 1989 and Cao 1989d].

The feasibility of the **Computational Burden**

The number of discrete physical states of a DEDS explodes in a combinatorial fashion. A brute-force enumeration of the states along the lines of a finite state machine or a Markov chain without additional structure quickly leads to infeasible computational requirements even for the largest computers envisioned. Successful performance-oriented models must acknowledge this computational burden. Thus we need to distinguish between conceptual and computational models. If the computational requirement of a model grows combinatorially with its size (e.g., NP-complete[5]), then the model cannot claim to have "solved" the general performance problem.

Since a DEDS is primarily a man-made rather than a physical / natural system, it tends to interact more with humans than with nature. A controlled but unmanned spacecraft to Mars is interacting mostly with nature (Newton's law of motion and gravitation, the electromagnetic spectrum, etc.). On the other hand, a flexible manufacturing system, even under computer control, is interacting more with human operators through display screens or actual materials handling. This has implications in practical implementation of any DEDS theory in two respects. First, AI interface and natural language-processing ability may be a much more important factor in DEDS than in CVDS.[6] In manufacturing automation, for example, DEDS are required to interface with personnel of lower skill levels than with those operating aerospace vehicles. Secondly, since DEDS are totally man-made, there are no invariant physical laws to constrain system configurations. Combinatorial

---

[5] NP-Complete, which stands for Nondeterministic Polynomial time Complete, is a technical term roughly indicating that the computational complexity of the problem is such that it cannot be bounded in computational time by any polynomial function of the size of the problem.

[6] As a prime example of the difficulties of natural language processing, consider the following sentence which appears on the side of a tube of a well-known brand of toothpaste: "for best results, squeeze tube from bottom and flatten as you go." Any child who can read can understand the meaning of this sentence. However, we are far from being able to devise an AI system with such a degree of comprehension.

explosion of complexity can be an easy occurrence. Consequently, a day when all solutions to all DEDS problems can be reduced to an algorithm is unforeseeable. Some form of rule-based solutions approach and/or fuzzy logic which tolerates imprecision, will probably be a necessary part of a general DEDS tool chest[7].

In summary, we see that DEDS research lies at the intersection of System Theory, Operations Research, and Artificial Intelligence - the STORAI triad in Fig. 4.

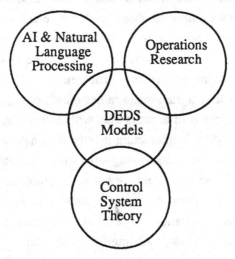

Fig. 4 The System Theory, Operations Research, and AI Union

Finally, as **scientific** disciplines must have both experimental and theoretical components, so must the study of DEDS. Carefully designed experiments accumulate evidence upon which a relevant theory can be built. Conversely, analytical reasoning pinpoints further experiments to be conducted for valida-tion. To give an example, consider the area of manufacturing automation which is often touted as a prime area for the application of DEDS research. Much of the performance-oriented literature using queueing network theory and discrete

---

7 For another interesting connection of AI with the subject of this book, see D.S. Weld [1990].

event simulation deals principally with performances which are related to "resource contention and allocation." This assumes that the individual operations of a manufacturing process are sufficiently well-understood to be totally described by timing considerations. Yet problems, such as the yield percentage in semi-conductor wafer manufacturing, have more to do with the physics and chemistry of materials involved in unit manufacturing operations than with resource contention. Direct experimentation and statistical quality control are integral parts of this problem. A second example is the subject of perturbation analysis itself. This book would not be possible if we did not have the experimental evidence in our early research to support our beliefs. Without this "observation - conjecture - experiment - theory - validation" cycle to inspire and to guide research, we might easily end with only a partial solution or a not so relevant model if a totally axiomatic mathematical development of DEDS is followed [Ho 1982, Hogarth 1986].

The purpose of this book is twofold: to advocate the rich opportunities in the field of DEDS, and to describe in some detail a newly developed performance evaluation technique called perturbation analysis (PA) of DEDS. DEDS is used here in the narrow sense as a parallel to CVDS in the form of dx/dt=f(x,u,t). We do not address issues such as protocol, implementation, integration, interfaces, and hardware in DEDS. By this we do not imply that such issues are unimportant. But our narrower definition of DEDS is consistent with our past use of the term as well as with the convention adopted in discussions of control theory for CVDS. Conceptually, we can visualize the entire sweep of classical control-theoretic problems such as controllability and observability, estimation and identification, information and control, awaiting formulation for DEDS. Our emphasis in this book is the subject of performance evaluation of DEDS, especially the sensitivity analysis of performance, and the use of perturbation analysis for this purpose.

Introduction to Perturbation Analysis

2.1 Basic Concepts and Terminology

To facilitate the discussion of Perturbation Analysis (PA), let us introduce some notations for DEDS. Let

θ = system parameter(s)

$x(t)$ = a time history of the evolution of the DEDS, i.e., the (state, holding time) sequence as illustrated in Fig.1.1. In more physical terms, this may consist of the content of all the queues as a function of time, durations of all service intervals, etc. Since DEDS are often stochastic, $x(t)$ will in general be dependent on the actual realized values of various random variables in the system.

ξ = a vector of random variables, defined on the underlying probability space, that represents all the random phenomena of the DEDS or a particular realization of all the random variables in the system.

We may write $x(t) = x(t ; \theta, \xi)$ to indicate the dependency on θ and ξ. $x(t ; \theta, \xi)$ is called a **trajectory**, or a **sample path**, of the DEDS. Also we let

τ_i = the ith transition of the state process $x(t)$, i.e., the ith time when $x(t)$ switches from one discrete state to another as a result of an occurrence of an event.

When we perturb the parameter θ to $\theta + \Delta\theta$, we obtain a perturbed sample path denoted by $x(t ; \theta + \Delta\theta, \xi)$. In the parlance of discrete event simulations, if

$x(t; \theta, \xi)$ is generated by a simulation, then $x(t; \theta+\Delta\theta, \xi)$ is the simulation generated by the same model using the same initial seed for the random number generator with only the value of θ changed to $\theta+\Delta\theta$. To colloquially differentiate between $x(t; \theta, \xi)$ and $x(t; \theta+\Delta\theta, \xi)$, we call the former the **nominal** path or trajectory, NP, the latter, the **perturbed** path or trajectory, PP.

Given a DEDS and its sample paths, a sample performance measure is a functional defined on that sample path, e.g., the time required to serve N customers Two common examples for the sample performance measure are

$$L(x(t; \theta, \xi)) \equiv \frac{1}{T} \int_0^T f(x(t, \theta, \xi))\, dt \equiv L(\theta, \xi) \tag{1}$$

and

$$L(x(t; \theta, \xi)) \equiv \frac{1}{N} \int_0^{\tau_N} f(x(t; \theta, \xi))dt \equiv L(\theta, \xi). \tag{2}$$

Note that the sample performance defined above depends implicitly on the length of the sample path, T or τ_N. Since the sample paths of a DEDS are piecewise constant, we can equivalently write Eqs.(1) and (2) as

$$L(\theta, \xi) \equiv \frac{1}{\tau_N} \sum_{i=0}^{N-1} f(x_i)(\tau_{i+1}-\tau_i) \tag{3}$$

$$L(\theta, \xi) \equiv \frac{1}{N} \sum_{i=0}^{N-1} f(x_i)(\tau_{i+1}-\tau_i), \tag{4}$$

where $x_i = x(\tau_i+)$ is the ith state and $(\tau_{i+1}-\tau_i)$ is the duration of the ith state, $\tau_0=0$. The expected value of $L(\theta, \xi)$ is $J(\theta) \equiv E[L(\theta, \xi)]$, which is the central item of interest in the performance evaluation of DEDS. Note that the above performance measure is defined in terms of finite time intervals. Often we are also interested in steady-state performance such as the average waiting time for a typical customer or the throughput rate. The steady-state performance can be expressed as the limiting value of the sample performance measures defined above on a finite interval. For example, let $f(x(t ; \theta, \xi))$ be the number of customers in a queue, $n(t)$, then the steady-state queue length of the queue can be expressed using (1):

$$J(\theta) = \lim_{T \to \infty} L(x(t;\theta,\xi)) = \lim_{T \to \infty} \frac{1}{T} \int_0^T n(t)dt. \tag{5}$$

Letting $f \equiv 1$ in (2) and taking the limit, we obtain the steady-state throughput of the system:

$$J(\theta) \equiv \lim_{\tau_N \to \infty} \frac{N}{\tau_N} = \frac{1}{\lim_{\tau_N \to \infty} (\tau_N/N)} = \frac{1}{\lim_{\tau_N \to \infty} \frac{1}{N} \int_0^{\tau_N} f(x(t;\theta,\xi))dt}. \tag{6}$$

The sample performance measure for the perturbed paths is defined in the same way.

Stated simply, Perturbation Analysis (PA), particularly Infinitesimal Perturbation Analysis (IPA), is a technique for the computation of the gradient of the expected or the limiting performance measure of a discrete event dynamic system with respect to its parameters, i.e., $\partial J(\theta)/\partial \theta$, using only one sample path or one Monte Carlo simulation of the system. A slightly more general definition of PA explains it as a technique to infer knowledge of $L(\theta+\Delta\theta, \xi)$ from observing $x(t ; \theta, \xi)$, the behavior of the system under θ alone. Thus, the basic goal of PA is to squeeze out as much information as possible from one sample path of a DEDS. Note that traditionally sensitivity information such as $dJ(\theta)/d\theta$ are estimated by doing two experiments with the DEDS, one at θ and the other at $\theta+\Delta\theta$. Taking the difference between the two performance measures and dividing through by $\Delta\theta$ yields an estimate for the derivative. In other words,

$$\frac{dE[L(\theta,\xi)]}{d\theta} = \lim_{\Delta\theta \to 0} \lim_{N \to \infty} \frac{1}{N} \sum_n \frac{L(\theta+\Delta\theta,\xi_n) - L(\theta,\xi_n)}{\Delta\theta}$$

for the case of Eq.(1) or

$$\frac{dJ}{d\theta} \equiv \lim_{\Delta\theta\to 0}\frac{1}{\Delta\theta}\{\lim_{T\to\infty}[L(x(t;\theta+\Delta\theta,\xi)) - L(x(t;\theta,\xi))]\}$$

for the case of Eq.(5). For multidimensional θ, the number of extra experiments required for gradient estimation equals the dimension of the θ vector. This is computationally intensive and numerically difficult since one is attempting to take the difference of two almost equal random numbers and dividing it by another small number. On the other hand, PA claims to be able to estimate the gradient from a single sample path or experiment. At first this may appear to be counter-intuitive stemming from the philosophical belief that "one cannot get something for nothing." A more concrete and equally intuitive objection is based on the obvious observation that sample paths of the DEDS under θ and $\theta+\Delta\theta$ will in general sooner or later become totally different as they evolve in time even for very small $\Delta\theta$. Thus, it does not seem reasonable at first glance that we can infer knowledge about $x(t ; \theta+\Delta\theta,\xi)$ from $x(t ; \theta,\xi)$. Finally, this disbelief may be developed into a more sophisticated and technical objection involving the legitimacy of interchanging differentiation with the expectation operators or taking the limit (for more about these see below and later in Chapters 3 and 4). That this is indeed possible constitutes a substantial portion of the development of PA and the content of this book[1].

Exercise Give at least one real world (or conceptually real-world) example for each of the DEDS performance measures mathematically defined below: Let $x(t)$ or x_i be the sample path which is implicitly dependent on the system parameter θ, and the sample realization ξ; L the performance criterion; f some cost function of the state, $x(t)$ or x_i; T the termination time which can be

---

[1] For readers who are electrical or system engineers, the following analog is enough to explain away the counter-intuition. We can obtain a complete characterization of a linear system via its impulse response by performing autocorrelation of the system output when the system is perturbed by white noise. The DEDS here is constantly being excited by the random occurrence of various events which serves to exercise the DEDS and to carry information to its outputs.

deterministic or random depending on the occurrence of an event, a state, a time instant or all three.   Some of the possibilities you should consider are utilization, throughput, probability of a given state, system (waiting) time, etc. Your task is to show how to model these by way of Eqs.(7-9).

$$L(\theta,\xi) = \frac{1}{T}\int_0^T L(x(t))dt \qquad (7)$$

$$L(\theta,\xi) = f(x(T)) \qquad (8)$$

$$L(\theta,\xi) = \frac{1}{N}\sum_{i=0}^{N-1} f(x_i,x_{i+1}) \qquad (9)$$

2.2 A Short History of the PA Development

In 1977, one of the authors (Ho) was presented with an interesting consulting problem [Ho, Eyler, and Chien 1979][2]. The FIAT Motor company in Torino, Italy had installed a production monitor system on one of their automobile engine production lines, which could be visualized as a simple serial queueing network with finite queue (buffer) capacity between servers (machines). The automatic line monitoring system recorded service initiations, completions, idlings and blockings of various machining stations as well as the movement of the engine parts among them, in short, a complete operating history of the DEDS. A tremendous amount of production information was being generated. The following questions were asked: "Besides the standard statistical information such as downtime, throughput, utilization, that were being generated by the monitoring system from the collected information, could this information be used further for control purposes? In particular, we (FIAT) were interested in whether or not the buffer spaces between machines are optimally

---

[2] M. Bello in a 1977 M.I.T. Masters degree thesis also had a version of the idea of perturbation analysis as applied to an M/D/1 queue, see [Bello 1977].

distributed for maximal throughput given a limited budget for buffer spaces."
The genesis of the PA idea came from the following paradigm, which was an
attempt to answer the above questions.

"Imagine you are standing next to a buffer between any two machines
of this automated production line. Your duty is to expedite the
production by helping out in case of blocking or idling of machines.
You soon realize that the only occasion in which you can be of help is
when the upstream machine is about to be blocked due to a full buffer
at its downstream receiving machine. At that time, you can behave as
an extra slot in the buffer space by unloading and holding the piece
from the upstream machine, thus permitting it to produce one extra
piece which will otherwise be blocked. Subsequently, the extra
production piece that you are holding will become useful when the
down-stream machine is about to become idle due to an empty buffer.
At that time the extra piece which you are holding can be fed to
the downstream machine allowing it to produce one extra piece. This
cycle from the initial blocking of the upstream machine to the next
idling of the downstream machine is pictorially illustrated in Fig.1.
The trajectory of the buffer content "bounces" from FULL to EMPTY
during the above described cycle. The key observation is that only
during an occurrence of such a "bounce" cycle can an extra buffer
space be of value, namely, it enables the production of one extra piece.
This is true regardless of the detail evolution of the system trajectory.
It is thus reasonable to suggest that we can infer the effect of one extra
buffer space on throughput by simply counting the number of such
"bounce" cycles along a nominal trajectory **without** the need of a
separate experiment. The next observation is that the gain in
production realized at the two machines does not necessarily mean
gain in production by the entire line. The production gains must be
PROPAGATED to all other machines further upstream or downstream. For example, if machine A has a gain in production, this gain
is propagated to the next downstream machine B only when an idling
period occurs on B. This is because the production gain will enable B
to shorten the idling period by starting production earlier than if the
gain were not there. Similarly, gains are propagated upstream when an

upstream machine is blocked due to a full buffer. Gains will release the blocking sooner. Finally, "gains" are relative; they can be cancelled by blockage downstream and no input from upstream corresponding to propagation of zero gains. The entire production line REALIZES the gain only when it is propagated to all the machines."

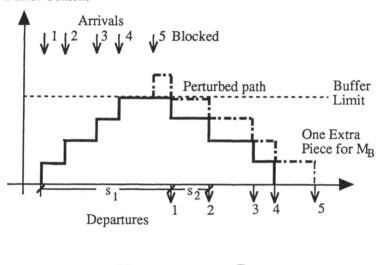

A 2-Machine 1-Buffer System

Fig. 1 The Nominal and Perturbed Paths of a "Bounce"

This rudimentary paradigm captures the essence of the PA technique:

(i) A parameter change (e.g., increasing the size of the buffer space by 1) can **generate** perturbations in the timing of events in the sample path of a DEDS.

(ii) Perturbations in the timing of one event (e.g., termination of a service period of a machine) can be **propagated** to other event (e.g., termination of an idling period at another machine).

(iii) Since all performance measures of a DEDS depend on the timing of events on its sample path, perturbations in the timing of events will induce perturbations in the sample performance measure,L.

Steps (i) - (iii) suggests a method to calculate efficiently the perturbed performance $L(\theta+\Delta\theta, \xi)$ of a sample path from $x(t; \theta, \xi)$ alone since all three steps only require information directly observable on $x(t; \theta, \xi)$. More specifically, for the case of continuous θ (see Chapters 3 and 4), the derivative $dL(\theta, \xi)/d\theta$ is calculated directly from observing $x(t; \theta, \xi)$ alone. Then by taking ensemble averages or letting the length of the sample path approach infinity, we can directly estimate $dJ(\theta)/d\theta$. This is known as Infinitesimal PA, or IPA. By late 1981, it was realized that the **generation** and **propagation** of perturbations in the timing of events of a sample path of a general DEDS are basic elements in the analysis of its dynamic behavior. A natural generalization of the work to arbitrary queueing networks was made and was experimentally verified to estimate $dE[L(\theta,\xi)]/d\theta$ accurately for a variety of single class non-product form queueing networks [Ho and Cao 1983][3].

Following this generalization, effort was initiated to study more rigorously the theoretical basis of PA. The basic mathematical question for PA can be posed as follows:

Question Infinitesimal PA (i.e., $\Delta\theta=d\theta$) estimates the expected value or the limiting value (as $T\rightarrow\infty$) of the derivative of $L(\theta,\xi)$. Yet one is really interested in the derivative of the expected or the limiting value of $L(\theta,\xi)$ assuming it is well defined and exists. When are these two quantities equal; i.e., when is the interchange between E and $d/d\theta$ or between $\lim_{T\rightarrow\infty}$ and $d/d\theta$ valid? That is,

---

[3] Independently, Woodside [1984] published a paper reporting the application of PA ideas to queueing networks.

$$E[\frac{dL(\theta,\xi)}{d\theta}] = ? = \frac{dE[L(\theta,\xi)]}{d\theta} \tag{10}$$

$$\lim_{t\to\infty} \frac{d[L(x(t;\theta,\xi)]}{d\theta} = ? = \frac{d}{d\theta}[\lim_{t\to\infty}L(x(t;\theta,\xi)] \tag{11}$$

The question for (10) is referred to as the "**unbiasedness**" question, while that of (11), the "**consistency**" question. This is the mathematical version of the same intuitive question concerning the inference of system behavior information under $\theta+\Delta\theta$ from that under θ.

Consider the issue of Eq.(10) first. Regardless of the size of $\Delta\theta$, an ensemble of nominal paths, $x(t ; \theta,\xi)$, will start to differ with the perturbed ensemble, $x(t ; \theta+\Delta\theta,\xi)$ for some value of ξ, if the ensemble is large enough. Once a path starts to differ, the future evolution of $x(t ; \theta,\xi)$ and $x(t ; \theta+\Delta\theta,\xi)$ may look entirely different. In other words, an arbitrarily small perturbation in θ will sooner or later lead to discontinuities in the ensemble of trajectories and hence in $L(\theta,\xi)$ for some ξ. Elementary analysis tells us that such discontinuities may rule out any possibility for the above-mentioned interchange to be valid; hence, the mathematical objection that PA cannot provide a consistent estimate of the derivative of expected L by computing the expectation of the derivatives. This key issue was addressed in general by [Cao 1985], and in specific by [Suri and Zazanis 1988] for the M/G/1 queue. The idea is that if for a small perturbation in θ, **both** the probability of encountering a discontinuity in $L(\theta, \xi)$ and the value of the discontinuity are sufficiently small then such sample paths do not contribute substantially to the averaging process and Eq.(10) is valid. In other words, the infinitesimal PA algorithm still gives the right answer even though on a small (of order $\Delta\theta$) percentage of the time the algorithm will make errors which also are small in magnitude (of order $\Delta\theta$). Consequently, the average error, being the product of the percentage of occurrence and the magnitude of the error, is of order $(\Delta\theta)^2$ and is insignificant for the calculation of the first derivatives. Furthermore, it also became apparent that the particular model used to represent a DEDS can be crucial in establishing the validity of (10). By proper reformulation or transformation, a seemingly discontinuous $L(\theta, \xi)$ can be made continuous, and hence, valid for IPA [Ho and Cao 1985, Glasserman 1988a, and Vakili 1989]. Similarly, for Eq.(11) the question regarding the IPA estimate is, "does the IPA estimate of the derivative of a performance, obtained on a sample path of a finite length, converge stochastically to the derivative of the steady-state performance as the length of the sample path goes to infinity?" To study this problem for closed Jackson networks, a concept called realization probability is defined, which measures the average final effect of a perturbation generated at a server when the system is in a particular state. Based on realization probability, it was shown in [Cao 1987a] that the IPA estimate for the throughput derivative with respect to a mean service time does converge with probability one as the length of the sample path goes to infinity. It was also shown that the limiting value in fact equals the derivative of the steady-state throughput with respect to the mean service time of the server. Furthermore, it was proved that this limiting value simply equals the expected value of the realization probability. This result justifies the use of a long sample path based IPA estimate instead of the average of N IPA estimates as indicated by (10).

Lastly, [Glasserman 1990b], [Hu 1991] and [Glasserman and Yao 1991] derived explicit conditions under which IPA is applicable (i.e., unbiased and consistent) to a large class of queueing networks and Markov chains. Simultaneously, IPA were shown to be extendible via proper re-formulation to various specific classes of queueing problems that do not satisfy the explicit conditions. Many of the above are treated in some detail in Chapters 4 and 5 which show that the domain of the applicability of IPA rather than being limited is actually under-explored.

At the same time, the issue of divergence between the nominal and the perturbed path were attacked directly by trying to modify the rules of perturbation generation an propagation. We recognized that as perturbations accumulate they eventually cause event order changes and hence the state sequence difference between the NP and PP. Both the generation rule and the propagation rule need to be modified when perturbations in timing are finite and state sequence changes are not negligible. By assuming that only the adjacent (or adjacent and next to adjacent) events in an NP may change order because of perturbations, we developed the first order Finite Perturbation Analysis (FPA) rules [Ho, Cao, and Cassandras 1983]. These are basically brute force thought experiments to generate the perturbed path in a short run. However, once a perturbation is generated and propagated according to the first order FPA rules, we do not extrapolate the evolution of the perturbation indefinitely into the future which would be equivalent to carrying out a separate brute force realization of the perturbed sample path. Instead we make the following heuristic assumption:

Statistical Similarity Assumption

Once a perturbation (finite or otherwise) has been introduced, the types of future interaction it may encounter and induce among customers and servers in the DEDS along the actual perturbed path is statistically similar to that of the nominal sample path.

In other words, we assume that for a small system parameter change, the change in the distribution of various interactions among customers and servers should be continuous from the nominal to the perturbed path. Even though finite or large perturbations may be generated along the sample path, the propagation of these perturbations along the nominal and the perturbed sample path will be essentially the same *on the average*. So long as we are only interested in the average performance measures, we can use the nominal path to calculate the propagations. So long as we implement accurately the short term effect of the perturbations on each and every interaction (state change), their long term accumulated average effects are essentially the same whether we use the perturbed or the nominal path for calculation. Although there is no theoretical justification for this totally heuristic belief, experimentally, the results of such finite PA are encouraging. In fact, a very large scale statistical experiment involving random generation of systems, routes, parameters, and initial conditions shows that such crude finite perturbation analysis rules do predict sensitivities more accurately than infinitesimal rules for systems for which Eq. (10) or (11) does not hold [Dille 1987]. Conceptually, this heuristic thinking can actually be viewed as the limit of approximations to an exact finite PA scheme to be discussed in Chapter 6.

More generally, PA is viewed as a first step and a general approach (time domain, dynamical, and performance oriented) towards the analysis and optimization of DEDS primarily within the queueing theoretic and the generalized semi-Markov process (GSMP) framework. The main tenets of PA are:

(i) A sample path of a DEDS (real or simulated) inherently contains information about the system far beyond the usual summary statistics, such as time or ensemble averages of variables of interest. If this information is collected in time and processed appropriately, it can yield gradient and other useful performance data. For example, by analyzing a long sample path of a GI/G/1 queue at one value of its service rate, one can deduce its performance at all other values of the service rate (see Chapters 6 and 7, and [Gong and Hu 1991]).

(ii) If the structure of the model of a DEDS did not change except for some parameter values, then the separate generation of sample paths in traditional simulation for "what if" studies entails a great deal of duplicated effort that can be and should be leveraged to improve computational efficiency. PA is one of the several possible ways to accomplish this. Likelihood ratio, (or score function,) change-of-measure, and discrete time conversion methods are other alternative ways.

Thus, the idea of cut-and-paste, or extended PA [Ho and Li 1988], standard clock [Ho, Li, and Vakili 1988], augmented chain [Cassandras and Strickland 1989c], aggregation of Markov chains via the likelihood ratio [Zhang and Ho 1991b], etc. are examples of this mind-set. Finally, it appears that a more or less coherent time domain model of DEDS along the lines of formal languages and generalized semi-Markov schemes and processes is slowly emerging [Lin and Yao 1989, Glasserman and Yao 1991]. A reasonably complete theoretical and computational framework now exists for the study of PA. These will be covered in succeeding Chapters 6, 7, and 8.

Lastly, it is worth stressing that PA algorithms, and particularly IPA algorithms, are applicable to sample paths of DEDS, *simulated as well as in real time*. Although most of our discussions and experiments pertain to simulated sample paths for the purpose of validation, we view the real time applications of PA to actual system trajectories as the ultimate important engineering payoff. Here, a PA or PA-related algorithm can perform **non-intrusive** performance analysis without the need of actually perturbing any parameter of the system. This is something that cannot be easily accomplished with simulation.

CHAPTER THREE

Informal Treatment of Infinitesimal Perturbation Analysis

3.1 The Basic Idea

As we sketched out in Appendices A and B, by and large there are two approaches to the performance oriented study of DEDS, namely, the analytical or the simulation- based approaches. The former has as its centerpiece the product-form formula of Jackson-Newell-Gordon and their various extensions (see Appendix A); the latter, simulation languages such as SIMSCRIPT, GPSS, SLAM, SIMAN, and their mathematical formalization, GSMP (see Appendix B). Perturbation Analysis in some sense attempts to combine the advantages of both the theoretical and the simulation approaches. PA is time-domain based. It views a queueing network as a stochastic dynamical system evolving in time and analyzes the sample realization of its state process. To this extent, it is no different from the viewpoint (and, hence, enjoys the same advantages) of the simulation approach to DEDS. However, by observing a sample path (trajectory) of the network trajectory, we use *analysis* to derive answers to the question: "what" will happen "if" we repeat the sample path exactly except for a small perturbation of the timing of some event at some time t? The efficiency of our approach lies in the fact that we can answer a multitude of such "what if" questions simultaneously, while a single sample path is being observed. Thus, compared with the brute-force simulation study, our approach has a computational advantage of N+1 : 1, where N is the number of "what if" questions asked (the brute-force simulation method requires one additional simulation experiment for each question asked). For pedagogical reasons, we shall introduce in this chapter the simplest form of PA in an intuitive and empirical manner deliberately overlooking some of the foundational issues alluded to in Chapter 2. They will be dealt with in succeeding Chapters 4 - 8.

Let us start by considering a sample path of a single class server, say the ith server, standing alone, with an infinite supply of waiting customers. This is nothing but a sequence of sample service time values $S_i(j)$, $j=1,2,...,$ where i denotes the ith server and j, the jth activation for service. When this server is placed in a network, additional idle periods, due to no waiting customers, and/or blocked outputs, due to the inability of customer to depart, are spliced into this sequence as a result of server-server interactions. We denote these periods as NI (for No Inputs and waiting for jobs) and FO (for Full Outputs)[1]. The entire event sequence including the NIs and FOs are denoted as $E_i(\xi)$, the ith server's event sequence where ξ represents the particular sample realization. A collection of such $E_i(\xi)$, $i=1,2,...,$ one for each server in a network, is called a "tableau" of event sequences. It should be emphasized that such a tableau represents rather completely a real or simulated sample path of a single class queueing network. This representation is "completely general," regardless of the complexity and/or the nonstandard nature of the underlying queueing network. Mathematically speaking, the tableau representation is equivalent to the state process (time history) x(t) introduced in Section 2.1; the tableau representation, however, also illustrates clearly the history of each individual server. A typical tableau from an open tandem queueing network with finite queues and an infinite number of arriving customers is illustrated in Fig. 1.

The special topology of the tandem network is manifested in the fact that the NI and FO intervals in $E_i(\xi)$ are always caused by the slow services in

---

[1] If a customer completes its service at a server and finds the queue of his next destination full, then he must remain at the current server and thus "blocks" it from serving another customer. We say, this server encounters a full output (FO) and is blocked. This terminology is from manufacturing. On the other hand, in communication technology, a message, instead of being blocked from advancing to the next queue, is simply "lost." We distinguish that two different forms of blocking by adding the modifier "manufacturing" or "communication" to the term. Unless otherwise indicated, blocking in this book refers to the manufacturing blocking.

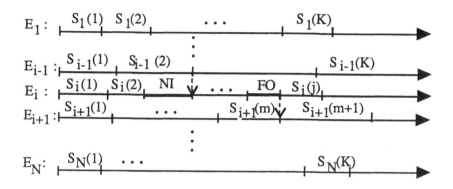

Fig. 1 Tableau Representation of Operating History of a Tandem Network

$E_{i-1}(\xi)$ and $E_{i+1}(\xi)$, respectively. For example, the NI interval after $S_i(2)$ in Fig.1. is caused by starving, since server i - 1 is not serving customers fast enough; similarly, the FO interval is caused by blocking since server i+1 is not serving fast enough. More generally, for arbitrary networks, the server-server interactions are time varying in the sense that an NI(FO) interval may be caused and terminated by different servers at different times. Fig. 2 illustrates the general situation where, for clarity, we have suppressed the notation for $S_i(j)$s and only display the FOs and NIs. It cannot be emphasized too strongly that these FO and NI intervals are the only means of interaction among servers in an ordinary single class network. Analysis of queueing network problems will be completely decomposed into individual queueing problems in the absence of these intervals (since each server will simply be serving at the maximal rate). One can also say that networks in which all servers are always busy present no opportunity for improvement or optimization from the viewpoint of utilization.

Now, suppose that one of the event times on the sequence $E_i(\xi)$ is perturbed by Δ (e.g., the kth customer's service time ends Δ seconds later or earlier with Δ arbitrarily small). We say that server i, or the customer being served at server i, gains a perturbation Δ. A perturbation of the event sequences

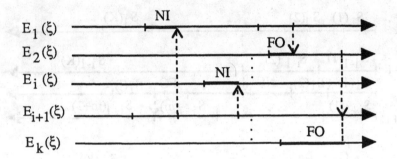

Fig. 2 Event Sequence Tableau for General Queueing Networks[2]

$E_i(\xi)$ will be propagated to other event sequences $E_j(\xi)$, $j \neq i$, only through these FO or NI intervals. This is clear, since only during these intervals can the event sequences of two servers be coupled together. The propagation of a perturbation follows some simple rules illustrated in Figs. 3.a and 3.b and described as follows.

Infinitesimal Perturbation Propagation Rules

(i)   A perturbation of a customer at server i will be propagated to the next customer in the same server unless the server meets an NI or an FO after the service completion of the customer.

(ii) If, after an NI, server i receives a customer from server j, then server i will have the same perturbation as server j after the NI. (We say that the NI of server i is terminated by server j and that the perturbation of server j is propagated to server i through an NI.)

(iii) If, after an FO of server j caused by server i, server i receives a customer from server j, then server j will have the same perturbation as server i

---

[2] The arrows in this figure are used to indicate the controlling event which terminates the NI or FO.

after the FO. (We say that the FO of server j is terminated by server i and that the perturbation of server i is propagated to server j through an FO.)

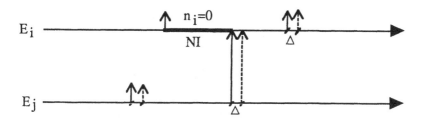

Fig. 3a Perturbation Propagation Through an NI

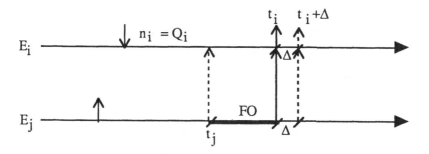

Fig. 3b Perturbation Propagation Through an FO

Propagation rule (i) has a very simple meaning: if the service completion time of a customer in a server is delayed (or advanced) by Δ, then the service starting time (hence the service completion time) of the next customer, if any, in the server will be delayed (or advanced) by the same amount of time. Fig. 3a shows that if after an NI server i receives a customer from server j and server j is perturbed by Δ, then the service starting/completion time of server i will be perturbed by the same amount. This explains the propagation rule (ii). In Fig. 3b, t_j is the service completion time of server j and Q_i the size of the buffer of server i. Since $n_i = Q_i$ at t_j, server j is blocked by server i. The service starting time of the next customer at server j is t_i instead of t_j, with t_i being the service completion time of the customer in server i. Therefore, if the service completion time of server i is perturbed by Δ, then the service starting time of server j will be perturbed by the same amount. This means that server j will have the same perturbation as server i after the FO.

The simplest case occurs in the tandem network of Fig. 1, where NI propagates perturbations downstream and FO upstream. Since the termination of NI is controlled by the upstream server, any perturbation of the upstream server will be propagated to the downstream server at the termination of the NI interval. Similar dual statements apply to the FO interval by following this simple rule as the system evolves in time. For more general queueing networks, the propagation rules are not much more difficult. We simply have to ascertain which server interacts with another server through an NI or FO interval at any given time and to propagate perturbations according to the propagation rules. It is always the server who terminates these NI and FO intervals that propagates its perturbation (if any) onto the server experiencing these intervals. Using the propagation rules, we can relate the effect of a small perturbation in the timing of an event on the trajectory of one server at some time τ to that of any other server at some later time t. Note that a perturbation in $E_i(\xi)$ can be either propagated or cancelled. For example, if server i experiences an NI(FO) interval, which is terminated by server j who has no perturbation, then this perturbation in $E_i(\xi)$ will be cancelled since, regardless of the starting time of the NI(FO) interval, the termination time is controlled by $E_j(\xi)$, which has no perturbation. Once a perturbation is introduced into the system, we can easily follow its propagation to the other servers via the rules described above. We shall show later that such a perturbation under appropriate conditions is eventually propagated to all other servers or is cancelled (see Chapter 4). In the former case, we say that the perturbation is "realized" by the system; in the latter case, it is "lost."

Now let us consider the effect of the parameter change on a sample path. Let $F(s, \theta)$ be the service time distribution of a server with θ being a parameter. Let $\xi = F(s, \theta)$, then ξ is a random variable uniformly distributed on $[0,1)$ and s

$= F^{-1}(\xi, \theta)$, where F^{-1} is the inverse function of F. In simulation, ξ is first generated and then converted to s via the above equation. This is illustrated in Fig.
4. If the service parameter changes from θ to $\theta + \Delta\theta$, then for the same ξ the
service time will be $s + \Delta s = F^{-1}(\xi, \theta + \Delta\theta)$ instead of $s = F^{-1}(\xi, \theta)$. Thus, the
service completion time of a customer will be delayed by

$$\Delta s = F^{-1}(\xi, \theta + \Delta\theta) - F^{-1}(\xi, \theta) = \left\{ \frac{\partial F^{-1}(\xi, \theta)}{\partial \theta} \right\}_{\xi = F(s, \theta)} \Delta\theta.$$

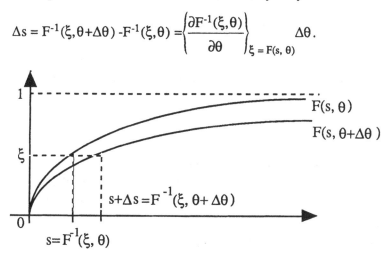

Fig. 4 Perturbation Generation Rule

That is, a parameter change $\Delta\theta$ will induce a perturbation Δs in each customer's
service time. Δs is called the perturbation generated in the service period of the
customer and the above formula is called the *perturbation generation rule*.
Fixing ξ and taking the derivative of both sides of $\xi = F(s, \theta)$, we get $[\partial F/\partial s]\Delta s$
$+ [\partial F/\partial \theta]\Delta\theta = 0$. From this, we can obtain another equivalent perturbation generation formula as follows.

$$\Delta s = - \left\{ \frac{\partial F/\partial \theta}{\partial F/\partial s} \right\}_{(s, \theta)} \Delta\theta.$$

The advantage of this formula is that ξ does not appear in it and that we do not
need the form for the inverse function F^{-1}.

Example 1 If $F(s)$ is exponentially distributed with mean θ, i.e.,

$$s_{i,k} = -\theta \ln(1-\xi_{i,k}),$$

then we get $ds_k/d\theta = s_k/\theta$.

Example 2 If $F(s)$ is uniform $[0, \theta]$, then $ds_k/d\theta = s_k/\theta$.

Example 3 If θ is a so-called scale parameter of a distribution function, i.e., the distribution of s_k/θ is independent of θ, then again $ds_k/d\theta = s_k/\theta$.

Example 4 If θ is the so-called location parameter of a distribution, i.e., the distribution function of $s_k-\theta$ is independent of θ, then $ds_k/d\theta = 1$.

Exercise Verify Examples 1 - 4.

By using the perturbation generation rule, we can determine the perturbations generated in each customer's service period because of the parameter perturbation in the service time distribution. These perturbations will be propagated along the nominal sample path according to the perturbation propagation rules. Therefore, by applying the perturbation generation and propagation rule on the nominal path we can construct a perturbed sample path. For performance studies, however, we are often interested in the effect of perturbing a parameter of a queueing system on a performance measure, such as throughput. This can be done by simply analyzing the perturbed path constructed from the nominal path.

To summarize, the problem of determining the effect of a parameter change on a performance measure can be decomposed into three subproblems. Let τ_i be the generic time of an event in a sequence; θ, the system parameter; and L, the performance measure of the system. We ask the following three questions:

(i) How does the change in the value of a system parameter, θ, change the timing of various system events, τ_i, i.e., $\partial\tau_i/\partial\theta=$? The answer to this question is often referred to as the perturbation generation rules.

Example: changing the mean service time of a resource will induce a series of changes in the service duration times of the resource.

(ii) How does the change in the timing of one event, τ_i, change the timing of another, τ_k, i.e., $\partial\tau_k/\partial\tau_i =$? This can be answered by the perturbation propagation rules described above .

(iii) How does the change in the timing of some events, e.g., τ_K, change the system performance, L, i.e., $\partial L/\partial\tau_K =$? Example: The change in the time required to finish 100,000 jobs will change the average throughput of the system.

Questions (i) and (iii) are parameter and performance measure specific. In the context of DEDS performance analysis, they can usually be resolved via ordinary calculus and probability. However, question (ii) is generic to the sample path of all DEDS. We have just illustrated this in the above figures and paragraphs. This notion of PA is also related to that of the **critical timing path** (CTP) of operations research, which couples the timing of one event to that of another. PA can be visualized as a very efficient method of keeping track of a large number of complex CTPs. In the parlance of discrete event simulation, the CTP is simply the "**Event Scheduler**" or "**Future Event List**" found in all general purpose simulation languages. Putting (i - iii) together, we get

$$\frac{\partial L}{\partial\theta} = \sum_i \frac{\partial L}{\partial\tau_K} \frac{\partial\tau_K}{\partial\tau_i} \frac{\partial\tau_i}{\partial\theta}. \tag{1}$$

The resemblance of (1) to the well-known equation of control theory (see Appendix C)

$$\frac{\partial J}{\partial\theta} = \int_{t_o}^{t_f} \frac{\partial\phi}{\partial x(t_f)} \Phi(t_f,\tau) \frac{\partial f}{\partial\theta} d\tau \tag{2}$$

for the system $dx/dt = f(x,\theta,t)$ and $J = \phi(x(t_f))$ is no coincidence.

The above discussion is based on the implicit assumption that for a small enough perturbation in the parameter value,θ, the CTPs of the nominal and the perturbed sample path are in fact the same. This is fine so long as we limit ourselves to trajectories of finite length and parameter values that are continuous (Chapters 6 and 7 will discuss the case when the parameters are integer-valued or discontinuous). In these continuous cases we can always visualize a conjectured perturbation of small enough size such that the nominal and the perturbed trajectory have the same CTP, or the same event sequence, or similar tableaus[3]. This condition is denoted as **deterministic similarity (DS)**.   (see also Definition 4.3.) Under DS we can reconstruct the perturbed event sequences E_i's or the perturbed trajectory $x(t; \theta+d\theta, \xi)$ from $x(t; \theta, \xi)$ easily. Section 3.2 below discusses one such algorithm explicitly. Furthermore, all the arguments so far used are completely non-stochastic,   nothing is said about the relationship of Eq.(1) to the derivative of the expected value of some performance measure. Intuitively, of course, we believe that if a sufficiently long sample path is observed the observed time average in   Eq.(1) may be related or equal to the expected value.

3.2 Single Class Queueing Networks

Now let us extend our idea to the classical closed queueing networks of the Jackson-Gordon-Newell type. These are closed queueing networks with K servers and N customers from a single class.   The servers have exponential service time distributions with mean service times $s_i = 1/\mu_i$. The customers are routed with constant routing frequency q_{ij}, which is the probability that a customer, upon completing service at server i, proceeds next to server j. First of all, such networks are particularly simple to simulate since in such networks no

---

[3] A fundamental assumption here is that no two events occur at the same time. They are separated by a finite time duration however small (see Appendix B Section 3 Assumption 1). For DEDSs with continuous random variables, this is not a serious restriction at all. However, for DEDSs with discrete random variables, such an assumption may not be warranted and can cause additional complications. See [Cao, Gong and Wardi 1990].

full outputs or blockings are present. More importantly, this provides an easy check of the validity of the perturbation analysis approach, since there exist explicit formulas of the derivative of throughput with respect to mean service time [William and Bhandiwad 1976]. During a simulation of such a network, we can consider a virtual perturbation of the mean service time of a particular server. Let $s_{i,k}$ be the kth customer's service time at server i. Then, $s_{i,k}$ can be obtained by using a uniform random variable $\xi_{i,k}$ as follows:

$$s_{i,k} = - s_i \ln(1-\xi_{i,k}).$$

Suppose that the mean service time is changed from s_i to $s_i + \Delta s_i$; then, according to the perturbation generation rule of the previous section, a perturbation

$$\Delta s_{i,k} = - (s_i + \Delta s_i)\ln(1- \xi_{i,k}) - [-s_i \ln(1- \xi_{i,k})] = s_{i,k}\frac{\Delta s_i}{s_i} \qquad (3)$$

will be generated during the kth customer's service period Note that this method of generating $\Delta s_{i,k}$ will result in a set of perturbations with the property that

$$\lim_{n \to \infty} \frac{1}{n}\sum_{k=1}^{n} (s_{i,k} + \Delta s_{i,k}) = s_i + \Delta s_i.$$

Exercise Suppose instead of the above method of generating $\Delta s_{i,k}$, we simply add a perturbation of Δs_i to each sample $s_{i,k}$ with the result that the samples will also have a mean equal to $s_i + \Delta s_i$. Is there anything wrong with this?

The following algorithm implements the perturbation generation and propagation rules and calculates the elasticity of the average throughput, TP, with respect to mean service time, s_i, $(s_i/TP)(\partial TP/\partial s_i)$ (i.e., the % change in TP for 1% change in s_i).

Step 0   Initialize all counters and accumulators (one accumulator per server).

Step 1   Each time server i is activated, add the sample service time, $s_{i,k}$, to an accumulator A_i.

Step 2 If, on the sample trajectory, server k terminates a No Input (NI) at server j, copy the content of A_k into A_j.

Step 3 At the end of the simulation/experiment, let $A* = \min (A_1, A_2, \ldots, A_K)$, $A^* = \max(A_1, A_2, \ldots, A_K)$ and let T be the total elapsed time of the experiment/simulation to serve C customers. Then

$$\frac{s_i}{TP} \frac{\partial TP}{\partial s_i} \approx - \frac{A*}{T - (A^* - A*)}, \tag{4}$$

where the approximation in Eq.(4) improves as the duration of the experiment increases.

We now provide an explanation of Eq.(4). According to Eq.(3), the perturbation generated in the kth customer's service period of server i because of the change in the mean service time, Δs_i, is proportional to $s_{i,k}$. Step 1 of the algorithm simply attempts to accumulate a number, A_i, which is proportional to the perturbations generated with a factor of $\Delta s_i / s_i$. (This factor is cancelled in the final calculation, so it does not appear in Eq.(4).) The number, in fact, is the sum of $s_{i,k}$ over all the service durations during which a perturbation is introduced. At any time, the difference between A_i and A_j represents the relative perturbations between the event times of the sequences E_i and E_j with respect to their nominal tableau. When an NI occurs at server j, which is terminated by server i, the perturbations accumulated at i are propagated to j. This explains Step 2. At the end of the experiment, $A*$ is a number proportional to the perturbations realized by all the servers (hence, by the network), i.e.,

$$\sum_{k \in R} s_{i,k} = A*,$$

where R represents the set of introduced perturbations that have been realized by the network. $A^* - A*$ corresponds to the sum of perturbations that have not been (but will be) realized or lost by the system at the end of the experiment. Thus, if we remove these intervals and take

$$T' = T - (A^* - A*)$$

as the effective total time duration of the experiment, then $A_*/[T-(A^*-A_*)]$ is more accurate than A_*/T for the proportion of the realized perturbations. For a long-duration experiment, the change in the termination time of the experiment due to Δs_i in the mean service time s_i is then

$$\Delta T' = A_* \frac{\Delta s_i}{s_i}.$$

From this, we have

$$\frac{s_i}{T'} \frac{\Delta T'}{\Delta s_i} = \frac{A_*}{T'}.$$

But, since $s_i = 1/\mu_i$, and $TP = C/T'$ where C is the total number of customers served in the time duration T', we have

$$\frac{s_i}{T'} \frac{\Delta T'}{\Delta s_i} \approx -\frac{\mu_i}{TP} \frac{\Delta TP}{\Delta \mu_i} \approx \frac{A_*}{T-(A^*-A_*)}. \tag{5}$$

On the other hand, $(\mu_i/TP)(dTP/d\mu_i)$ is given by the formula [William and Bhandiwad 1976]

$$\frac{\mu_i}{TP} \frac{\partial TP}{\partial \mu_i} = Q_N(i) - Q_{N-1}(i), \tag{6}$$

where $Q_N(i)$ is defined as the average queue length of server i when there are N customers in the network. Thus, we may directly verify our approach by numerically comparing the calculation using Eqs.(4) - (6). Numerical results are reported in [Ho and Cao 1983], which indicate that the estimates obtained from the IPA algorithm are very accurate.

One cannot emphasize too strongly the simplicity of the above IPA algorithm. Step 2, which is the heart of the PA process, can be carried out with minimal extra coding whether we are programming a simulation or modifying a real time data collection system (see Appendix E for an explicit implementation). The only requirement is that we must be able to perform step 2 *at the time* of the occurrence of the event, NI termination. Step 2 does not depend on the particular exponential distribution we have assumed. Only step 1 is affected by it. To illustrate the ease with which a sample-path based technique like PA can be modified to take care of other classes of problems, let us assume the service time distribution of the ith server is uniform in $[0, 2S_i)$ and one or more buffers is less than a finite number N^4. This will immediately turn the problem into one for which no analytical solutions are known, much less closed form derivative information for the TP. However, to accommodate these changes, we only need to modify the above IPA algorithm ever so slightly by adding the words "or Full Output (FO)" to the words "... terminates a No Input (NI)" in Step 2.

Exercise Convince yourself this modification is correct in the spirit of this section.

3.3   A GSMP Formulation of IPA

The informal description of the IPA algorithm above can be formalized using a GSMP framework (see Appendix B)[5]. It is instructive to look at the formalization. We now present a development due to [Glasserman 1990c]. We define

τ_n =   the epoch of the nth state transition

a_n =   the nth event

x_n =   the nth state visited by the process: $x_n = x(\tau_n^+)$

c_n =   the vector of clock readings at τ_n^+

$c_n(\alpha)$ =   at τ_n, the time remaining until α occurs, provided $\alpha \in \Gamma(x_n)$, the feasible event list associated with x_n

---

[4] With finite buffer size, blocking of one server by another can occur. However, for simplicity, let us assume every finite buffer in the network can only be fed by a single source, i.e., simultaneous blocking of two servers by one queue is ruled out. The reason for this restriction will be clear in Chapters 4 and 5.

[5] For readers who are not interested in mastering a great deal of notations, the text below to the end of the section can be skipped on a first reading.

$N(\alpha,n)$ = number of instances of α among $a_1, \ldots, a_n$ [6]

Also let $\{Y(\alpha,k), \alpha \in \Gamma, k=1,2, ..\}$ and $\{U(\alpha,k), \alpha \in \Gamma, k=1,2, ..\}$ be two doubly indexed sequences of independent random variables. For each k, $Y(\alpha,k)$ is distributed according to ϕ_α and represents the kth clock sample for α. The *routing indicator*, $U(\alpha,k)$, is uniformly distributed on $[0,1)$ and will be used to determine the state transition at the kth occurrence of α. State transitions are determined by a mapping ψ: $X \times \Gamma \times [0,1) \to X$: if α occurs in state x with routing indicator $u=U(\alpha,k)$, then $x(t)$ jumps to $x'=\psi(x,\alpha,u)$. The requirement for ψ is that for all x,α,x'

$$\text{Prob}(\psi(x,\alpha,u)=x') = p(x';x,\alpha) \qquad (7)$$

whenever u is uniformly distributed on $[0,1)$. It is straightforward to show that with these definitions, the evolution of the trajectory $x(t)$ of the GSMP is governed by

$$\tau_{n+1} = \tau_n + \min\{c_n(\alpha): \alpha \in \Gamma(x_n)\} \qquad (8)$$

$$a_{n+1} = \text{Arg}\{\alpha \in \Gamma(x_n): c_n(\alpha) = \min\{c_n(\alpha'): \alpha' \in \Gamma(x_n)\}\} \qquad (9)$$

$$N(\alpha,n+1) = N(\alpha,n)+1, \qquad \alpha=a_{n+1}$$

$$= N(\alpha,n), \qquad\qquad \text{otherwise} \qquad (10)$$

$$x_{n+1} = \psi(x_n, a_{n+1}, U(a_{n+1}, N(a_{n+1},n+1))) \qquad (11)$$

At each state transition the clock readings are adjusted by setting clocks for any new events and reducing the time left on any surviving "old" clocks by the time since the last transition, i.e., the duration of the state x_n or the triggering event's life time. Thus,

$$c_{n+1}(\alpha) = c_n(\alpha) - (\tau_{n+1} - \tau_n) \quad \text{if } \alpha \in \Gamma(x_n) \text{ and } \alpha \neq a_{n+1} \qquad (12)$$

$$c_{n+1}(\alpha) = Y(\alpha, N(\alpha,n+1)+1) \quad \text{if } \alpha \in \Gamma(x_{n+1}) \text{ and}$$

$$\text{either } \alpha \notin \Gamma(x_n) \text{ or } \alpha=a_{n+1} \qquad (13)$$

---

[6] By convention, an event can be scheduled but does not occur until its clock reading runs down to zero. Thus $N(a,n)$ does not include those α which have been scheduled but are still alive.

The sample path x(t) is finally generated by setting $x(t)=x_n$ on $[\tau_n,\tau_{n+1})$.

Exercise Using a typical busy period of a G/G/1 queue, illustrate all the variables in Eqs.(8 - 13).

To define IPA estimates associated with x(t) we further introduce $T(\alpha,k)$ to be the epoch of the kth occurrence of event α, i.e., $T(\alpha,k)$ is equal to τ_n* where

$$n* = \min\ \{n{\geq}0: N(\alpha, n) = k\}.$$

If the event α does not occur k times in any finite time period, then $T(\alpha,k)=\infty$. Every τ_n is equal to some $T(\alpha,k)$ - in particular, with $\alpha = a_n$ and $k = N(a_n, n)$. Thus, if we define $r_n \equiv (a_n, N(a_n, n))$, we may write $\tau_n=T(r_n)$. We call r_n defined in this way the sequence of *event-order pairs*. Next, we introduce the notions of the *triggering sequence* for (α, k) and the *triggering indicator* $\eta(\alpha,k; \beta,j)$, which formalize the notion of "critical timing path" or "event schedule" defined in the previous section. Note that the clock that runs out and triggers the jump at τ_{n+1} was set at some earlier transition τ_m, $m{\leq}n$, with some clock sample $Y(\beta,j)$. In particular, with $\beta=a_m$ and $j=N(a_m, m)$ we have $\tau_n=\tau_m+Y(\beta,j)$. Tracing backward this way from any τ_n, we can find a sequence of events and indices $(\beta_1,j_1), \ldots, (\beta_{m_n},j_{m_n})$ such that

$$\tau_n = Y(\beta_1,j_1)+ \ldots +Y(\beta_{m_n},j_{m_n}) \tag{14}$$

with $j_1=1$. Now we define the triggering indicators as follows: $\eta(\alpha,k ; \beta,j)$ is equal to zero except

(i)   for every α and k, $\eta(\alpha,k;\alpha,k)=1$,

(ii) if the kth clock for α is set at the jth occurrence of β, then $\eta(\alpha,k;\beta,j)=1$,

(iii) if $\eta(\alpha,k;\beta,j)=1$ and $\eta(\beta,j;\beta',j')=1$, then $\eta(\alpha,k;\beta',j')=1$.

An immediate consequence of the above is that for every $\alpha{\in}\Gamma$ and every $k>0$, if $T(\alpha,k)<\infty$, then

$$T(\alpha,k) = \sum_{\beta,j} Y(\beta,j)\eta(\alpha,k;\beta,j) \tag{15}$$

and

$$\tau_n = \sum_{i=1}^{n} Y(r_i)\eta(r_n;r_i). \tag{16}$$

From these equations, for any finite $T(\alpha,k)$, we have

$$\frac{d\,T(\alpha,k)}{d\theta} = \sum_{\beta,j}\frac{dY(\beta,j)}{d\theta}\eta(\alpha,k;\beta,j) \quad \text{and} \quad \frac{d\tau_n}{d\theta} = \sum_{i=1}^{n}\frac{dY(r_i)}{d\theta}\eta(r_n;r_i). \tag{17}$$

Sample performance derivatives can then be related to the derivative of τ_n according to the simple calculus of $dL/d\tau_n$. Eq.(17) is given in the batch calculation form. If a simulation or a real sample path of $x(t)$ is available, e.g., from realizing the recursions (8 - 13), then we can state a recursive form of the IPA algorithm by associating an accumulator A_α for every event $\alpha \in \Gamma$ and adding to Eqs.(8 - 13) the following:

At τ_{n+1} set

$$k_{n+1} := N(a_{n+1},n+1); \tag{18}$$

$$A_{a_{n+1}} := A_{a_{n+1}} + dY(a_{n+1}, k_{n+1})/d\theta \tag{19}$$

for every $\beta \in \Gamma(x_{n+1})\text{-}\Gamma(x_n)$, $A\beta := A_{a_{n+1}}$

$d\tau_i/d\theta$ is just the content of A_{a_i} at τ_i^+. Eq.(18) corresponds to the perturbation generation: a perturbation $[dY(a_{n+1}, k_{n+1})/d\theta]\Delta\theta$ is generated during the event lifetime. Eq.(19) describes the perturbation propagation: if an event β is scheduled by an event a_{n+1}, then the perturbation of a_{n+1} is propagated to the event β. It is worth emphasizing that the algorithm of Eqs.(18 - 19), that of Eqs.(3 - 4), and that of Appendix E are different embodiments of basically the same thing.

3.4 Realization Ratios

An old fashioned exhibit, called the great probability machine, is often found in science museums.   The device consists of a vertical wood board with neatly spaced rows of nails anchored at each imaginary rectangular grid point on the board. Small marbles are released at the top center of the board. As they fall down and hit the nails by chance, they are diverted randomly either to the left or to the right at each encounter with a nail. At the end, each marble falls into one of the many equally spaced slots at the bottom of the board. The path taken by each marble is not predictable.   But the ensemble of marbles eventually forms a Gaussian histogram at the bottom of the board convincingly demonstrating the idea of statistical determinism. A similar visualization can be made with the tableau   diagrams of the perturbation   generation and propagation process of Figs. 1 and 2. We can imagine each perturbation generation as the creation of a ball which rolls along the event sequence E_i to the right in the direction of increasing time.   Whenever one of these balls reaches a service completion which terminates an NI or an FO of another event sequence E_j, an additional ball is "created" on E which will also continue to roll to the right. On the other hand, if these balls encounter an NI or an FO along its own event sequence, then they are simply stopped or annihilated. As the sample path evolves in time, this mental picture of an ensemble of balls being generated, created, and annihilated as NIs and FOs occur corresponds to the processes of perturbation generation and propagation described in Section 3.1. If we measure the performance of the system at a finite time at a particular server, k, then when a ball reaches E_k it will affect the performance. It is intuitively reasonable to conjecture that in the limit of a long sample path, a given percentage of the balls generated along E_i will eventually reach E_K. This percentage can be called the "realization ratio" of the network. Pictorially, this is illustrated in Fig. 5. This concept of the "realization ratio" is a probabilistic notion similar to the great probability machine described earlier. In other words, in statistical equilibrium, we expect "realization ratio" to be a deterministic number for each server generating perturbations.   Through this realization ratio we can obviously evaluate the impact of the parameter perturbation on the steady-state system performance. We shall formalize this notion of "realization ratio" in Chapter 4.

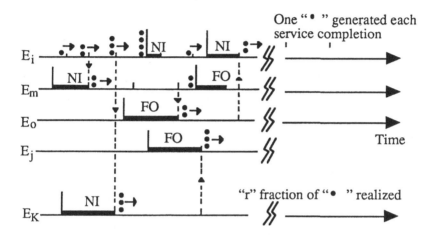

Fig. 5 Marble Ball Analogy of Perturbation Generation (at Server i) and Propagation (to K)

3.5 The GI/G/1 Queue

In Sections 3.1 - 3.4 we are interested primarily in conveying the basic conceptual and intuitive idea behind the perturbation analysis of queueing networks or general DEDS and its potential general applicability. In this section, we plan to develop more algorithmically the PA for a particularly simple DEDS, i.e., a queue-server combination with general arrival and service distributions. Let us assume that the system is in steady state and denote

$$J(\theta,\lambda) \equiv E[T(\theta,\lambda,\xi)] \equiv \text{mean system time for a customer}$$

where λ and θ are parameters of the arrival and service distribution, respectively. If we further define M as the number of busy periods, N as the total number of customers served along the sample path, and ST_i as the system time of the ith customer, then

$$J(\theta,\lambda) \equiv \lim_{M\to\infty} T(\theta,\lambda,\xi,M) \equiv \lim_{N\to\infty} \frac{1}{N}\sum_{i=1}^{N} ST_i, \qquad (20)$$

where $T(\lambda,\theta,\xi,M)$ is the average system time of a customer over M busy periods or N total customers on a sample path. Now we wish to develop a formula for $dJ/d\theta$. To do this, we shall first develop some relationships between θ and the service time samples, s_i, $i=1, 2, \ldots$, of the server. Recall that the perturbation of each service time can be generated according to the formula

$$\Delta s = - \left\{ \frac{\partial F/\partial \theta}{\partial F/\partial s} \right\}_{(s,\,\theta)} \Delta\theta.$$

We now investigate how a given perturbation Δs_i in the service time s_i affects the mean system time $T(\theta,\lambda)$. Fig. 6 illustrates the situation in a typical busy period of the GI/G/1 queue. In the figure, a_i and d_i denote the ith arrival and departure, respectively.

We have tacitly assumed the case of $\Delta s_i > 0$ ($\Delta s_i < 0$). Under the deterministic similarity assumption, no two busy periods will coalesce into one (or no busy period will split into two), and the order of all events remains the same after perturbations are introduced to the sample path. We assume that the function F is smooth enough (the probability density function does not contain atoms) so that we can always choose $\Delta\theta$ small enough such that deterministic similarity holds. Now let C_{k+i} be the ith customer in a busy period, with C_{k+1} being the first customer in the busy period. In the perturbed path,

additional time for $C_{k+i} = \sum_{j=1}^{i} \Delta s_{k+j}.$

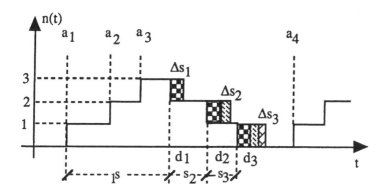

Fig. 6 A Busy Period in a GI/G/1 Queue

Suppose there are M busy periods observed during a total time t, and n_m is the number of customers served in the mth busy period, then to get the effect of $\Delta\theta$ on T as in Eq.(20), we have

$$\Delta T = \frac{1}{N}\sum_{m=1}^{M}\sum_{i=1}^{n_m}\sum_{j=1}^{i}\Delta s_{k_m+j} = \frac{1}{N}\sum_{m=1}^{M}\sum_{i=1}^{n_m}\sum_{j=1}^{i}\frac{ds_{k_m+j}}{d\theta}\Delta\theta \qquad (21)$$

where $k_m \equiv n_1 + \ldots + n_m$ is the number of customers served in the first m busy periods and $N = n_1 + \ldots + n_M \equiv k_M$. Note from the generation formula everything on the right-hand side of (21) can be calculated using knowledge of the sample path. The computation is extremely simple. Dividing Eq.(21) by $\Delta\theta$, we get an estimate of $dT/d\theta$. In Chapter 4, we shall show that this estimate in fact converges to the correct analytically known formula in the case of the M/G/1 queue. For more details as well as empirical result on Eq.(21), see [Suri and Zazanis 1988].

Eq.(21) requires the computation of $ds/d\theta$. As mentioned in Examples 1 - 4 earlier in Section 3.1, for a class of random variables, the computation is in fact very simple. A random variable, S, is said to be parameterized by a location parameter, θ, if $S-\theta$ is independent of θ. In terms of probability density function, $p(S)=f(S-\theta)$ where $f(S)$ is an arbitrary density function. Similarly,

S is said to be parameterized by a scale parameter if S/θ is independent of θ or $p(S)=f(S/\theta)$. It is directly verified that $(S=F^{-1}(\xi, \theta)$ as shown in Fig. 4)

$$\frac{dS}{d\theta} = \text{constant}, \quad \frac{d^2S}{d\theta^2} = 0$$

and

$$\frac{dE[S]}{d\theta} = E\frac{dS}{d\theta} \text{ and } \frac{dE[S^2]}{d\theta} = 2E\left[S\frac{dS}{d\theta}\right] = 2E(S) \qquad (22)$$

for location parameterized random variables, and

$$\frac{dS}{d\theta} = \frac{S}{\theta}$$

and

$$\frac{dE[S]}{d\theta} = E\left[\frac{dS}{d\theta}\right] = \frac{E[S]}{\theta}, \quad \frac{dE[S^2]}{d\theta} = 2E\left[S\frac{dS}{d\theta}\right] = \frac{2E[S^2]}{\theta} \qquad (23)$$

for scale parameterized ones.

Exercise Use Eqs.(21-23) to verify directly the known formula for $dE[T]/d\theta$ for an M/M/1 queue for the case of θ =mean of service time distribution (see Section 4.4).

Exercise Derive the analog of Eq.(21) for case of θ =mean of arrival distribution.

Exercise Consider a cyclic queueing network consisting of two servers and one customer. The service distribution of the two servers are exponentially distributed with means s_1 and s_2, respectively. What is the PA estimate of the elasticity of the system throughput TP with respect to s_1, $(s_1/TP)(\partial TP/\partial s_1)$?

The developments in Section 3.2 and this section suggest that the method of IPA is basically distribution independent. We shall now be more explicit and demonstrate some robustness property of IPA with respect to the lack of knowledge of the specific distributions of a server. In particular, IPA turns out to be robust with respect to the unknown probability distributions of the arrival and service times in a class of distributions governed by a scale parameter.

As an example, consider the Erlang family of probability distributions given by

$$F_{er}(x;\lambda,R) = 1 - e^{-R\lambda x} \sum_{n=0}^{R-1} \frac{(R\lambda x)^n}{n!}, \quad 0 \leq x \tag{24}$$

It is well-known that this probability distribution characterizes a random variable x which describes the time spent in a service facility consisting of R identical "stages" (see Appendix A.2). Each stage in turn represents a server with an exponentially distributed service time with a rate $R\lambda$. The Erlang family is a rich and useful two parameter class of distributions, since it covers a broad range from R=1, when x is exponentially distributed, to R $\rightarrow\infty$, when x approaches the constant $1/\lambda$ with probability one. The mean and variance of $F_{er}(x; \lambda,R)$ are given by $E[x]=1/\lambda$ and $Var[x]=1/R\lambda^2$. Now in order to utilize Eq.(21), we need to evaluate $ds/d\theta$ for $s\equiv x$ and $\theta=\lambda$. Using the generation formula, we find

$$\frac{\partial F_{er}}{\partial x} = \frac{\partial F_{er}}{\partial(\lambda x)} \frac{\partial(\lambda x)}{\partial x}, \quad \frac{\partial F_{er}}{\partial \lambda} = \frac{\partial F_{er}}{\partial(\lambda x)} \frac{\partial(\lambda x)}{\partial \lambda} \Longrightarrow \frac{\partial x}{\partial \lambda} = -\frac{x}{\lambda} \tag{25}$$

which is independent of "R." In fact, λ is a scale parameter of the Erlang distribution. Consequently, regardless of the actual number of stages of the Erlang distribution, for the purpose of the PA estimation we could simply use Eq.(25) which is just the same as that for the exponential distribution with the same mean. Furthermore, for any distributions characterized by a scale parameter, we know from Eq.(23) that

$$\Delta x = \frac{\Delta\theta}{\theta} x \tag{26}$$

then we can generate, for every sample of x, a Δx which is a pre-specified fraction, $\Delta\theta/\theta$, of x. No knowledge of the value of θ is required. Using finite perturbation analysis (FPA)[7], we can propagate such perturbations and analyze their effect on system performance. This is often useful in real world experimental situations.

Finally, if we erroneously assumed the distribution is characterized by a scale parameter when in fact it is not, i.e., we use $(s_j/\theta)\Delta\theta$ instead of $(\partial s_j/\partial\theta)\Delta\theta$ as the perturbation generated, then the error is given by

$$\text{Error}|_{\text{IPA}} = \frac{1}{N}\sum_{m=1}^{M}\sum_{i=1}^{n_m}\sum_{j=1}^{i}\left(\frac{\partial s_j}{\partial\theta} - \frac{s_j}{\theta}\right) \tag{27}$$

But any sample service time satisfies $E[\partial s/\partial\theta] = \partial E[s]/\partial\theta = 1$, which implies $E[\partial s/\partial\theta - s/\theta] = 0$. Consequently, the expectation of each term of the triple sum in Eq.(27) is always zero. And it is not unreasonable to expect that the error of IPA and the scale parameter assumption will be small particularly when n_m, the number of customers served in busy periods is small so that no particular term is repeated often in the triple sum. For experimental data and further analysis of the robustness of IPA as compared with the "likelihood ratio method" for derivative estimation, see [Cassandras, Gong, and Lee 1991] and Chapter 7.

3.6 Load-Dependent Servers and IPA

Beyond the general service time distribution, the next order of complications arises when the service time at a server is dependent on the load, i.e., the number of customers in the queue. Load-dependent servers arise naturally both in the study of queueing networks as well as in the course of aggregating a complex network (Appendix A.3.3 and Chapter 8). The usual specification of

---

[7] See Chapters 6 and 7.

such a server is to define a modifying coefficient, $a_i(n_i)$, depending on the load, for the mean service rate, μ_i, i.e.,

The mean service rate $\equiv a_i(n_i)\,\mu_i$

where n_i is the number of customers (waiting and being served) at server i. More generally, we can simply write $a_i(n_i)\mu_i$ as $\mu_i(n_i)$. Such a situation arises naturally when we view an M/M/m server as an M/M*/1 server, where M* denotes an exponential server with $a(n)=n\mu$ for $n\leq m$ and $a(n)=m\mu$ for $n>m$, and μ is the mean service rate of the server in the original M/M/m system. Another situation where load dependent servers arise are with problems of aggregation of queueing networks (see Ch.8). Our primary interest in this section is to modify the IPA algorithm described above in Section 3.2 to accommodate load-dependent servers. We do his both as an extension of IPA and as a preparation for later chapters when the results of this section are needed.

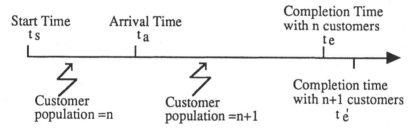

Fig. 7 A Typical Service Period of a Load-Dependent Server

The major difference between the perturbation analysis in the case of load-dependent servers and that of Sections 3.2 is the fact that whenever a customer goes from one server to another, it can propagate the perturbations of one server to the other regardless of whether or not the servers are blocked or idle (as in the case of Section 3.2). This is because the arrival of a customer to a server changes the load at the server, and hence its mean service rate. In other words, the coupling among servers is much more immediate in the load dependent case. Fig. 7 illustrates a typical case. Let

$$r = \frac{\text{mean service rate with load } n}{\text{mean service rate with load } n+1} = \frac{\mu_i(n)}{\mu_i(n+1)},$$

$$t'_e = t_a + r(t_e - t_a), \tag{28}$$

and

$$dt'_e = (1-r)dt_a + r dt_e + dr(t_e - t_a) \tag{29}$$

which shows how perturbations in the arrival time of a customer, dt_a, or in the scheduled completion of the server with customer load n, dt_e, or in the mean load-dependent service rate, dr, can affect the completion time of the server with n+1 customer load, dt'_e. Otherwise, there is no change in the perturbation analysis algorithm described in Section 3.3.

3.7 Remarks About PA

In a narrow sense, perturbation analysis (PA) is a computational technique that calculates the sensitivities of performance measures of a DEDS with respect to system parameters by analyzing a single sample path. PA in effect is a method of reconstructing the perturbed performance value from the nominal (original) experiment or sample path of a DEDS without the need of actually carrying out the perturbed experiment. This point deserves emphasis. PA is simply an analytical means to process information inherent in a sample path of an experiment. It makes no difference whether or not the experiment deals with steady-state or transient performance. To the extent PA is applicable to the particular experiment, it can be used for the determination of either steady state or transient performance gradients or sensitivities.

The important thing to point out is the ease with which we can keep track of the propagation of any perturbation on the system trajectory as it evolves. Since the tableau representation is general, perturbation analysis is thus in principle distribution-free and applicable to any system for which we can simulate or display a tableau. When this perturbation analysis is used to predict the system performance, we need to distinguish among three kinds of predictions.

The first kind, called the deterministic prediction, is the original basis of PA. We ask the question: What if we repeat this simulation or experiment or sample path "exactly," except for this small perturbation? The implied assumption here is that the nominal and perturbed paths of the system will be essentially similar, except for the duration of certain event intervals due to the perturbation. In particular, the assumption is that the order of occurrence of all the events in an event sequence for both the nominal and the perturbed tableaus (paths) remains invariant. Perturbation analysis will answer the "what if" question correctly if this assumption of "deterministic similarity" of the sample paths is valid. The second kind of prediction using PA is statistical in nature and usually represents the more practical questions asked. A parameter change (e.g., the change in the me an service time of a server) induces an ensemble of perturbations (each service activation) in a tableau. We wish to predict the average number of such perturbations actually realized at (i.e., propagated to) the output server. Note that, in this case, we are not necessarily interested in repeating the experiment or the sample path exactly. We use the sensitivity derived from one nominal path to predict results from an entirely different sample path (in other words, a new simulation with a new random seed). An implied assumption here is that the two sample paths or tableaus are "stochastically similar" or the nominal tableau is statistically representative. The intuitive idea behind this assumption is clear and is the basis for all simulation experiments. In any tableau representing sufficiently long operating history of the network, we expect the FOs and NIs to be distributed essentially in a random manner. Two equally long tableaus should have the same distribution of FOs and NIs and, hence, the same realization percentage of perturbations (the marble analog and the notion of "realization ratio" in Fig. 5). This assumption of "stochastic similarity" is behind all applications of Monte Carlo experiments. Our approach cannot escape it either. Many more or less standard techniques to ensure the "statistical represen-tativeness" of a sample tableau exist. Peculiar to PA, we have also the fundamental question of "is the expectation or limit of the sample PA 'derivatives' equal to the derivative of the expectation or limit" mentioned in Chapter 2. PA computes the former while we are interested in the latter. This question will be treated in great detail in Chapters 4 and 5. Lastly, the third kind of prediction is related to optimization. When we use many sensitivities (e.g., the sensitivities of the average throughput to the mean service times of all servers) to estimate gradient in any kind of optimization algorithm, we are assuming a certain amount of "linearity." In other words, if an increase of $\Delta_i\%$ in the mean service rate μ_i, $i = 1, 2$, causes an increase of $\varepsilon_i\%$ in the average throughput, respectively, we postulate that a simultaneous increase of $\Delta_1\%$ and $\Delta_2\%$ in μ_1 and μ_2 will result in an increase of $(\varepsilon_1+\varepsilon_2)\%$ in the average throughput provided the perturbations are sufficiently small. Whether or not this kind of superposition holds, of course, depends on the particular system. Some systems or response surfaces are more linear than others. To predict the "degree of linearity" of a queueing network or a DEDS requires knowledge about the response curve, or the surface $J(\theta)\equiv E[L(\theta,\xi)]$, which is unknown. Note also that this problem of the "degree of linearity" of a system is not peculiar to the particular method of the PA approach. It is an inherent issue that must be addressed by all optimization problems.

Finally, it is worthwhile to emphasize the mind-set behind PA. Because of the man-made nature of DEDS and queueing networks, the possible different kinds of systems that one may encounter in DEDS study is literally infinite. PA has only been developed so far to deal explicitly with a small fraction of such systems. However, we submit PA is nevertheless a general approach for the time domain study of these systems. Succeeding chapters will develop more elaborate algorithms to illustrate this mind-set.

CHAPTER FOUR

Foundations of Perturbation Analysis

The infinitesimal perturbation generation and propagation rules were explained in the last chapter. Following these rules, we can obtain a perturbed sample path by analyzing a nominal sample path and then obtain the performance sensitivity. In this chapter, we shall formulate and study the problem in a more formal manner. Our purpose is to provide a mathematical foundation for IPA. To this end, we first choose the closed Jackson network and the GI/G/1 queue as the basic cases for the discussion. The same principle is then applied to the networks with general service time distributions, for which the so-called product-form solution does not exist. The important point is that the principle illustrated in this chapter applies to many other DEDS[1]. The process of establishing these results will also illustrate the limitation of the infinitesimal perturbation analysis; the limitation leads to the extensions of IPA, which will be discussed in the next several chapters. For technical integrity, this chapter contains more details, many of which can be passed over in a first reading, than other chapters. This chapter is not crucial for the understanding of the other chapters. However, we shall, wherever appropriate, provide a "forest view" for readers not interested in the technical details.

There are two fundamental issues regarding the IPA estimates. One is related to the transient behavior, or the IPA estimate in a finite period; the other to the steady-state quantity, or the convergence property of the IPA estimate as the length of the sample path goes to infinity. In mathematical language, these two issues can be stated as two questions: (1) Is the estimate obtained by

---

[1]PA is a sample-path based approach and, as such, has the potential of being distribution-free.

applying the IPA algorithm to a sample path with a finite length an unbiased estimate of the derivative of the expected value of the performance measure in the same finite period? (2) Does the IPA estimate converge to the derivative of the steady-state performance with probability one (w.p.1) as the length of the sample path goes to infinity? (equivalently, is the IPA estimate a strongly consistent estimate (see Appendix A) of the derivative of the steady-state performance?) These two questions for closed Jackson networks are addressed in Sections 4.1, 4.2 and Sections 4.2, 4.3, respectively.

We start with a discussion on the sensitivity analysis of a general stochastic system with a parameter θ in Section 4.1. We first propose a formal description of a DEDS using a random vector ξ (or a sequence of random variables). Next, we define the "sample derivative" in which the common random vector ξ is used for both systems with parameters θ and $\theta+\Delta\theta$. The perturbation analysis sensitivity estimate can be formally described as the sample derivative. The issue of the unbiasedness of the perturbation analysis estimate becomes that of the interchangeability of the expectation and the differentiation. Sufficient conditions are then developed for the interchangeability in a stochastic environment. The study of the IPA for closed Jackson networks is in Sections 4.2 and 4.3. In Section 4.2, we first give a set of very simple equations formulating the perturbation generation and propagation rules for the closed Jackson networks. Using these equations, we show that the sample performance function (the performance as a function of θ for a fixed ξ, to be defined) of the average time required for one service completion is a continuous and piecewise linear function of the mean service times. Applying the sufficient condition developed in Section 4.1, we prove the unbiasedness of the IPA estimate in a finite period, which corresponds to the transient behavior of the system. Based on these results, the steady-state performance is further studied. We prove that the IPA estimate of the sensitivity of the system throughput is an asymptotically unbiased estimate of the sensitivity of the steady-state throughput. (In other words, the IPA estimate converges in mean to the sensitivity of the steady-state throughput as the length of the sample path goes to infinity.) In Section 4.3, the concept called the realization probability first introduced in Section 3.3 is formalized. A set of equations specifying the realization probabilities is derived. Using realization probability, we show that as the length of a sample path goes to infinity the IPA estimate of the sensitivity of the system throughput based on a sample path converges with probability one to the sensitivity of the steady-state throughput of the system which, in fact, equals the expected value of the realization probability. The concept of realization probability provides a new theoretical method of calculating the performance sensitivity. The results in Sections 4.2 and 4.3 justify the application of IPA and provide a clear insight to IPA. In Section 4.4, we extend these results to closed networks with general service time distributions. We define a perturbation generation function which specifies the perturbation generated in an infinitesimal interval. The equations for realization probabilities become differential equa-tions, and the sensitivity of the steady-state throughput of the system can be expressed as the expected value of the product of the generation function and the realization probability. In Section 4.5, we study the M/G/1 queue and the GI/G/1 queue. The purpose of this section is to introduce two different approaches for proving the strong consistency of the IPA estimates. In the case of the M/G/1 queue, the strong consistency is validated by computing directly the limiting value of the IPA estimate and comparing it with the derivative obtained by existing queueing theory formulas. In the case of the GI/G/1 queue, the result is established by using the convexity of the sample performance function. Finally, in Section 4.6 we prove some technical details Since one picture is worth a thousand words, we let the following diagram to summarize the results of this chapter.

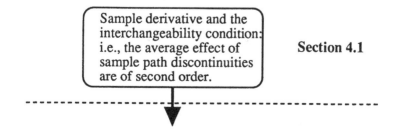

Sample derivative and the interchangeability condition: i.e., the average effect of sample path discontinuities are of second order.

Section 4.1

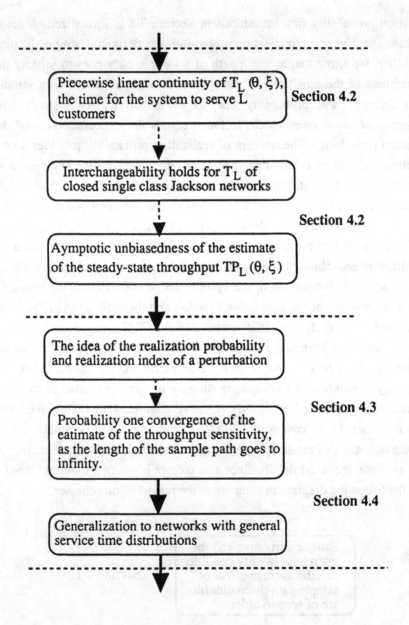

Piecewise linear continuity of $T_L(\theta, \xi)$, the time for the system to serve L customers — **Section 4.2**

Interchangeability holds for T_L of closed single class Jackson networks

Section 4.2

Aymptotic unbiasedness of the estimate of the steady-state throughput $TP_L(\theta, \xi)$

The idea of the realization probability and realization index of a perturbation

Section 4.3

Probability one convergence of the eatimate of the throughput sensitivity, as the length of the sample path goes to infinity.

Section 4.4

Generalization to networks with general service time distributions

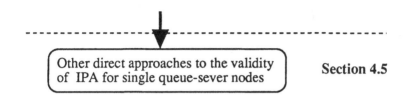

Other direct approaches to the validity
of IPA for single queue-sever nodes   **Section 4.5**

Three Different Approaches to IPA Validity:
Sample path continuity, realization probability,
and direct calculation

4.1 Sample Derivative and Interchangeability

In this section, we give a mathematical formulation of the perturbation analysis estimation problem. We start with a general description of a DEDS.

A DEDS can be represented by the pair (θ, ξ), where θ is a parameter of the DEDS and ξ is a random vector representing all the random effects involved in the system. Typically, the components of ξ are random variables u.i.i.d. (uniformly, independently, and identically distributed) on $[0, 1)$. ξ is defined on a probability space (Ω, F, P). That is, ξ is a measurable mapping $\Omega \to \Xi$, where $\Xi = [0,1)^{\infty}$ is an infinite dimensional hypercube. In a closed queueing network consisting of M servers, θ may be the mean service time of a server, denoted as s_i, and ξ is a random vector whose components contain the random variables $r_{i,k}$, i=1,2,...,M, k=1,2,..., that determine the customer service times of every server, and some other random variables that determine the destinations of customers and the initial state of the system. Each (θ,ξ) determines a sample path of the DEDS. A performance measure obtained from the path (θ,ξ) can be denoted by $L(\theta,\xi)$. For any realization of ξ, $L(\theta,\xi)$ is a function of θ. This function is called a *sample performance function,* or simply a *sample function* if there is no confusion.

The nominal and perturbed paths studied in perturbation analysis correspond to the same realization of the random vector and can be denoted as (θ,ξ) and $(\theta+\Delta\theta, \xi)$, respectively. The performance based on these two paths are denoted as $L(\theta,\xi)$ and $L(\theta+\Delta\theta, \xi)$, respectively. Thus, the derivative obtained by perturbation analysis is

$$\frac{\partial L(\theta, \xi)}{\partial\theta} = \lim_{\Delta\theta\to 0} \frac{L(\theta+\Delta\theta, \xi) - L(\theta, \xi)}{\Delta\theta}.$$

We call this derivative a *sample derivative*. In other words, the perturbation analysis estimate is in fact the sample derivative of the performance of a DEDS. For many queueing networks, this sample derivative can be obtained based on a single sample path. The concept of a sample derivative for a DEDS is first explained in [Cao 1985] and [Cao and Ho 1987b]. We assume that for the DEDS studied the sample derivative exists w.p.1. (i.e., for almost every (θ,ξ) the sample derivative exists.) This is true for most DEDS[2]. A proof of this statement for queueing networks is in the next section.

In most cases, we are interested in the expected value of the performance, $J(\theta)=E[L(\theta,\xi)]$. The sample derivative obtained by perturbation analysis, $\partial L(\theta,\xi)/\partial\theta$, is just an estimate of the derivative of the expected performance. This estimate is unbiased if and only if

$$E\{\frac{\partial L(\theta,\xi)}{\partial\theta}\} = \frac{\partial J(\theta)}{\partial\theta} = \frac{\partial}{\partial\theta}\{E[L(\theta,\xi)]\}, \tag{1}$$

i.e., if and only if the two operators "E" and "$\partial/\partial\theta$" are interchangeable [Cao 1985]. We shall called Eq.(1) as the interchangeability (IC) condition. Thus, in the formulation described above, the unbiasedness of the IPA estimate is equivalent to the interchangeability of "E" and "$\partial/\partial\theta$" for the sample function $L(\theta,\xi)$.

Eq.(1) requires that the function $L(\theta, \xi)$ satisfies some conditions. The most common one can be specified by using the Lebesgue dominated convergence theorem. However, for practical systems it is usually difficult to verify

---

[2] [Wardi 1989] and [Cao, Gong, Wardi 1990] discuss the case where sample derivatives are discontinuous due to the possibility of two events' occurring at the same time.

that the condition required by the Lebesgue theorem is satisfied. In the following, we shall provide some sufficient conditions for Eq.(1), which have an intuitive explanation. More explicit and easily checked conditions are further developed in Chapter 5.

First, we observe that the functions for which Eq.(1) does not hold are usually discontinuous with respect to θ. The following example explains the situation. Let $0<\theta<1$, $\xi \in [0,1)$ be a uniformly distributed random variable, and

$$L(\theta, \xi) = \begin{cases} 1 & \text{if } \xi \leq \theta, \\ 0 & \text{if } \xi > \theta. \end{cases} \tag{2}$$

Then

$$J(\theta) = E[L(\theta, \xi)] = \theta, \qquad \frac{\partial J}{\partial \theta} = 1.$$

However,

$$\frac{\partial}{\partial \theta}[L(\theta, \xi)] = 0, \qquad \text{w.p.1.}$$

Thus, Eq.(1) does not hold for this function. Note that for each realization of ξ, the sample function $L(\theta, \xi)$ jumps from 0 to 1 at $\theta = \xi$. As shown in this example, the sample derivative $[\partial L(\theta, \xi)]/\partial \theta$ at θ does not reflect the jumps of the function $L(\theta, \xi)$. On the other hand, the derivative of the expected value, $[\partial J(\theta)]/\partial \theta$, reflects the effect of the jumps in $[\theta, \theta + \Delta \theta]$.

Exercise   Let θ be any value in [0, 1). Now plot $L(\theta, \xi)$ for five or six randomly but uniformly chosen values of ξ. Graph the average value of $L(\theta, \xi)$.

Although most functions for which the interchangeability does not hold are discontinuous, the continuity of L is neither a sufficient nor a necessary condition for Eq.(1) to be valid. One of the sufficient condition for Eq.(1) is called the *uniform* (or *dominated*) *differentiability* defined below.

Assume that $\partial L(\theta,\xi)/\partial\theta$ exists at θ with probability one, then

$$L(\theta+\Delta\theta, \xi) - L(\theta,\xi) = \frac{\partial L(\theta,\xi)}{\partial\theta}\Delta\theta + r(\Delta\theta,\xi), \qquad \text{w.p.1,} \qquad (3)$$

and

$$\lim_{\Delta\theta\to 0}\frac{r(\Delta\theta, \xi)}{\Delta\theta} = 0, \qquad \text{w.p.1.} \tag{4}$$

Based on (3) and (4), we can introduce some new concepts of the differentiability:

Definition 4.1 A function $L(\theta,\xi) : R\times\Omega \to R$ is said to be *uniformly differentiable* w.p.1 on Ω at θ, if and only if for any $\varepsilon > 0$, there exists a $\delta > 0$ such that if $|\Delta\theta| < \delta$, then $|r(\Delta\theta,\xi)/\Delta\theta| < \varepsilon$ holds w.p.1 on Ω.

Definition 4.2 A function $L(\theta,\xi) : R\times\Omega \to R$ is said to be *dominantly differentiable* w.p.1 on Ω at θ, if and only if there exists an integrable function $f(\theta,\Delta\theta,\xi)$ and a $\delta > 0$ such that if $|\Delta\theta| < \delta$, then $|r(\Delta\theta,\xi)/\Delta\theta| < f(\theta,\Delta\theta,\xi)$ holds w.p.1 on Ω.

By the definitions of $\partial/\partial\theta$ and the expectation and Eq.(3), we have

$$\frac{\partial J(\theta)}{\partial\theta} = \frac{\partial E[L(\theta, \xi)]}{\partial\theta}$$

$$= \lim_{\Delta\theta\to 0}\frac{E[L(\theta+\Delta\theta,\xi)] - E[L(\theta,\xi)]}{\Delta\theta} = \lim_{\Delta\theta\to 0}E[\frac{\partial L(\theta,\xi)}{\partial\theta}] + \lim_{\Delta\theta\to 0}E[\frac{r(\Delta\theta,\xi)}{\Delta\theta}].$$

The first term on the most right-hand side does not depend on $\Delta\theta$. By the Lebesgue dominated convergence theorem and Eq.(4), if $L(\theta,\xi)$ is uniformly (or dominantly) differentiable, we have

$$\lim_{\Delta\theta\to 0}E[\frac{r(\Delta\theta,\xi)}{\Delta\theta}] = E\{\lim_{\Delta\theta\to 0}\frac{r(\Delta\theta,\xi)}{\Delta\theta}\} = 0 .$$

From the above two equations, we can obtain Eq.(1). Therefore, if $L(\theta,\xi)$ is uniformly (or dominated) differentiable at θ w.p.1, then the interchangeability of the expectation and differentiation holds.

Now we assume that the function $L(\theta,\xi)$ is discontinuous, and we give a technical analysis of the condition of Eq.(4). A typical example of such functions is one which can be decomposed into two parts:

$$L(\theta,\xi) = L_1(\theta,\xi) + L_2(\theta,\xi), \tag{5}$$

where $L_1(\theta,\xi)$ is a uniformly (or dominated) differentiable function and $L_2(\theta,\xi)$ is a piecewise constant function of θ. We assume that the probability that $L_2(\theta,\xi)$ jumps at any particular value of θ is zero and that the probability that $L_2(\theta,\xi)$ jumps in $[\theta, \theta+\Delta\theta]$ is, however, positive. Under this assumption, we have

$$\frac{\partial L_2(\theta, \xi)}{\partial \theta} = 0, \qquad \text{w.p.1.}$$

Thus,

$$\frac{\partial L(\theta, \xi)}{\partial \theta} = \frac{\partial L_1(\theta, \xi)}{\partial \theta}, \qquad \text{w.p.1,}$$

and

$$E\left\{\frac{\partial L(\theta, \xi)}{\partial \theta}\right\} = E\left\{\frac{\partial L_1(\theta, \xi)}{\partial \theta}\right\}.$$

Since $L_1(\theta, \xi)$ is uniformly (or dominated) differentiable, Eq.(1) holds for $L_1(\theta,\xi)$. Thus, we have

$$E\left\{\frac{\partial L(\theta, \xi)}{\partial \theta}\right\} = \frac{\partial E[L_1(\theta, \xi)]}{\partial \theta} = \frac{\partial E[L(\theta, \xi)]}{\partial \theta} - \frac{\partial E[L_2(\theta, \xi)]}{\partial \theta}.$$

From this, Eq.(1) holds for $L(\theta,\xi)$ if and only if

$$\frac{\partial E[L_2(\theta, \xi)]}{\partial \theta} = 0.$$

We shall derive some conditions for this equation. Let $h(\theta, \Delta\theta, \xi)$ be the total height of the jumps of $L_2(\theta, \xi)$ in $[\theta, \theta + \Delta\theta]$, and

$$c(\theta, \Delta\theta, h) = \Pr\{ \xi \mid \xi : h(\theta, \Delta\theta, \xi) \le h \} \qquad (6)$$

be the probability that the height of jumps in $[\theta, \theta + \Delta\theta]$ is less than h. Then

$$E\{h(\theta, \Delta\theta, \xi)\} = \int_{-\infty}^{\infty} h \, d_h c(\theta, \Delta\theta, h),$$

where d_h denotes that the differentiation is with respect to the variable h. Furthermore, we assume that $c(\theta, \Delta\theta, h)$ in (6) can be decomposed as follows:

$$c(\theta, \Delta\theta, h) = b(\theta, \Delta\theta, h) \times e(\theta, \Delta\theta), \qquad (7)$$

where $b(\theta, \Delta\theta, h)$ is the conditional probability that the jump is less than h given that a jump occurs (i.e., the conditional distribution of h), and $e(\theta, \Delta\theta)$ is the probability that the jump occurs in $[\theta, \theta + \Delta\theta]$. We also assume that the first and the second derivatives of $e(\theta, \Delta\theta)$ with respect to $\Delta\theta$ exist at $\Delta\theta = 0$, i.e., the probability that $L(\theta, \xi)$ jumps in $[\theta, \theta + \Delta\theta]$ does not change too rapidly. Since the probability that $L(\theta, \xi)$ jumps at θ is zero, we can write $e(\theta, \Delta\theta)$ in the following form:

$$e(\theta, \Delta\theta) = g(\theta)\Delta\theta + o(\Delta\theta) \qquad (8)$$

From Eqs.(6) - (8), we have

$$\lim_{\Delta\theta \to 0} \frac{E\{h(\theta, \Delta\theta, \xi)\}}{\Delta\theta}$$

$$
\begin{aligned}
&= \lim_{\Delta\theta \to 0} \frac{\displaystyle\int_{-\infty}^{\infty} h \, d_h \{b(\theta, \Delta\theta, h) \times [g(\theta)\Delta\theta + o(\Delta\theta)]\}}{\Delta\theta} \\[2em]
&= \lim_{\Delta\theta \to 0} \int_{-\infty}^{\infty} h \, d_h \{b(\theta, \Delta\theta, h) \times [g(\theta) + \frac{o(\Delta\theta)}{\Delta\theta}]\} \\[2em]
&= g(\theta) \lim_{\Delta\theta \to 0} \int_{-\infty}^{\infty} h \, d_h [b(\theta, \Delta\theta, h)] .
\end{aligned}
\tag{9}
$$

Let

$$
b(\theta, h) = \lim_{\Delta\theta \to 0} b(\theta, \Delta\theta, h)
$$

be the conditional distribution of the height of the jump at θ. Then the distribution in (9), $b(\theta, \Delta\theta, h)$, converges weakly to the distribution $b(\theta, h)$ as $\Delta\theta$ goes to 0. Thus, if the jump height h is bounded, then the limit and the integration in Eq.(9) are interchangeable [Billingsley 1979, Theorem 25.8]. We get

$$
\lim_{\Delta\theta \to 0} \frac{E\{h(\theta, \Delta\theta, \xi)\}}{\Delta\theta} = g(\theta) \int_{-\infty}^{\infty} h \, d_h \, b(\theta, h).
\tag{10}
$$

The bounded condition for h can be replaced by some milder restrictions. In fact, by Theorem 25.6 of [Billingsley 1979], for any sequence $(\Delta\theta)_n \to 0$, n=1,2,..., there exist random variables κ_n and κ on a common probability space such that κ_n has distribution $b[\theta, (\Delta\theta)_n, h]$, κ has distribution $b(\theta, h)$, and $\kappa_n \to \kappa$ w.p.1. By changing variables, (10) becomes equivalent to $E(\kappa_n) \to E(\kappa)$. This holds if κ_n are bounded or dominated by another integrable function. Particularly, in most DEDS, the distribution of the jump height does not depend on $\Delta\theta$. In this case, (10) follows from (9) directly.

From Eq.(10), we have:

$$\frac{\partial E[L_2(\theta,\xi)]}{\partial\theta} = \lim_{\Delta\theta\to 0} E\{\frac{L_2(\theta+\Delta\theta,\ \xi) - L_2(\theta,\xi)}{\Delta\theta}\}$$

$$= \lim_{\Delta\theta\to 0}\frac{E\{h(\theta,\ \Delta\theta,\ \xi)\}}{\Delta\theta} = g(\theta)\int_{-\infty}^{\infty} h\ d_h\ b(\theta,\ h). \qquad (11)$$

Thus, Eq.(1) holds for functions of form (5) if and only if

$$g(\theta)\int_{-\infty}^{\infty} h\ d_h\ b(\theta,\ h)=0.$$

i.e., either

$$g(\theta) = 0,$$

or

$$\int_{-\infty}^{\infty} h\ d_h\ b(\theta,\ h)=0.$$

This is stated formally as follows: If function $L(\theta,\xi)$ is of the form (5), then Eq.(1) holds if and only if the probability that $L(\theta,\xi)$ jumps in $[\theta,\ \theta+\Delta\theta]$ is $o(\Delta\theta)$, or the average height of the jump at θ equals zero[3].

This result provides an intuitive explanation for why in some cases the sample derivative may be a biased estimate of the derivative of the expected value of the performance measure. If the performance function has jumps with respect to θ, and the effect of the jumps is not negligible (either because the jump probability is not $o(\Delta\theta)$, or the average height is not zero), then the two operators "E" and "$\partial/\partial\theta$" are not interchangeable. In this case, $\partial/\partial\theta\{E[L(\theta,\xi)]\}$ contains information about the average jump heights of $L(\theta,\xi)$ in the neighborhood of θ, while the sample derivative $\partial/\partial\theta\{L(\theta,\xi)\}$ does not.

---

[3] This could result from a probability of jump of order $O(\Delta\theta)$ and a jump height of order $O(\Delta\theta)$ given rise to average height of jump of order $o(\Delta\theta)$ which is negligible for first derivative calculations.

For the function defined in the Eq.(2), the probability that $L(\theta,\xi)$ jumps in $[\theta,\theta+\Delta\theta]$ is $P(\xi \in [\theta,\theta+\Delta\theta]) = \Delta\theta$, which is not $o(\Delta\theta)$, and the height of jumps is always 1. This explains why Eq.(1) does not hold for this function.

For DEDS, changes in any parameter θ will induce perturbations to the event times. These perturbations will be propagated along any sample path (θ, ξ) and will be accumulated. Thus, as time t increases, the sizes of the perturbations of events will increase. Eventually, the perturbations of some events will become so big that the order of events changes. For example, in a system with multiclass customers, a server may serve a class 1 customer before it serves a class 2 customer in the nominal path (e.g., because the class 1 customer arrives just before the class 2 customer does); however, because of the accumulated perturbation in the class 1 customer's arrival, the server may serve the class 2 customer first in the perturbed system. This kind of order change of events causes the discontinuity of the sample performance function. (Examples will be given in Chapter 5.) Therefore, for these systems, as t increases, the sample function based on any sample path will eventually become discontinuous. If, on the other hand, one is only concerned with the average finite-time perfor-mance, this same problem appears in a different way. No matter how small $\Delta\theta$ is, there may always exist, with a non-zero probability, some sample paths (θ,ξ) from the ensemble of paths to be averaged over, such that on these paths the perturbations induced by a parameter change $\Delta\theta$ cause event changes. Thus, the sample performance functions $L(\theta,\xi)$ based on these sample paths are discon-tinuous; the interchangeability defined in Eq.(1) may or may not hold.

In the next section, we shall study closed Jackson networks with single-server nodes and single-class customers. We shall prove that the interchange-ability holds for the derivative of the average time required for the system to serve one customer with respect to a mean service time of a server. Using this, we shall prove that the estimate of the derivative of the system throughput obtained by perturbation analysis is strongly consistent [Cao 1988a]. Both the transient and steady-state performance will be discussed.

4.2 Perturbation Analysis Estimates for Closed Jackson Networks

As discussed in Section 4.1, the perturbation analysis estimate is the derivative of the sample function $L(\theta,\xi)$. In this section, we shall study in detail the closed Jackson network. To prove the interchangeability of the expectation and the differentiation on $L(\theta,\xi)$, we first investigate the shape of the function. This is the main topic of Section 4.2.1.

4.2.1 Sample Performance Functions

Consider a class of closed queueing networks containing M single-server nodes and N single-class customers. Each server has a buffer with an infinite size. The service discipline is first come first served. The customers circulate among servers, and at the completion of the service at server i a customer goes to server j with probability $q_{i,j}$. The service time required by a customer at server i is exponentially distributed with mean $s_i = 1/\mu_i$.

First we make some mild restrictions on the system structure. We do not consider networks that can be decomposed into two or more subnetworks that never interact with each other. For such networks, we simply consider each individual sub-network. To be more precise, we introduce the notion of irreducible networks (this is closely related to the idea of irreducibility of a Markov chain). A closed queueing network is said to be *reducible* if, by reordering the indices of servers, the matrix $Q=[q_{i,j}]$ can be written in the following form:

$$Q = \begin{pmatrix} Q_1 & O \\ P & Q_2 \end{pmatrix}$$

where Q_1 is an $r \times r$ matrix, $0<r<M$, O is a matrix whose entries are all zeros. Otherwise, the network and the matrix Q are said to be *irreducible*.

It is easy to see that in an irreducible queueing network a customer in any server i has a positive probability of going to any other server in the network either directly or through other servers. The matrices Q_1 and Q_2 may also be

reducible. For ease of discussion, we assume that these two matrices are irreducible. If P =0, then the network can be decomposed into two subnetworks. A
customer in each of the sub-networks can never reach the servers in the other.
If P $\neq$0, then the customers in servers r+1 to M have a positive probability of
going to servers 1 to r. The customers in servers 1 to r, however, cannot go
to servers r+1 to M. Thus, in steady state there are no customers in servers r+1
to M. In this case, we need only study the sub-network with the transition
matrix Q_1, which is irreducible. A thorough study of irreducible matrices can
be found in [Seneta 1981].

Let the sample path start at time t=0. Let $s_{i,k}$ be the service time of the
kth customer served by server i since the start of the sample path. Let

$$r_{i,k} = 1 - \exp\{ -\frac{s_{i,k}}{s_i}\}, \qquad i=1,2,...,M, \; k=1,2,....$$

Then $r_{i,k}$ are independent random variables uniformly distributed on [0,1), and

$$s_{i,k} = - s_i \ln (1 - r_{i,k}), \qquad i=1,2,...,M, \; k=1,2,... \qquad (12)$$

The parameters of this system are s_i and $q_{i,j}$, i,j = 1,2,...,M. In the following,
we choose the mean service time of one server, say s_u, as the parameter θ to
which the derivative is taken. The random effect involved in the system can be
represented by the random vector $\xi = (\eta_1, \eta_2,..., \eta_M; \zeta_1, \zeta_2, ..., \zeta_M, \phi)$,
where $\eta_i =(r_{i,1}, r_{i,2}, ...)$, ζ_i is a random vector; its kth component determines
the destination of the kth customer in server i, and ϕ is a random vector which
represents the initial condition, i.e., the number of customers in each server at t
=0. Note that the initial state can also be written as a function of a random variable ξ_0, which is uniformly distributed on [0,1). For example, let n_i, i=1,2, ...,
H, be the possible initial states of the system, and $P(n_i) = p_i$ be the probabilities
of these states. Then we can let the initial state be n_i if

$$\sum_{j=0}^{i-1} p_j \leq \xi_0 < \sum_{j=0}^{i} p_j, \qquad p_0 = 0.$$

However, this representation of the initial state is not essential for the discussion of this book. We shall simply consider ϕ as the initial state of the system rather than a uniformly distributed random variable.

We observe the system until it completes a total of L services. We denote this time as $T_L = T_L(\theta, \xi)$. Let g_i be the number of service completions of server i in the time period $[0, T_L]$. Then

$$L = \sum_{i=1}^{M} g_i.$$

We choose the length of the period in which the system completes L services, $T_L(\theta, \xi)$, as the performance measure concerned.

The system state can be represented as $\mathbf{n} = (n_1, n_2, ..., n_M)$, where n_i is the number of customers in server i. The state evolution of the system can be described as a pure-jump right-continuous Markov process $N(t, \xi)$. Each time a customer transfers from one server to the other, the system changes its state. The state transition instants of the Markov process are denoted as T_d, d=1,2,. ..,L. In $[0, T_L]$, the sample path goes through a total of L states. Let $S_d = T_d - T_{d-1}$ be the sojourn time of the system in its dth state.

From the theory of the irreducible matrix [Seneta 1981], it can be shown that the visit ratios $v_1, v_2, ..., v_M$, determined by the equation $vQ = v$, $v = (v_1, v_2,...,v_M)$, are all positive for an irreducible matrix Q. By the Gordon-Newell formula [Gordon and Newell 1967], the steady-state probabilities $P(\mathbf{n})$ are positive for all state $\mathbf{n}$. This implies that the Markov process $N(t, \xi)$ of an irreducible network and its imbedded Markov chain are both irreducible.

Let $t_{i,k}$ be the kth service completion time of server i. Note that the set $\{T_d, d = 1,2, ...\}$ is just the same as the set $\{t_{i,k}, i=1,2,...,M, k = 1,2,...\}$. A sample path of the system is completely determined by the set $\{t_{i,k}, i=1,2,...,M, k = 1,2,...\}$, the initial state, and the customer transitions.

To study the derivative of the performance function $T_L(\theta, \xi)$, we first investigate the shape of the function. We shall see, for any realization of ξ, $T_L(\theta, \xi)$ is a continuous and piecewise linear function of θ. The proof of this statement is based on perturbation analysis, which describes the behavior of

$T_L(\theta,\xi)$ in a small neighborhood of θ. We shall first derive a mathematical expression of the perturbation analysis rules for the closed Jackson network.

As discussed in Chapter 3, perturbation analysis depends heavily on server-to-server interaction. In this respect, idle (NI) periods play a dominant role. A server is said to be in an *idle period* if the number of customers in the server is zero (i.e., $n_i = 0$). In fact, if there were no idle periods, each server would work as if it stood alone, and there would be no perturbation propagated from one server to others. Now, suppose that server i experiences an idle period after its (k-1)th customer leaves it, then the number of customers at server i at $t_{i,k-1}+$, $n_i(t_{i,k-1}+)$, is 0. If after this idle period, server i receives a customer from server j, who was the hth customer served by server j in $[0, T_L]$, and starts a new busy period, then we say that the idle period of server i is *terminated* by the hth customer of server j. Obviously, the following recursive formulas hold:

$$t_{i,k} = \begin{cases} t_{i,k-1} + s_{i,k} & \text{if } n_i(t_{i,k-1}+) \neq 0, \\ t_{i,h} + s_{i,k} & \text{if } n_i(t_{i,k-1}+) = 0. \end{cases} \tag{13}$$

This equation simply says that if a customer starts a new busy period, then its service starting time is decided by the arrival time to the server $(t_{j,h})$; otherwise it is decided by the service completion time of the previous customer. The set $\{t_{i,k}, i=1,2,...,M, k=1,2,...\}$ is completely determined by Eqs. (12) and (13).

Now, suppose that a parameter, $\theta = s_u$, changes by a small amount $\Delta\theta = \Delta s_u$. We call the system with parameter $\theta+\Delta\theta$ the *perturbed system*. A sample path of the perturbed system is called a *perturbed path*. Let $s'_{i,k}$ be the service time of the kth customer of server i in the perturbed system, and $t'_{i,k}$ be the service completion time of this customer. The equations for $t'_{i,k}$ are similar to Eqs.(12) and (13):

$$s'_{i,k} = -s'_i \ln(1-r_{i,k}), \qquad i=1,2,...,M, k=1,2,... \tag{14}$$

and

$$t'_{i,k} = \begin{cases} t'_{i,k-1} + s'_{i,k} & \text{if } n_i(t'_{i,k-1}+) \neq 0, \\ t'_{i,h} + s'_{i,k} & \text{if } n_i(t'_{i,k-1}+) = 0. \end{cases} \qquad (15)$$

Let $\Delta s_{i,k} = s'_{i,k} - s_{i,k}$. $\Delta s_{i,k}$ is called the *perturbation generated during the corresponding service period*. From (12) and (14), we have

$$\Delta s_{i,k} = \begin{cases} 0 & \text{if } i \neq u, \\ -\Delta s_u \ln(1 - r_{u,k}) = \dfrac{\Delta s_u}{s_u} s_{u,k} & \text{if } i = u. \end{cases} \qquad (16)$$

Note that the perturbation generated in the kth customer's service period, $\Delta s_{u,k}$, is proportional to both the change in the mean service time Δs_u and the service time $s_{u,k}$. Although the form of Eq.(15) looks the same as that of Eq.(13), the equations for a particular pair (i,k) may be different for both perturbed and the nominal systems. For instance, server i may meet an idle period after it serves its (k-1)th customer in the nominal path (hence the second equation in (13) holds for this (i,k)); this idle period may not exist in the perturbed path if the kth customer in the perturbed system enters server i earlier than that in the nominal system (hence the first equation in (15) holds for this (i,k)). This is because as perturbations are generated, not only the $t_{i,k}$ and $s_{i,k}$ change but the qualifying conditions of Eqs.(13) and (15) also may change. Thus, before we can derive a simple equation for $\Delta t_{i,k}$, we have to make some restrictions on Eqs.(13) and (15).

Let $X_d = N(T_d, \xi)$, i.e., the dth state of the DEDS during its evolution. Then $X = \{X_d, d=1,2,..\}$ is the Markov chain imbedded in $N(t, \xi)$. We define:

Definition 4.3 Two sample paths are said to be *deterministically similar* in $[0, T_L]$, if and only if the state sequences $\{X_d, d=1,2,...,L\}$ are the same for both paths, i.e., the nominal and the perturbed state sequences are identical.

If the nominal and the perturbed sample paths are similar, then for any (i,k), Eqs.(13) and (15) are exactly the same. Thus, by taking the difference between the two sides of these two equations, we obtain:

$$\Delta_{i,k} = \begin{cases} \Delta_{i,k-1} + \Delta s_{i,k} & \text{if } n_i(t_{i,k-1}+) \neq 0, \\ \Delta_{i,h} + \Delta s_{i,k} & \text{if } n_i(t_{i,k-1}+) = 0. \end{cases} \tag{17}$$

$\Delta_{i,k} = t'_{i,k} - t_{i,k}$ is called the *perturbation of server i at $t_{i,k}$*, or the *perturbation of the kth customer of server i..* (Colloquially, we have also referred to this perturbation as the "GAIN" of the server or the customer in literature.) For convenience, we also say that server i continuously possesses this perturbation, $\Delta_{i,k}$, after $t_{i,k}$ until the perturbation is changed because of generation or propagation. Eqs.(16) and (17) are the rules describing the evolution of perturbation for two deterministically similar sample paths. Eq.(16) is the perturbation generation rule for a server with exponential distributed service time: The perturbation generated in a customer's service period is proportional to the service time of the customer. Eq.(17) describes the perturbation propagation rules. Suppose that $\Delta s_{i,k} = 0$ in (17), i.e., there is no perturbation generated in $s_{i,k}$. Then Eq.(17) becomes

$$\Delta_{i,k} = \begin{cases} \Delta_{i,k-1} & \text{if } n_i(t_{i,k-1}+) \neq 0, \\ \Delta_{i,h} & \text{if } n_i(t_{i,k-1}+) = 0. \end{cases}$$

This corresponds to the infinitesimal perturbation propagation (IPA) rules described in Chapter 3: (i) A perturbation will be propagated to the next customer in the same busy period ($\Delta_{i,k} = \Delta_{i,k-1}$); (ii) If a customer coming from server j starts a new busy period at server i, then the perturbation of server j will be propagated to server i ($\Delta_{i,k} = \Delta_{j,h}$).

As pointed above, IPA rules only apply to two deterministically similar sample paths. In the following, we shall prove that for the closed Jackson network studied in this chapter, given any finite L, there exists, with probability one, a small positive number δ (which may depend on the sample path) such that if $|\Delta s_u| < \delta$ then the perturbed path is similar to the nominal one in $[0, T_L]$.

The intuitive idea behind this is that on a DEDS trajectory under Assumption 1 of Appendix B.3 any two events are always separated by a finite time interval however small. Hence we can always arrange to have the accumulated perturbations on the perturbed trajectory to be smaller than such finite time intervals. To prove this mathematically, we need some preliminary results.

First, since $0 < \mu_i < \infty$ and $0 < \mu(\mathbf{n}) < \infty$ for all $\mathbf{n}$, it is easy to show that the following result holds

$$S_{min}(\theta, \xi) \equiv \min \{S_d, d = 1, 2, ..., L\} > 0, \quad \text{w.p.1.}$$

The next result needed is about the bound of the perturbation generated in a finite period $[0, T_L]$. Let $t_{u,h}$ be the last service completion time of server u in $[0, T_L]$. By the perturbation propagation rule, the perturbation acquired by any server in the network at time t is the accumulation of the perturbations that have been propagated to the server from server u. On the other hand, according to the perturbation generation rule, the total perturbation generated in $[0, t]$ at server u is at most $(\Delta\theta/\theta)t_{u,h}$ (there is no perturbation generated in an idle period of server u). Therefore, the perturbation of any server at time t should be less than or equal to $(\Delta\theta/\theta)t_{u,h}$. That is, if the nominal and the perturbed paths are similar in $[0, T_L]$, then the perturbation of any server in the system at time t is bounded by $(\Delta\theta/\theta)t_{u,h} < (\Delta\theta/\theta)T_L$.

Now, if we choose $(\Delta\theta/\theta) T_L < S_{min}(\theta, \xi)$, then the perturbation of any server in $[0, T_L]$ will be at most S_{min}. In this case, the perturbation in θ will not change the order of the service completions. In other words, for almost every sample path in $[0, T_L]$, we can choose $\Delta\theta$ small enough (say, $\Delta\theta < \theta S_{min}/T_L$) such that both the perturbed and the nominal paths are deterministically similar in $[0, T_L]$. Of course, the size of $\Delta\theta$ depends on ξ.

Another important feature of the perturbation analysis equations (16) and (17) is the linearity. That is, the perturbation at any $t_{i,k}$ is a linear function of $\Delta\theta$ in a small neighborhood of θ. In fact, if the perturbed sample paths corresponding to $\Delta\theta = \Delta\theta_1$, $\Delta\theta_2$, and $\alpha\Delta\theta_1 + \beta\Delta\theta_2$ (α, $\beta >0$) are similar to the nomi-nal one, then we have

$$\Delta_{i,k}(\alpha\,\Delta\theta_1 + \beta\,\Delta\theta_2) = \alpha\,\Delta_{i,k}(\Delta\theta_1) + \beta\,\Delta_{i,k}(\Delta\theta_2) \tag{18}$$

where $\Delta_{i,k}(\delta)$ is the perturbation of server i at $t_{i,k}$ corresponding to $\Delta\theta = \delta$. The linearity is due to the fact that $\Delta s_{i,k}$ is proportional to $\Delta\theta = \Delta s_u$. By the perturbation generation rule (Eq. (3.22) and Examples 1-3 in Section 3.1), this linearity also holds for many other systems including those for which the cumulative service distribution has θ as a scale parameter.

From the above discussion, if we choose

$$|\theta_i - \theta| < \frac{S_{min}}{T_L(\theta,\xi)}\,\theta, \qquad i=1,2,$$

then all the sample paths with $\theta \in [\theta_1(\xi), \theta_2(\xi)]$ are deterministically similar. Moreover, from Eq.(18), $T_L(\theta,\xi)$ is linear in $[\theta_1(\xi), \theta_2(\xi)]$ with respect to θ. Therefore, for almost any sample path (θ,ξ), there exists w.p.1 an $\theta_1(\xi)$ and $\theta_2(\xi)$, $\theta_1(\xi) < \theta < \theta_2(\xi)$, such that $T_L(\theta,\xi)$ is linear in $[\theta_1(\xi), \theta_2(\xi)]$.

As $\theta + \Delta\theta$, $\Delta\theta>0$, increases, this linearity may be violated. The following two cases may cause this violation: (i) An idle period at one server is eliminated, or (ii) A new idle period at one server is created. Figs. 1 and 2 illustrate these two cases. In the figures, the integers above each line are the numbers of customers in the corresponding servers in the nominal path, and the integers in the parentheses below each line are the numbers of customers in the perturbed path. In Fig.1, if $\Delta_{i,k-1} - \Delta_{j,h}$ is greater than the length of the idle period, then the idle period would not appear in the perturbed path. In this case, $t'_{i,k} = t'_{i,k-1} + s_{i,k}$, while $t_{i,k} = t_{j,h} + s_{i,k}$. We say that an idle period is eliminated in the perturbed path. In Fig. 2, if $\Delta_{j,h} - \Delta_{i,k-1}$ is greater than the length of the period in which the system stays in the state with $n_i=2$, then the customer in server j would come to server i after server i completes the service to its last customer in the busy period; thus a new idle period would be created at server i in the perturbed path. In this case, $t'_{i,k} = t'_{j,h} + s_{i,k}$ while $t_{i,k} = t_{i,k-1} + s_{i,k}$. We say that an idle period is created in the perturbed path. In both cases the propagation rules in (17) no longer hold for $\Delta_{i,k}$. Let $\theta_2(\xi)$ be the smallest value such that if $\theta + \Delta\theta < \theta_2(\xi)$, the linearity still holds. Then at $\theta_2(\xi)$ an idle period is about to be either eliminated or created. Thus, on the sample path (θ_2, ξ), two customer transitions occur at the same time (e.g., $t'_{i,k-1} = t'_{j,h}$ in Figs. 1 and 2). When θ increases further from $\theta_2(\xi)$, the sample paths of the system form another set of similar paths. Similar discussion holds for $\Delta\theta < 0$. Thus, we conclude that $T_L(\theta,\xi)$ is a piecewise linear function of θ.

Furthermore, we can show that the function $T_L(\theta,\xi)$ is continuous with respect to θ. In fact, the linearity of $T_L(\theta,\xi)$ in the open interval $(\theta_1(\xi), \theta_2(\xi))$ implies its continuity in the same interval. Now, suppose that θ increases to $\theta_2(\xi)$. On the sample path $(\theta_2(\xi), \xi)$, two customer transitions happen at the same time. The following three cases may occur: (i) An idle period in the nominal path is just eliminated, e.g., $t'_{i,k-1} = t'_{j,h}$ in Fig. 1. In this case $t'_{i,k} = t'_{i,k-1} + s_{i,k} = t'_{j,h} + s_{i,k}$. Thus $\Delta_{i,k} = \Delta_{j,h} + \Delta s_{i,k}$ still holds at $\theta_2(\xi)$. Since the same equation holds for $(\theta_2(\xi)-, \xi)$, $T_L(\theta,\xi)$ is left continuous at $\theta_2(\xi)$. (ii) A new idle period is about to be created, e.g., $t'_{i,k-1} = t'_{j,h}$ in Fig. 2. (iii) None of the above two cases happens. It is ready to see that $\Delta_{i,k} = \Delta_{i,k-1} + \Delta s_{i,k}$ holds

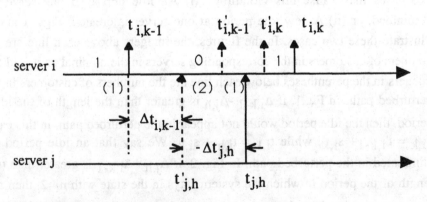

Fig. 1 An Idle Period Is Eliminated

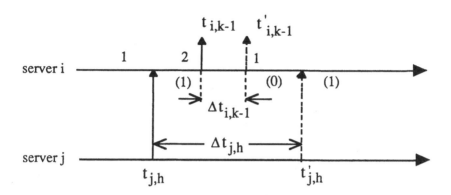

Fig. 2 An Idle Period Is Created

in both cases (ii) and (iii), which is the same for sample path $(\theta_2(\xi)-, \xi)$. Therefore, $T_L(\theta,\xi)$ is left continuous at $\theta_2(\theta)$. By the same reason, it is also right continuous. Finally, we conclude: The sample function $T_L(\theta,\xi)$ is a continuous, piecewise linear function of θ for any ξ. This indicates the following fact: If we fix a realization of ξ then we can obtain a continuous and piecewise linear function as an estimate of the curve $J(\theta) = E[T_L(\theta,\xi)]$.

However, the continuity or piecewise linearity may not hold for other sample functions and other parameters. For example, let θ be one of the routing probabilities. For simplicity, let $\theta = q_{1,2}$, and assume that $q_{1,3}=1-\theta$ and $q_{1,i}=0$ for $i \neq 2,3$. In computer simulation, for the kth customer in server 1, we draw a number $r_{1,k}$ from a random number generator generating uniformly distributed numbers on $[0,1)$. If $r_{1,k}<\theta$, then this customer goes to server 2; otherwise, the customer goes to server 3. For any finite L and any θ there exists, with probability one, a $\Delta\theta$ small enough such that there is no customer whose $r_{i,k}$ lies between θ and $\theta + \Delta\theta$. This means that the corresponding customers in both the nominal and the perturbed sample paths have the same destinations. In other words, the sample paths with θ and $\theta + \Delta\theta$ are exactly the same. Therefore, as θ increases, $T_L(\theta,\xi)$ does not change as long as $\theta + \Delta\theta$ does not hit any value of $r_{1,k}$, k=1,2,..., g_1. Here, the sample function $T_L(\theta,\xi)$ is a piecewise constant function with jumps at certain points.

By definition, $E[T_L(\theta,\xi)] = J(\theta)$ holds for any sample function, i.e., $T_L(\theta,\xi)$ is an unbiased estimate of $J(\theta)$. The sample derivatives $(\partial/\partial\theta)T_L(\theta,\xi)$, however, may behave differently. For example, if θ is one of the routing probabilities, then, as discussed above, $(\partial/\partial\theta)T_L(\theta,\xi) =0$. Thus, $E\{(\partial/\partial\theta)\ T_L(\theta,\xi)\}$ $= 0$. On the other hand, $E[J(\theta)]$ may be a smooth function with $(\partial/\partial\theta)\ E[J(\theta)] \neq$ 0. We can easily see this by noting that the average of an infinite series of piece-wise constant functions may in fact be a smooth curve with well-defined slopes. Fig.3 illustrates the average of three piecewise constant functions. Each of these individual functions has a larger jump, while the average curve has three smaller jumps. As the number of averaged curves becomes larger, the average curve becomes smoother. The expected value of $T_L(\theta,\xi)$ corresponds to the average of infinitely many piecewise constant functions; therefore, it has infinitely small jumps and may be a differentiable function. (The function defined in Eq.(2) is a good example for this. See also the exercise following Eq.(2)).

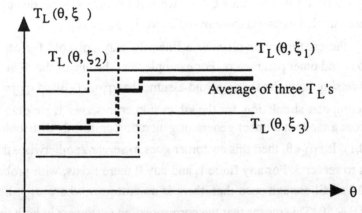

Fig. 3 The Average of Three Piecewise Constant Functions

We have proved that the sample performance function $T_L(\theta,\xi)$ is a continuous (i.e., no jumps or discontinuities) and piecewise linear function of the mean service time θ. Using this result, in the next section, we shall prove that the derivative of the sample performance function with respect to θ is an unbiased estimate of $(\partial/\partial\theta)[J(\theta)]$.

4.2.2 Sample Derivative of Performance in Transient Periods

The sample derivative of $T_L(\theta,\xi)$ with respect to θ is defined as

$$\frac{\partial T_L(\theta,\ \xi)}{\partial\theta} = \lim_{\Delta\theta\to 0} \frac{T_L(\theta+\Delta\theta,\ \xi) - T_L(\theta,\ \xi)}{\Delta\theta}.$$

Since $T_L(\theta,\xi)$ is defined on a finite time period, its expected value depends on the initial state of the time period, n_0. Thus, an important property of the sample derivative is

$$\frac{\partial}{\partial\theta}\{E[T_L(\theta,\ \xi)\mid n_0]\} = E\{\frac{\partial}{\partial\theta}[T_L(\theta,\ \xi)]\mid n_0\} \tag{19}$$

In the above equation, we assume that the both sides of the equation exist. In fact, because $T_L(\theta,\xi)$ is continuous and piecewise linear, it is differentiable at any θ unless θ is a kink point on the curve of $T_L(\theta,\xi)$, or, equivalently, on the corresponding sample path $(\theta,\ \xi)$ two customer transitions occur at the same time. Thus, for any θ, the probability that $T_L(\theta,\xi)$ is not differentiable is less than or equal to the probability of the sample paths on which two customer transitions overlap, i.e., $S_{min} = 0$. Since $P(S_{min}=0)=0$, $T_L(\theta,\xi)$ is differentiable w.p.1 at any θ. Let

$$r(\theta, h,\ \xi) = \frac{T_L(\theta+h,\ \xi) - T_L(\theta,\ \xi)}{h} - \frac{\partial T_L(\theta,\ \xi)}{\partial\theta}, \qquad h\in [0,\ \Delta\theta].$$

The last term on the right-hand side exists w.p.1. Thus we have

$$\lim_{h\to 0} r(\theta,\ h,\ \xi) = 0, \qquad\qquad \text{w.p.1.} \tag{20}$$

Eq.(19) can be written in the following form:

$$\lim_{h \to 0} E\{r(\theta, h, \xi) \mid n_0\} = 0.$$

(21)

Recall that $\xi = (\eta_1, \eta_2, ..., \eta_M; \zeta_1, \zeta_2, ..., \zeta_M, \phi)$. We denote $\rho = (\eta_1, \eta_2, ..., \eta_M; \zeta_1, \zeta_2, ..., \zeta_M)$. Since $\phi = n_0$ is given, we can write $\xi = (\rho, n_0)$. The two random vectors ρ and n_0 are independent. Thus, the underlying probability space Ω can be represented by the decomposition $\Omega = \Omega_1 \times \Omega_2$, where the random vectors ρ and n_0 are defined in Ω_1 and Ω_2, respectively. We shall see that $r(\theta, h, \xi)$ is dominated by a differentiable function w.p.1 on Ω_1; therefore, $T_L(\theta, \xi)$ is dominated differentiable w.p.1 on Ω_1 at θ. According to the results in Section 4.1 and Eqs.(1) and (3), Eq.(19) holds.

Using the results in Section 4.2.1, we can find the bound of the slopes of the curve of $T_L(\theta, \xi)$. First of all, any two points on the same linear segment of the curve represent two similar sample paths. Let us take one point as the performance of the nominal path and the other as that of the perturbed one. We have shown that the maximum perturbation of any server in $[0, T_L]$ is less than $(\Delta\theta/\theta)T_L(\theta, \xi)$. (The range of $\Delta\theta$ in which this statement holds depends on ξ.) Thus, we have

$$\Delta T_L(\theta, \xi) \leq \frac{\Delta\theta}{\theta} T_L(\theta, \xi).$$

Therefore, the bound of the slopes at θ can be determined by

$$\frac{\Delta T_L(\theta, \xi)}{\Delta\theta} \leq \frac{T_L(\theta, \xi)}{\theta}.$$

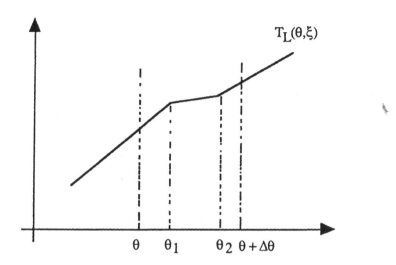

Fig. 4 An Example of $T_L(\theta,\xi)$

Fig. 4 shows an example of the curve $T_L(\theta,\xi)$ with three linear segments and two kink points in a neighborhood of θ. Since $T_L(\theta,\xi)$ is a non-decreasing function of θ, we conclude that all the derivatives of $T_L(\theta,\xi)$ in $[\theta, \theta+\Delta\theta]$ are bounded by $T_L(\theta+\Delta\theta,\xi)/\theta$. From this, we have

$$\frac{T_L(\theta + h, \xi) - T_L(\theta, \xi)}{h} \leq \frac{T_L(\theta+\Delta\theta, \xi)}{\theta}, \qquad h \in [0, \Delta\theta]. \quad (22)$$

On the other hand, since $(\partial/\partial\theta)T_L(\theta,\xi)$ is the slope of $T_L(\theta,\xi)$ at θ, we have

$$\frac{\partial[T_L(\theta, \xi)]}{\partial\theta} \leq \frac{T_L(\theta, \xi)}{\theta} \leq \frac{T_L(\theta+\Delta\theta, \xi)}{\theta}. \quad (23)$$

From (22)-(23), we get

$$|r(\theta, h, \xi)| \leq \frac{T_L(\theta+\Delta\theta, \xi)}{\theta}, \qquad h \in [\theta, \theta+\Delta\theta].$$

Therefore,

$$f(\theta, \xi) = \frac{T_L(\theta+\Delta\theta, \xi)}{\theta}$$

is a dominating function of $r(\theta, h, \xi)$ in $[\theta, \theta+\Delta\theta]$.

Next, since the service rates of all the servers in a closed Jackson network are finite, the state transition rates of the Markov process $N(t, \xi)$ for all the states are finite. This means that the mean sojourn times of the system in each state are finite. Therefore, the mean of the time in which the system goes through L states, $E[T_L(\theta+\Delta\theta,\xi)]$, is finite. Namely, $f(\theta, \xi) = T_L(\theta+\Delta\theta,\xi)/\theta$ is integrable. $T_L(\theta,\xi)$ is dominated differentiable w.p.1 on Ω_1. Thus, Eq. (21) can be verified by interchanging the order of $\lim_{\Delta\theta\to0}$ and "E;" Using the result in Section 4.1, we obtain the following important result about the interchangeability.

For a closed Jackson network with single-class customers and single-server nodes, given any initial state n_0, the sample derivative $(\partial/\partial\theta)T_L(\theta,\xi)$ is an unbiased estimate of the derivative of the conditional mean of $T_L(\theta,\xi)$, i.e.,

$$E\{\frac{\partial}{\partial\theta}[T_L(\theta, \xi) \mid n_0]\} = \frac{\partial}{\partial\theta}\{E[T_L(\theta, \xi) \mid n_0]\}.$$

The sample derivative $(\partial/\partial\theta)T_L(\theta,\xi)$ can be obtained by applying perturbation generation and propagation rules to a single sample path of the network. This result shows that the perturbation analysis estimate in a finite period $[0, T_L]$ is unbiased.

Now, let us turn to the problem of estimating the throughput sensitivity. The conditional mean throughput of the system in a finite period in which all the servers in the system complete L services, given that the initial state is n_0, is defined as

$$TP_L(\theta \mid n_0) = \frac{L}{E[T_L(\theta,\xi) \mid n_0]} .$$

Its derivative with respect to θ is

$$\frac{\partial TP_L(\theta \mid n_0)}{\partial \theta} = - \frac{L}{\{E[T_L(\theta,\xi) \mid n_0]\}^2} \{\frac{\partial}{\partial \theta} E[T_L(\theta,\xi) \mid n_0]\} .$$

Using perturbation analysis, we can obtain both $T_L(\theta,\xi)$ and $(\partial/\partial\theta)T_L(\theta,\xi)$ based on a single sample path of the system. Suppose that we get K such estimates corresponding to K random vectors ξ_k, k=1,2,...,K, with the same initial state n_0. By the law of large numbers and the unbiasedness of the perturbation analysis estimate, we have

$$\lim_{K \to \infty} \{ \frac{1}{K} \sum_{k=1}^{K} \frac{\partial}{\partial \theta} [T_L(\theta, \xi)] \} = \frac{\partial}{\partial \theta} \{E[T_L(\theta, \xi) \mid n_0]\} \qquad \text{w.p.1,}$$

and

$$\lim_{K \to \infty} \{ \frac{1}{K} \sum_{k=1}^{K} T_L(\theta, \xi) \} = E[T_L(\theta, \xi) \mid n_0] \qquad \text{w.p.1,}$$

Therefore,

$$\lim_{K \to \infty} \{ - L \times \frac{\frac{1}{K} \sum_{k=1}^{K} \frac{\partial}{\partial \theta} [T_L(\theta, \xi)]}{[\frac{1}{K} \sum_{k=1}^{K} T_L(\theta, \xi)]^2} \} = - L \times \frac{\frac{\partial}{\partial \theta} \{E[T_L(\theta, \xi)] \mid n_0\}}{\{E[T_L(\theta, \xi)] \mid n_0\}^2}$$

$$= \frac{\partial[TP_L(\theta \mid n_0)]}{\partial \theta} , \qquad \text{w.p.1.} \qquad (24)$$

This means that the term in the braces on the most left side, which can be obtained by infinitesimal perturbation analysis, is a strongly consistent estimate (as K goes to infinity) of $(\partial/\partial\theta)[TP_L(\theta \mid n_0)]$.

Summarizing, we have studied the properties of the perturbation analysis estimates in a finite period for a closed Jackson network with single-class customers and single-server nodes. We showed that, for any realization of ξ, the time required by the network to complete a finite number of services, $T_L(\theta,\xi)$, is a continuous and piecewise linear function of the mean service time of any server in the network, θ. The slope of the curve of this function (i.e., the derivative of this sample performance function) is an unbiased estimate of the derivative of its conditional expected value under a given initial state. We also showed that perturbation analysis provides a strongly consistent estimate for the conditional expected throughput of the system with any given initial state.

It is worthwhile to note that these properties do not hold for all performance measures and parameters. For example, if θ is one of the routing probabilities, then the sample derivative $(\partial/\partial\theta)[T_L(\theta,\xi)]$ is zero w.p.1. In this case, the sample derivative does not contain any information about the derivative of the expected value of $T_L(\theta,\xi)$. Other means have to be used to estimate such sensitivities. See Chapters 5 and 6 below.

4.2.3 Derivative of Steady-State Throughput
In Section 4.2.2, we studied the performance measure $T_L(\theta,\xi)$ and the sample derivative $(\partial/\partial\theta)T_L(\theta,\xi)$. Since L is finite , both $T_L(\theta,\xi)$ and $(\partial/\partial\theta)T_L(\theta,\xi)$ depend on the initial state n_0 and reflect the transient behavior of the system. In this section, we shall discuss the steady-state performance. We shall show that, as $L\rightarrow\infty$, the elasticity (i.e., the percentage change in a function divided by the percentage change in a parameter) of the sample function $T_L(\theta,\xi)$, $[\theta/T_L(\theta,\xi)]$ $\times [\partial T_L(\theta,\xi)/\partial\theta]$, converges in mean to the elasticity of the mean throughput in steady-state. This means that the sample elasticity of T_L given by perturbation analysis is an asymptotically unbiased estimate of the elasticity of the steady-state throughput. This result reveals a new property of closed Jackson networks and justifies the application of perturbation analysis to the estimation of the derivative of the steady-state throughput.

The proof of the convergence property is quite lengthy. To help to understand the idea, we sketch here the main procedure of the proof. First, we observe that $T_L(\theta,\xi)$ is unbounded as L→∞. Thus, we study the convergence property of $R_L(\theta, \xi) = T_L(\theta, \xi)/L$ instead of $T_L(\theta,\xi)$. Next, we prove that $R_L(\theta,\xi)$ converges with probability one as L→∞ and also converges in mean if we choose the steady-state probability as the initial state probability. Then, we study the derivative of $R_L(\theta,\xi)$ with respect to θ and prove that it converges in mean as L→∞. Finally, using these results, we prove that the elasticity of $R_L(\theta, \xi)$ with respect to θ converges in mean as L→∞, which immediately implies the main result of this section, i.e., the elasticity of the sample throughput $TP_L(\theta,\xi)$ converges in mean as L→∞.

The proof of the results in Section 4.2.3 depends on the ergodicity of the imbedded Markov chain **X**. Note that for some special networks **X** may be periodic, hence not ergodic. A simple example of such networks is a cyclic network consisting of two tandem nodes and one customer. The state sequence of the imbeded Markov chain of this network is ... (0,1), (1,0), (0,1), (1,0),. .. However, as will be explained in Section 4.6, for studying the IPA estimate we can always assume that **X** is ergodic, since otherwise we can construct an equivalent network whose imbedded Markov chain is ergodic. For example, in the two-server cyclic network we can construct the equivalent network as follows. Let s be the mean service time of one server in the network. We replace the server with a server having a mean service time of s/2 and a feedback loop with a probability of 1/2. (i.e., upon service completion, a customer will leave the server with probability 1/2 and will come back to the same server for another service with probability 1/2.) The resulting network is equivalent to the original one in the sense that the two networks have the same steady-state probabilities (see Section 4.6), and the imbedded Markov chain of the equivalent network is aperiodic and ergodic.

We assume that at time t =0 one server completes a service, i.e., t =0 is a transition instant of the Markov process N(t, ξ). Let

$$R_L(\theta, \xi) = \frac{T_L(\theta, \xi)}{L} = \frac{1}{L} \sum_{d=1}^{L} S_d, \tag{25}$$

where $S_d = T_d - T_{d-1}$. $R_L(\theta,\xi)$ is the average time between two successive transitions. Let

$$R_L(\theta \mid n_0) = E\{R_L(\theta, \xi) \mid n_0\}$$

be the conditional expectation of $R_L(\theta,\xi)$, given that the initial state is n_0. The steady-state probability of the system state at transition instants T_d is that of the imbedded Markov chain, denoted as $\pi(n)$. Since we are interested in the steady-state properties, we assume that the probability distribution of the initial state is $\pi(n_0)$, i.e., the observation period starts after the system reaches its steady state. Taking expectation of both sides of Eq.(25), we have

$$E\{R_L(\theta, \xi)\} = \frac{1}{L} \sum_{d=1}^{L} E(S_d) = E(S_d). \tag{26}$$

The last equation is due to the fact that in steady state $E(S_d)$s are equal for all d. Eq.(26) shows that $E\{R_L(\theta,\xi)\}$ is independent of L. Therefore, we can define

$$R(\theta) = E\{R_L(\theta, \xi)\} = E\{E[R_L(\theta, \xi) \mid n_0]\}.$$

The first expectation on the right-hand side is taken over the probability distribution $\pi(n)$.

Let

$$\mu(n) = \sum_{i=1}^{M} \varepsilon(n_i)\mu_i$$

be the transition rate of the Markov process $N(t, \xi)$, and $1/\mu(n)$ is the mean sojourn time that the Markov process stays in state n. By Eq.(26), we have

$$R(\theta) = E(S_d) = E\{E(S_d \mid n)\} = \sum_{\text{all } n} \frac{\pi(n)}{\mu(n)}.$$

Now, let us study the convergence property of $R_L(\theta,\xi)$. First we define:

$$\chi_n(X_d) = \begin{cases} 1 & \text{if } X_d = n, \\ 0 & \text{if } X_d \neq n. \end{cases}$$

Then,

$$R_L(\theta, \xi) = \frac{1}{L}\sum_{d=1}^{L}S_d = \frac{1}{L}\sum_{\text{all }n}\sum_{d=1}^{L}\chi_n(X_d)S_d \tag{27}$$

By the ergodicity of the Markov chain X (cf. Eq. (7.15) in [Breiman 1968]),

$$\lim_{L\to\infty}\frac{1}{L}\sum_{d=1}^{L}\chi_n(X_d) = \pi(n) \qquad\qquad \text{w.p.1.}$$

By the strong law of large numbers, we have

$$\lim_{L\to\infty}\frac{\sum_{d=1}^{L}\chi_n(X_d)S_d}{\sum_{d=1}^{L}\chi_n(X_d)} = E\{S_d|n\} \qquad\qquad \text{w.p.1.}$$

The denominator of the left-hand side is the number of state n in $\{X_d, d=1,2, ...,L\}$, and the numerator is the total time that the Markov process is in state n in $[0, T_L]$. Therefore,

$$\lim_{L\to\infty}\frac{1}{L}\sum_{d=1}^{L}\chi_n(X_d)S_d = \lim_{L\to\infty}\{\frac{1}{L}\sum_{d=1}^{L}\chi_n(X_d)\}\times \lim_{L\to\infty}\frac{\sum_{d=1}^{L}\chi_n(X_d)S_d}{\sum_{d=1}^{L}\chi_n(X_d)}$$

$$= \pi(n)\,E\{S_d|n\} = \frac{\pi(n)}{\mu(n)}, \qquad\qquad \text{w.p.1.}$$

Thus, we have

$$\lim_{L \to \infty} R_L(\theta, \xi) = \sum_{\text{all } n} \{\lim_{L \to \infty} \frac{1}{L} \sum_{d=1}^{L} \chi_n(X_d) S_d\}$$

$$= \sum_{\text{all } n} \frac{\pi(n)}{\mu(n)} = R(\theta), \qquad \text{w.p.1.} \qquad (28)$$

Finally, since $E\{R_L(\theta,\xi)\} = R(\theta)$ for all L, we have

$$\lim_{L \to \infty} E[R_L(\theta, \xi)] = R(\theta). \qquad (29)$$

Eqs.(28) and (29) and the fact that $R_L(\theta,\xi) > 0$ imply that $R_L(\theta,\xi)$ is uniformly integrable (see e.g., Problem 16.13 in [Billingsley 1979]).

Next, we study the derivative of $R(\theta)$. We have

$$\frac{\partial[R(\theta)]}{\partial \theta} = \frac{\partial}{\partial \theta} E\{E[R_L(\theta, \xi)|n]\} = \frac{\partial}{\partial \theta} \{\sum_{\text{all } n} E[R_L(\theta, \xi)|n] \pi(n)\}$$

$$= \sum_{\text{all } n} \frac{\partial}{\partial \theta} \{E[R_L(\theta, \xi)|n] \pi(n)\}.$$

The last equation is due to the fact that the number of states is finite. Note that the steady-state probability distribution $\pi(n)$ also depends on $\theta = s_u$, we have

$$\frac{\partial[R(\theta)]}{\partial \theta} = \sum_{\text{all } n} \frac{\partial}{\partial \theta} \{E[R_L(\theta, \xi)|n]\}\pi(n) + \sum_{\text{all } n} E\{R_L(\theta, \xi)|n\}\frac{\partial \pi(n)}{\partial \theta}.$$

Applying the interchangeability to the first term on the right-hand side of the above equation, we get

$$\frac{\partial[R(\theta)]}{\partial \theta} = \sum_{\text{all } n} E\{\frac{\partial}{\partial \theta} R_L(\theta, \xi)|n\}\pi(n) + \sum_{\text{all } n} E\{R_L(\theta, \xi)|n\}\frac{\partial \pi(n)}{\partial \theta}$$

$$= E\{E[\frac{\partial R_L(\theta, \xi)}{\partial \theta} \mid n]\} + \sum_{\text{all } n} \{R_L(\theta \mid n)\} \frac{\partial \pi(n)}{\partial \theta}$$

$$= E\{\frac{\partial [R_L(\theta, \xi)]}{\partial \theta}\} + \sum_{\text{all } n} \{R_L(\theta \mid n)\} \frac{\partial \pi(n)}{\partial \theta}. \tag{30}$$

The second term on the right-hand side of (30) is the bias caused by the change in the probability of the initial state. Letting $L \to \infty$ on both sides of (30), we get

$$\frac{\partial [R_L(\theta)]}{\partial \theta} = \lim_{L \to \infty} E\{\frac{\partial [R_L(\theta, \xi)]}{\partial \theta}\} + \sum_{\text{all } n} \{\lim_{L \to \infty} R_L(\theta \mid n)\} \frac{\partial \pi(n)}{\partial \theta}. \tag{31}$$

Using the uniform integrability of $R_L(\theta, \xi)$ as shown above, we have

$$\lim_{L \to \infty} R_L(\theta \mid n) = \lim_{L \to \infty} E[R_L(\theta, \xi) \mid n_0]$$

$$= E[\lim_{L \to \infty} R_L(\theta, \xi) \mid n_0] = R(\theta)$$

From this equation, the second term on the right-hand side of (31) is

$$\sum_{\text{all } n} \{\lim_{L \to \infty} E[R_L(\theta, \xi) \mid n]\} \frac{\partial \pi(n)}{\partial \theta} = R(\theta) \sum_{\text{all } n} \frac{\partial \pi(n)}{\partial \theta}$$

$$= R(\theta) \frac{\partial}{\partial \theta} \{\sum_{\text{all } n} \pi(n)\} = 0.$$

From this and Eq.(31), we obtain the following conclusion:

In an irreducible single class closed Jackson network,

$$\lim_{L \to \infty} E\{\frac{\partial [R_L(\theta, \xi)]}{\partial \theta}\} = \frac{\partial [R_L(\theta)]}{\partial \theta}. \tag{32}$$

> That is, the sample derivative of $R_L(\theta, \xi)$ is an asymptotically unbiased
> estimate of the derivative of its steady-state mean.

Now we turn to the convergence property of the derivative of the system throughput. The steady-state throughput is defined as

$$TP(\theta) = \sum_{\text{all } n} P(n)\mu(n),$$

where $P(n)$ is the steady-state probability of state n of the Markov processes describing the trajectory of the queueing network,[4] and $\mu(n)$ is the transition rate of the Markov process $N(t, \xi)$ when the system is in state n. We have

$$TP(\theta) = \frac{1}{R(\theta)}.$$

To insure continuity of exposition, we will postpone the proof of this equation to Section 4.6

As will be clear later, it is convenient to study the property of the elasticity. The elasticity of the steady-state throughput is defined as (% change in throughput per unit % change in θ)

$$\frac{\theta}{TP(\theta)} \frac{\partial TP(\theta)}{\partial \theta} = - \frac{\theta}{R(\theta)} \frac{\partial R(\theta)}{\partial \theta}.$$

The throughput based on a single sample path with a finite number of states is

$$TP_L(\theta, \xi) = \frac{L}{T_L(\theta, \xi)} = \frac{1}{R_L(\theta, \xi)}.$$

The sample elasticity of the system throughput is then

$$\frac{\theta}{TP_L(\theta, \xi)} \frac{\partial TP_L(\theta, \xi)}{\partial \theta} = - \frac{\theta}{R_L(\theta, \xi)} \frac{\partial R_L(\theta, \xi)}{\partial \theta} = - \frac{\theta}{T_L(\theta, \xi)} \frac{\partial T_L(\theta, \xi)}{\partial \theta}.$$

---

[4] $P(n)$ is to be distinguished from $\pi(n)$ which is the steady-state probability of n of the embedded Markov chain. They are related, of course. See Section 4.5.

As explained above, the sample elasticity of $T_L(\theta, \xi)$ can be obtained by perturbation analysis. We take this sample elasticity as an estimate of the elasticity of the steady-state throughput. To study the convergence property of this estimate, we need the following technical result (the proof of the result is again postponed to Section 4.6 to avoid disruptions):

> If f_n, n=1,2,... are uniformly integrable and $|g_n| \le |f_n|$ for all n, then g_n, n=1,2,... are also uniformly integrable.

Now, let us study the convergence property of the sample elasticity. Consider

$$f_L(\theta, \xi) \equiv \frac{\theta}{R_L(\theta, \xi)} \frac{\partial R_L(\theta, \xi)}{\partial \theta} - \frac{\theta}{R(\theta)} \frac{\partial R_L(\theta, \xi)}{\partial \theta}$$

$$= \{1 - \frac{R_L(\theta, \xi)}{R(\theta)}\} \times \frac{\theta}{R_L(\theta, \xi)} \frac{\partial R_L(\theta, \xi)}{\partial \theta}.$$

From Eq.(28), we have

$$\lim_{L \to \infty} \frac{R_L(\theta, \xi)}{R(\theta)} = 1, \qquad \text{w.p.1}$$

By Eq.(25) and the fact that the slope of $T_L(\theta, \xi)$ is bounded by $T_L(\theta, \xi)/\theta$, we have

$$0 \le \frac{\theta}{R_L(\theta, \xi)} \frac{\partial R_L(\theta, \xi)}{\partial \theta} = \frac{\theta}{T_L(\theta, \xi)} \frac{\partial T_L(\theta, \xi)}{\partial \theta} \le 1, \qquad \text{for all L>0.}$$

Thus,

$$\lim_{L \to \infty} f_L(\theta, \xi) = 0, \qquad \text{w.p.1.}$$

and

$$|f_L(\theta, \xi)| \le 1 + \frac{R_L(\theta, \xi)}{R(\theta)}.$$

Since $R_L(\theta,\xi)$ are uniformly integrable on $L \in \{1,2,...\}$, $1+R_L(\theta,\xi)/R(\theta)$ are also uniformly integrable (cf. e.g., Problem 16.17 in [Billingsley 1979]). Next, since all the $f_L(\theta,\xi)$'s are bounded by uniformly integrable functions, they are also uniformly integrable. Finally, (cf, Theorem 16.13 in [Billingsley 1979])

$$\lim_{L \to \infty} E\{f_L(\theta, \xi)\} = E\{\lim_{L \to \infty} f_L(\theta, \xi)\} = 0.$$

This is equivalent to

$$\lim_{L \to \infty} E\{\frac{\theta}{R_L(\theta,\xi)} \frac{\partial R_L(\theta,\xi)}{\partial \theta}\} = \frac{\theta}{R(\theta)} \lim_{L \to \infty} E\{\frac{\partial R_L(\theta,\xi)}{\partial \theta}\} = \frac{1}{R(\theta)} \frac{\partial R(\theta)}{\partial \theta}.$$

The last equation is due to Eq. (32). From this, we obtain:

> The sample elasticity of the system throughput is an asymptotically unbiased estimate of the elasticity of the steady-state throughput:
>
> $$\lim_{L \to \infty} E\{\frac{\theta}{TP_L(\theta, \xi)} \frac{\partial[TP_L(\theta, \xi)]}{\partial \theta}\} = \frac{\theta}{TP(\theta)} \frac{\partial TP(\theta)}{\partial \theta}. \qquad (33)$$

4.3 Realization and Sensitivity Analysis

We have shown that, for closed Jackson queueing networks, IPA provides a strongly consistent estimate for the derivative of the transient throughput (based on a sample path with a finite length) with respect to a mean service time (see Section 4.2.2). IPA also provides an asymptotically unbiased estimate for the elasticity of the steady-state throughput with respect to a mean service time. The implementation of perturbation analysis is based on the simple rules of perturbation generation and perturbation propagation.

In this section, we shall introduce a new quantity, called *realization probability*, which measures the ultimate effect of a perturbation on the system and was first intuitively introduced in Section 3.3. The notion of realization probability is closely related to the strong interaction between servers in a closed queueing network. We shall prove that if in a closed network a server's service completion time is somehow perturbed by a small amount, then finally either all the servers in the network will be perturbed by the same amount or all the servers will return back to their normal schedules as if nothing had happened. This gives a clear picture of the evolution of a perturbation in a closed queueing network. The probability that a perturbation is realized measures the average of the final effect of the perturbation on the service completion times of the servers in the network. Next, the effect of an infinitesimal change in a mean service time can be decomposed into the effects of the changes in each individual service time. Using this approach, we can obtain the sensitivity with respect to a mean service time by summing up the final effects of each perturbation induced by the change of the mean service time. Based on this, it is easy to prove that the elasticity of the steady-state throughput with respect to a mean service time can be expressed in terms of the realization probability and the steady-state proba-bility in a simple form. Finally, since realization probability describes the asymptotical behavior of a perturbation, it provides a vehicle of proving convergence theorems. Specifically, we shall prove that the sample elasticity of the steady-state throughput provided by perturbation analysis in a finite period converges to the elasticity of the steady-state throughput with probability one as the length of the period goes to infinity [Cao 1987a]. Note that the result about convergence with probability one is different from, and practically more useful than, that about convergence in mean, which was proved in the previous sections.

In Section 4.3.1, we introduce the concept of realization probability and derive some basic formulas for the realization probability in a closed Jackson network. In Section 4.3.2, we prove some convergence theorems of the perturbation analysis estimate of the elasticity of throughput. Using these theorems, we derive in Section 4.3.3 equations for the elasticity of the steady-state throughput using the realization probability. Although the throughput sensitivity can be calculated by taking derivatives from the Jackson formula, these equations in terms of realization probability provide an alternative way of calculating the throughput sensitivity; the perturbation analysis algorithms can be viewed as the practical implementations of these equations. Finally, Section 4.3.4 presents some other sensitivity calculations which are unique to the perturbation analysis method.

4.3.1 Realization Probability

In this section, we shall define the realization probability and derive a set of equations specifying the realization probability. Consider the class of closed queueing networks described in Section 4.2.1. The network consists of M single-server nodes and N single-class customers. The service time of server i is exponentially distributed with mean $s_i = 1/\mu_i$, and the routing probabilities are $q_{i,j}$, $i,j = 1,2,...,M$. Let $\Gamma = \{1,2,...,M\}$ and $\varnothing$ be the null set.

Suppose that at some time a server in this network obtains a perturbation Δ, which is so small that the deterministic similarity holds during its propagation (we shall specify the size of this perturbation later). Thus, the perturbation will be propagated to other servers according to the propagation rules described in Chapter 3 and Section 4.2 (Eq.(17)). Note that, through propagation, some servers may obtain the perturbation and others may lose the perturbation after obtaining it as a result of the interaction with other servers which have or do not have the perturbation. We use $V(t, \xi)$ to denote the set of integers representing the servers that have the perturbation at time t. The random variable ξ indicates that the set V depends on the sample path of the system. The state of a perturbed system with a perturbation Δ can be denoted as $(N(t, \xi), V(t, \xi))$. Note that both $N(t, \xi)$ and $V(t, \xi)$ change their values only at the transition instants T_d, $d=1,2, ..., L$. Set V is called the *perturbation set* of the system.

Definition 4.4 For any sample path (θ, ξ), the perturbation in $V(t, \xi)$ is said to be realized along the path, if $\lim_{d \to \infty} V(T_d, \xi) = \Gamma$; to be lost, if $\lim_{d \to \infty} V(T_d, \xi) = \varnothing$.

Let

$$\Xi_0 = \{ \xi \in \Xi : \lim_{d \to \infty} V(T_d, \xi) = \emptyset \},$$

$$\Xi_1 = \{ \xi \in \Xi : \lim_{d \to \infty} V(T_d, \xi) = \Gamma \}.$$

Note that sets Γ and $\emptyset$ are two "absorbing" sets with respect to the propagation rules; i.e., if $V(t_1, \xi) = \Gamma$ (or $\emptyset$) at some time t_1, then $V(t, \xi) = \Gamma$ (or $\emptyset$) for all $t > t_1$.

In the following, we shall derive a set of equations which characterize the probabilities of $\xi \in \Xi_1$ and $\xi \in \Xi_0$, denoted as $P(\Xi_1)$ and $P(\Xi_0)$, respectively, and the conditional probabilities of $\xi \in \Xi_1$ and $\xi \in \Xi_0$ given that the initial state is $\mathbf{n}$, denoted as $P(\Xi_1| X_0 = \mathbf{n})$ and $P(\Xi_0| X_0 = \mathbf{n})$ (or simply $P(\Xi_1| \mathbf{n})$ and $P(\Xi_0| \mathbf{n})$), respectively. First, in an irreducible closed queueing network, we have:

$$\lim_{L \to \infty} V(T_d, \xi) = \{\Gamma \text{ or } \emptyset\}, \qquad \text{w.p.1.} \tag{34}$$

or

$$\lim_{t \to \infty} V(t, \xi) = \{\Gamma \text{ or } \emptyset\}, \qquad \text{w.p.1.} \tag{35}$$

The reason for these two equations is simple: The state process of an irreducible closed queueing network is an ergodic Markov process. Thus, every sample path will reach the state $(N, 0, 0,...,0)$ with probability one. At this state, all the servers except server 1 are idle. According to the perturbation propagation rules, the perturbation set will always be either Γ or $\emptyset$ afterwards, depending on whether or not server 1 has the perturbation when the system reaches the state $(N, 0, 0,...,0)$.

Eqs. (34) and (35) can be stated formally as follows: In an irreducible closed queueing network with single-class customers and exponential service times, a perturbation will, with probability one, be realized or lost, i.e.,

$$P(\Xi_0 \cup \Xi_1) = P(\Xi_0) + P(\Xi_1) = 1.$$

From Eqs. (34) and (35), we also have

$P(\Xi_0 \cup \Xi_1 \mid X_0 = n) = P(\Xi_0 \mid X_0 = n) + P(\Xi_1 \mid X_0 = n) = 1.$

Using this result, we can make the following definition:

Definition 4.5 $P(\Xi_1 \mid X_0 = n) = P(\xi \in \Xi : \lim_{t \to \infty} V(t, \xi) = \Gamma \mid X_0 = n)$ is called the *realization probability* of the perturbation in set $V(T_0, \xi)$ at state n.

Recall that the size of the perturbation Δ must be small enough so that the deterministic similarity holds and the simple propagation rules described in Eq. (17) can be applied. Now, since the perturbation will be realized with probability one in a finite number of transitions, e.g., the perturbation will be realized before the state reaches $(N, 0,...,0)$, we only have to choose the size of the perturbation to be so small that the deterministic similarity holds within this finite number of transitions. Therefore, if the perturbation is smaller than the smallest value of S_d, $d = 1,2,..., L^*$, with L^* being the number of transitions before the perturbation is realized, then the deterministic similarity holds and the propa-gation rules for infinitesimal perturbation can be applied before the perturbation is realized. Thus, with probability one there exists a $\Delta > 0$ such that Eqs. (34) and (35) are well-defined.

By the Markov property, the realization probability depends only on the initial condition $V(T_0, \xi)$ and $N(T_0, \xi)$. Thus, we use $f(n, V)$ to denote the realization probability of the perturbation in (n, V). If $V = \{i\}$, it is denoted as $f(n, i)$. The Markov chain $X = \{X_d, d = 0,1,2...\}$ with a perturbation set V can be regarded as a new Markov chain with states (n, V). Eq. (34) indicates that, in this new chain with the augmented states, there are two absorbing states $\{(n, \Gamma): \text{all } n\}$ and $\{(n, \varnothing) : \text{all } n\}$. All other states are transient.

The realization probability satisfies the following properties:

(i)   If $n_i = 0$, then $f(n, i) = 0$.   (36)

(ii) $f(n, \Gamma) = 1$.   (37)

(iii) If $V_1 \cap V_2 = \varnothing$, $V_1 \cup V_2 = V_3$, then $f(\mathbf{n}, V_1) + f(\mathbf{n}, V_2) = f(\mathbf{n}, V_3)$. $\hspace{3em}$ (38)

(iv) $f(\mathbf{n}, 1) + f(\mathbf{n}, 2) + ... + f(\mathbf{n}, M) = 1$. $\hspace{3em}$ (39)

Eq.(36) involves a convention, i.e., we may consider an idle server as having a perturbation. This perturbation will be lost immediately after the server receives a customer from any other server which does not have the perturbation. Eq.(37) follows directly from the "absorbing" property of set Γ. Eq.(38) represents superposition. Eq.(39) is a consequence of (37) and (38). The proof of (38) is postponed to the next subsection.

The realization probability in a closed Jackson network satisfies the following set of linear equations:

$$\{\sum_{i=1}^{M} \epsilon(n_i)\,\mu_i\}\, f(\mathbf{n}, k) = \sum_{i=1}^{M}\sum_{j=1}^{M}\epsilon(n_i)\,\mu_i\,q_{i,j}\,f(\mathbf{n}_{i,j}, k)$$

$$+\sum_{j=1}^{M}\{1 - \epsilon(n_j)\}\,\mu_k\,q_{k,j}\,f(\mathbf{n}_{k,j}, j), \hspace{2em} n_k>0,\ k\in\Gamma. \hspace{1em} (40)$$

These equations are conceptually similar to the flow balance equations of the network (see Appendix A.4.1 Eq.(15)). To prove these equations, recall that a state $\mathbf{n}$ has a probability $[\epsilon(n_j)\mu_j / \Sigma\,\epsilon(n_k)\mu_k]\,q_{i,j}$ of transferring to state $\mathbf{n}_{i,j}$. The terms in the last summation in Eq.(40) are equal to 0 except when $n_j=0$. These terms represent the perturbation propagation effect. The superposition property $f(\mathbf{n}, \{k,j\}) = f(\mathbf{n}, k) + f(\mathbf{n}, j)$ is employed in the derivation of (40).

Solving Eqs. (36) - (40), we can obtain the values of all $f(\mathbf{n}, V)$. For symmetric closed Jackson networks with $\mu_i\,q_{i,j} = \mu_j\,q_{j,i}$, we have a closed form solution for these equations.

Example 4.1 Consider a closed Jackson network in which $\mu_i\,q_{i,j} = \mu_j\,q_{j,i}$. We can prove that the solution to Eqs.(36) - (40) is

$$f(\mathbf{n}, V) = \frac{\displaystyle\sum_{i \in V} n_i}{N}.$$ (41)

For $V = \{k\}$, we have

$$f(\mathbf{n}, k) = \frac{n_k}{N}, \qquad k \in V.$$ (42)

Obviously, (41) satisfies (36) - (40) by direct substitution and summation. To check that (41) also satisfies (40), we assume that there are M_0 idle servers in state $\mathbf{n}$, $0 \le M_0 < M$. Without loss of generality, we label these servers 1,2, ..., M_0, and consider $k > M_0$. Thus, $\varepsilon(n_i) = 0$, if $i \le M_0$; $\varepsilon(n_i) = 1$, if $i > M_0$. Now the right-hand side of Eq.(40) is

$$\sum_{i=M_0+1, i \neq k}^{M} \sum_{j \neq k}^{M} \mu_i q_{i,j} f(\mathbf{n}_{i,j}, k) + \mu_k q_{k,k} f(\mathbf{n}, k) + \sum_{i=M_0+1, i \neq k}^{M} \mu_i q_{i,k} f(\mathbf{n}_{i,k}, k)$$

$$+ \sum_{j \neq k}^{M} \mu_k q_{k,j} f(\mathbf{n}_{k,j}, k) + \sum_{j=1}^{M_0} \mu_k q_{k,j} f(\mathbf{n}_{k,j}, k)$$

$$= \sum_{i=M_0+1, i \neq k}^{M} \sum_{j \neq k}^{M} \mu_i q_{i,j} \frac{n_k}{N} + \mu_k q_{k,k} \frac{n_k}{N} + \sum_{i=M_0+1, i \neq k}^{M} \mu_i q_{i,k} \frac{n_k+1}{N}$$

$$+ \sum_{j \neq k}^{M} \mu_k q_{k,j} \frac{n_k-1}{N} + \sum_{j=1}^{M} \mu_k q_{k,j} \frac{1}{N}$$

$$= \sum_{i=M_0+1}^{M} \sum_{j=1}^{M} \mu_i q_{i,j} \frac{n_k}{N} + \Big\{ \sum_{i=M_0+1, i \neq k}^{M} \mu_i q_{i,k} \frac{1}{N} - \sum_{j \neq k}^{M} \mu_k q_{k,j} \frac{1}{N}$$

$$+ \sum_{j=1}^{M_0} \mu_k q_{k,j} \frac{1}{N} \Big\}$$

$$= \frac{n_k}{N} [\sum_{i=M_\sigma+1}^{M} \mu_i],$$

which equals the left-hand side of (40). By $\mu_k q_{k,j} = \mu_j q_{j,k}$, it is easy to check that the terms in the braces sum up to zero.

4.3.2 Realization Index and the Convergence Theorems

In Section 4.3.1, we introduced the concept of realization probability and derived some equations which specify the realization probability in a closed Jackson network. As mentioned earlier, one of the purpose of introducing this concept is to prove the convergence w.p.1 property of the perturbation analysis sensitivity estimate. The perturbation analysis estimate is a sample derivative based on a single sample path of a network. In order to study the perturbation realization on a sample path, we need a sample-path version of the realization probability, called the *realization index.*

Definition 4.6   Let V be the perturbation set at the dth state of a sample path $N(t, \xi)$ and $X_d = n$. The realization index of the perturbation on the sample path is

$$RI(n, V, d) = \begin{cases} 1 & \text{if the perturbation is realized on } N(t, \xi), \\ 0 & \text{otherwise.} \end{cases}$$

By the Markov property, for any V and n, the expected value of $RI(n, V, d)$ does not depend on d. By definition, we have

$$E\{RI(n, V, d)\} = f(n, V), \qquad \text{for all d.} \tag{43}$$

First of all, four properties similar to (36) - (39) hold for realization indices. In particular, for all d, the following equation holds:

If $V_1 \cap V_2 = \varnothing$, $V_1 \cup V_2 = V_3$,

then $RI(n, V_1, d) + RI(n, V_2, d) = RI(n, V_3, d)$ \qquad (44)

To prove this equation, it is convenient to use an M-dimensional 0-1 vector to denote the set V. As an illustrative example, assume that at the dth state there are three perturbed sets V_1, V_2, and V_3 corresponding to the following three vectors, respectively:

$g_1 = (0, 1, 0, 0, 1, 1, 0, 0, 0)$;

$g_2 = (0, 0, 1, 0, 0, 0, 0, 1, 1)$;

$g_3 = (0, 1, 1, 0, 1, 1, 0, 1, 1)$.

For instance, the vector g_1 shows that $V = \{2, 5, 6\}$. Note that $V_1 \cap V_2 = \varnothing$ and $V_1 \cup V_2 = V_3$ is equivalent to

$$g_{3,k} = g_{1,k} + g_{2,k} \qquad\qquad 1 \leq k \leq M \qquad\qquad\qquad (45)$$

where $g_{i,k}$ is the kth component of vector g_i, i=1,2,3. By the perturbation propagation rules, if no server meets an idle period, then g_1, g_2, and g_3 remain the same; if server j receives a customer from server i after an idle period, then server j will possess the same perturbation as server i after this customer transition. This is simply copying the ith column of the above array to its jth column. Thus, Eq.(45) always holds after the dth state. In particular, as d goes to infinity, (45) still holds. However, by Eq. (35), g_1, g_2, and g_3 will converge to a vector whose components are either all zeros or all ones. Therefore, only three cases are possible: (i) all g_1, g_2, and g_3 converge to vectors with all zeros; in this case, $RI(n,V_1,d) = RI(n,V_2,d) = RI(n,V_3,d) = 0$. (ii) g_1 and g_3 converge to vectors with all ones and g_2 converges to a vector with all zeros; in this case, $RI(n,V_1,d) = RI(n,V_3,d) = 1$ and $RI(n,V_2,d) = 0$. (iii) g_2 and g_3 converge to vectors with all ones and g_1 converges to a vector with all zeros; in this case $RI(n,V_2,d) = RI(n,V_3,d) = 1$ and $RI(n,V_1,d) = 0$. For all these three cases, (44) holds.

Taking the expectation of the both sides of (44) and using (43), we obtain (38).

In the rest of this subsection, we shall prove the following equation:

$$\lim_{L \to \infty} \frac{1}{T_L} \sum_{d=1}^{L} S_d RI(n, i, d) = \sum_{\text{all } n} P(n) f(n, i) \qquad \text{w.p. 1.}$$

In the next subsection, we shall see that the left-hand side of this equation is, in fact, equal to the sample elasticity of the throughput with respect to s_i.

The proof of the above equation depends heavily on the ergodicity of the state process. First, we note that whether a perturbation is realized or lost depends on the initial state n and the customer transitions afterwards. It is independent of the sojourn times in each state, S_d, S_{d+1},..... Therefore, $RI(n, V, d)$ is a function of X_d, X_{d+1},..... We denote this relation by

$$RI(n, V, d) = \Phi(X_d, X_{d+1}, \ldots), \qquad \text{for } X_d = n.$$

Φ is a function which depends on V and is independent of d.

Recall that $\chi_n(X_d) = 1$, if $X_d = n$; 0, otherwise. Note that for all d such that $\chi_n(X_d) = 1$, S_d are independent and identically distributed. By the law of large numbers, we have

$$\lim_{L \to \infty} \frac{\sum_{d=1}^{L} S_d RI(n, i, d) \chi_n(X_d)}{\sum_{d=1}^{L} RI(n, i, d) \chi_n(X_d)} = E\{S_d \mid X_d = n, RI(n, i, d) = 1\}, \quad \text{w.p.1,}$$

where the value of the denominator is the number of non-zero terms in the numerator. From this, we have

$$\lim_{L \to \infty} \frac{1}{L} \sum_{d=1}^{L} S_d RI(n, i, d) \chi_n(X_d)$$

$$= E\{S_d \mid X_d = n, RI(n, i, d) = 1\} \times \lim_{L \to \infty} \frac{1}{L} \sum_{d=1}^{L} RI(n, i, d) \chi_n(X_d)$$

$$= E\{S_d \mid X_d = n, RI(n, i, d) = 1\} \times \lim_{L \to \infty} \frac{1}{L} \sum_{d=1}^{L} Y_d,$$

where $Y_d = RI(n, i, d)\chi_n(X_d) = \Phi(X_d, X_{d+1}, ...)\chi_n(X_d) \equiv \Psi(X_d, X_{d+1},...)$. For the ergodic process $X_d, X_{d+1}, ...,$ the process $Y_d, Y_{d+1},...$ is also ergodic (see Proposition 6.3 in [Breiman 1968]). Thus,

$$\lim_{L\to\infty}\frac{1}{L}\sum_{d=1}^{L} Y_d = E(Y_d) \qquad\qquad\qquad\qquad \text{w.p.}1$$

$$= E\{RI(n, i, d)\,\chi_n(X_d)\} = P\{\,RI(n, i, d)\,\chi_n(X_d) = 1\} \qquad\qquad \text{w.p.}1.$$

Therefore,

$$\lim_{L\to\infty}\frac{1}{L}\sum_{d=1}^{L} S_d RI(n, i, d)\chi_n(X_d)$$

$$= E\{S_d \mid X_d = n,\ RI(n, i, d) = 1\,\} \times P\{\,X_d = n,\ RI(n, i, d) = 1\}$$

$$= E\{S_d\,\chi_n(X_d)\,RI(n, i, d)\} = E\{\,S_d\,RI(n, i, d)\mid X_d = n\,\} \times \pi(n)$$

where $\pi(n) = \pi(X_d=n)$ is the steady-state probability of state n in the imbedded chain $X = \{X_d, d=1,2,...\}$. By the Markov property of $N(t, \xi)$, for any given $X_d = n$, two random variables S_d and $RI(n,i,d)$ are independent. Thus,

$$E\{S_d\,RI(n, i, d)\mid X_d = n\,\}$$

$$= E\{S_d \mid X_d = n\}E\{RI(n, i, d)\mid X_d = n\} = \frac{f(n, i)}{\mu(n)}.$$

The independence can also be proved by using a reversibility argument (see [Kelly 1979]). The time reversed process $N(-t, \xi)$ of a closed Jackson queueing network is also a stationary Markov process. Hence the length S_d in the time reversed process does not depend on all previous states, $X_{d+1}, X_{d+2},...,$ which determine $RI(n, i, d)$. Finally, we have

$$\lim_{L \to \infty} \frac{1}{L} \sum_{d=1}^{L} S_d RI(n, i, d) \chi_n(X_d) = f(n, i) \frac{\pi(n)}{\mu(n)} \qquad \text{w.p.1.}$$

It is shown in Section 4.6.B that

$$P(n) = \alpha \frac{\pi(n)}{\mu(n)},$$

where

$$\alpha = \frac{1}{\displaystyle\sum_{\text{all } n} \frac{\pi(n)}{\mu(n)}}$$

is a normalizing constant. From this,

$$\lim_{L \to \infty} \frac{1}{L} \sum_{d=1}^{L} S_d RI(n, i, d) \chi_n(X_d) = \frac{1}{\alpha} P(n) f(n, i) \qquad \text{w.p.1.}$$

Note that $\sum_{\text{all } n} \chi_n(X_d) = 1$. Summing both sides of the above equation over all n, we get

$$\lim_{L \to \infty} \frac{1}{L} \sum_{d=1}^{L} S_d RI(n, i, d) = \frac{1}{\alpha} \sum_{\text{all } n} P(n) f(n, i), \qquad \text{w.p.1.} \qquad (46)$$

On the other hand, using the same approach, we can prove

$$\lim_{L \to \infty} \frac{T_L}{L} = \lim_{L \to \infty} \frac{1}{L} \sum_{d=1}^{L} S_d = \lim_{L \to \infty} \frac{1}{L} \sum_{\text{all } n} \sum_{d=1}^{L} S_d \chi_n(X_d)$$

$$= \sum_{\text{all } n} \{ \lim_{L \to \infty} \frac{1}{L} \sum_{d=1}^{L} S_d \chi_n(X_d) \}$$

$$= \sum_{\text{all } n} E[\chi_n(X_d)] \times E(S_d \mid X_d = n) \qquad \text{w.p.1}$$

$$= \sum_{\text{all } n} \frac{\pi(n)}{\mu(n)} = \frac{1}{\alpha}, \qquad \text{w.p.1.} \qquad (47)$$

Note that

$$\lim_{L\to\infty}\frac{1}{T_L}\sum_{d=1}^{L}S_d RI(\mathbf{n}, i, d) = \lim_{L\to\infty}\frac{L}{T_L}\times\lim_{L\to\infty}\frac{1}{L}\sum_{d=1}^{L}S_d RI(\mathbf{n}, i, d).$$

From this and Eqs. (46) and (47), we obtain the following result:

For closed Jackson networks,

$$\lim_{L\to\infty}\frac{1}{T_L}\sum_{d=1}^{L}S_d RI(\mathbf{n}, i, d) = \sum_{\text{all }\mathbf{n}} P(\mathbf{n})f(\mathbf{n}, i), \qquad \text{w.p.1.} \qquad (48)$$

The left-hand side of Eq.(48) is based on a single sample path of a net-work. $RI(\mathbf{n},i,d)$ describes the final effect of a perturbation generated in the dth state of the Markov process. The perturbation generated in the dth state is pro-portional to S_d, and $S_d RI(\mathbf{n},i,d)$ is proportional to the finally realized perturba-tion, which can be determined by propagating the perturbation along the sample path. Thus, the left-hand side of Eq.(48) can be considered as the limiting value of the perturbation analysis estimate. Eq.(48) provides an analytical formula for calculating this limiting value. However, to implement the left-hand side of (48) in practice is somewhat inconvenient. After the time period in which L service completions occur, one has to continue observing the system until all perturbations generated in that period have been either realized or lost. Sometimes this additional observation period may be too long. No practical algorithms currently in use are written in this way. Besides, the left-hand side of (48) does not correspond to the sample derivative which is based on the sample path in $[0, T_L]$. Below we shall provide a more practical result.

To express precisely the final effect of the perturbation in a finite period, we need a new definition. Let T_L be a service completion time of server v, i.e., $T_L = t_{v,k}$ for some v and k . (We assume that the last service completion hap-pens at server v.) We define:

$$
RI_L(n, V, d) = \begin{cases} 1 & \text{if the perturbation in } (n, V, d) \text{ is propagated} \\ & \text{to server } v \text{ at } T_L, \\ 0 & \text{if not.} \end{cases}
$$

A more practical version of Eq. (48) is stated as follows.

For closed Jackson networks,

$$
\lim_{L \to \infty} \frac{1}{T_L} \sum_{d=1}^{L} S_d RI_L(n, i, d) = \sum_{\text{all } n} P(n)f(n, i), \qquad \text{w.p.1.} \quad (49)
$$

The proof of this result is lengthy and purely technical, and is provided in Section 4.6.D. The result can be intuitively explained. By definition, we have

$$
\lim_{L \to \infty} RI_L(n, i, d) = RI(n, i, d).
$$

Thus,

$$
RI_L(n, i, d) - RI(n, i, d) \neq 0
$$

holds only if L-d is relatively small. As L becomes bigger, the difference between the both sides of (48) and (49) becomes smaller.

4.3.3 Sensitivity of Throughputs

In this section, we shall prove that the left-hand side of Eq.(49) is just the sample elasticity of the throughput with respect to the mean service time of server i. This elasticity can be easily obtained by applying the perturbation generation and propagation rules to a sample path of the system. Hence, Eq.(49) gives the limiting value of the sample elasticity estimate provided by perturbation analysis. We shall also show that this value i s the same as the elasticity of the steady-state throughput with respect to the mean service time. That is, the sample elasticity given by perturbation analysis converges to the elasticity of the steady-state throughput with probability one as the length of the sample path goes to infinity.

Let $TP_L(s_u, \xi)$ be the sample system throughput based on the sample path (s_u, ξ) in $[0, T_L]$. $TP_L(s_u, \xi)$ is defined as

$$TP_L(s_u, \xi) = \frac{L}{T_L(s_u, \xi)}.$$

The sample elasticity of the system throughput is

$$\frac{s_u}{TP_L(s_u, \xi)} \frac{\partial[TP_L(s_u, \xi)]}{\partial s_u} = - \frac{s_u}{T_L(s_u, \xi)} \frac{\partial T_L(s_u, \xi)}{\partial s_u}$$

$$= \lim_{\Delta s_u \to 0} \{ - \frac{s_u}{T_L(s_u, \xi)} \frac{\Delta T_L(s_u, \xi)}{\Delta s_u} \}.$$

where $\Delta T_L(s_u, \xi)$ is the perturbation of $T_L(s_u, \xi)$, which is assumed to be a service completion time of server v, because of the change in server u's service time s_u, Δs_u. By the results in Section 4.2.1, if $|\Delta s_u| < s_u (S_{min} / T_L)$, then $T_L(s_u, \xi)$ is a linear function of s_u. Therefore, if $|\Delta s_u| < s_u (S_{min} / T_L)$, we have

$$\frac{s_u}{TP_L(s_u, \xi)} \frac{\partial[TP_L(s_u, \xi)]}{\partial s_u} = - \frac{s_u}{T_L(s_u, \xi)} \frac{\Delta T_L(s_u, \xi)}{\Delta s_u}. \tag{50}$$

According to perturbation analysis rules, $\Delta T_L(s_u, \xi)$ can be determined in two steps: (i) At each service completion time of server u, generate a perturbation at server u; (ii) propagate all the perturbations generated at server u's service completion times along the sample path. $\Delta T_L(s_u, \xi)$ is the sum of all the perturbations propagated to server v at T_L. By Eq. (16), the perturbation generated at service completion time $t_{u,k}$ is

$$\Delta s_{u,k} = \frac{\Delta s_u}{s_u} s_{u,k} = \lambda s_u, \qquad\qquad \lambda = \frac{\Delta s_u}{s_u}.$$

We assume that $t_{u,k}$ is the d_kth transition time of the Markov process $N(t, \xi)$, i.e.,

$$t_{u,k} = T_{d_k}.$$

Let $n(d_k)$ be the system state before the kth service completion time of server u. Then by the definition of $RI_L(n(d_k), u, d_k)$, we have

$$\Delta T_L = \sum_{k\,:\,t_{u,k}\leq T_L} \Delta s_{u,k}\, RI_L(n(d_k),u,\ d_k)$$

$$= \lambda \sum_{k\,:\,t_{u,k}\leq T_L} s_{u,k}\, RI_L(n(d_k),u,\ d_k). \tag{51}$$

Note that during a customer's service time at server u, customers in other servers may transfer from one server to another, i.e., the system may traverse several states. Let $n(d_{k-1}+1)$, $n(d_{k-1}+2)$,..., $n(d_k)$ be the sequence of states that the system visits during the service time of the kth customer at server u. The state transition times at the end of these states are

$$T_{d_{k-1}+1},\ T_{d_{k-1}+2},...,\ T_{d_k}.$$

Thus, $s_{u,k}$ can be decomposed as

$$s_{u,k} = S_{d_{k-1}+1} + S_{d_{k-1}+2} +... + S_{d_k}. \tag{52}$$

During a customer's service time at server u, there is no customer transition from server u to any other server. Thus, the perturbation at server u can not be propagated to any other server. Therefore, for any sample path it holds that

$$RI_L(n(d_{k-1}+1),\ u,\ d_{k-1}+1) = RI_L(n(d_{k-1}+1),\ u,\ d_{k-1}+1)$$

$$=... =\ RI_L(n(d_k),\ u,\ d_k).$$

From this equation and Eq. (52), we get

$$\lambda s_{u,k}\, RI_L(n(d_k),\ u,\ d_k) = \lambda S_{d_{k-1}+1}\, RI_L(n(d_{k-1}+1),\ u,\ d_{k-1}+1)$$

$$+ \lambda S_{d_{k-1}+2}\, RI_L(n(d_{k-1}+2),\ u,\ d_{k-1}+2)+...+\ \lambda S_{d_k}\, RI_L(n(d_k),\ u,\ d_k).$$

Substituting the above equation to Eq. (51), we have

$$\Delta T_L = \lambda \sum_{\substack{0 \le d \le L, \\ (X_d)_u \ne 0}} S_d \, RI_L(n, u, d),$$

where $(X_d)_u$ is the uth component of X_d, i.e., the number of customers in server u at T_d. The indices d for which $(X_d)_u=0$ are excluded in the above expression because there is no perturbation generated in idle periods of server u. However, if $(X_d)_u=0$, then $RI_L(n, u, d) = RI_L(X_d, u, d)=0$. Thus, we can rewrite the above equation as

$$\Delta T_L = \lambda \sum_{d=1}^{L} S_d RI_L(n, u, d).$$

Therefore, from Eq. (50), we obtain

$$\frac{s_u}{TP_L(s_u, \xi)} \frac{\partial[TP_L(s_u, \xi)]}{\partial s_u} = -\frac{1}{T_L} \sum_{d=1}^{L} S_d \, RI_L(n, u, d). \tag{53}$$

This equation expresses the sample elasticity in terms of realization indices. Using (49) and (53), we can obtain the following convergence property of the sample elasticity of the system throughput.

For closed Jackson networks, the sample elasticity obtained by perturbation analysis satisfies

$$\lim_{L \to \infty} \left\{ \frac{s_u}{TP_L(s_u, \xi)} \frac{\partial[TP_L(s_u, \xi)]}{\partial s_u} \right\} = -\sum_{\text{all } n} P(n)f(n, u), \quad \text{w.p.1.} \tag{54}$$

Next, since $| RI_L(n, u, d) | \le 1$, from (53), we have

$$\frac{s_u}{TP_L(s_u, \xi)} \frac{\partial TP_L(s_u, \xi)}{\partial s_u} | \le \frac{1}{T_L} \sum_{d=1}^{L} S_d \le 1.$$

By the bounded convergence theorem [Billingsley 1979], we have

$$\lim_{L \to \infty} E\{\frac{s_u}{TP_L(s_u, \xi)} \frac{\partial[TP_L(s_u, \xi)]}{\partial s_u}\} = - \sum_{\text{all } n} P(n)f(n, u), \quad \text{w.p.1.} \quad (55)$$

Finally, the steady-state throughput is defined as

$$TP(s_u) = \frac{L}{E[T_L(s_u, \xi)]} .$$

Recall that the sample elasticity of the system throughput is an asymptotically unbiased estimate of the steady-state throughput (Eq.(33)). Thus, from Eq.(55), we conclude

For a closed Jackson network, it holds

$$\frac{s_u}{TP(s_u)} \frac{\partial TP(s_u)}{\partial s_u} = - \sum_{\text{all } n} P(n)f(n, u), \quad u \in \Gamma. \quad (56)$$

Eq.(56) provides an alternative way of calculating the sensitivity of the steady-state throughput in a closed Jackson network. Although the sensitivity can be obtained by directly taking the derivative from the well-known Jackson formula, this equation provides a new insight into the problem. The sensitivity expressed in the equation is based on only one sample path, rather than based on the difference of two steady-state throughputs. In fact, Eq.(56) can be viewed as the theoretical basis of the infinitesimal perturbation analysis of closed Jackson network. Combining Eqs.(48), (49), and (56), we obtain the following equations for closed Jackson networks:

$$\lim_{L \to \infty} \frac{1}{T_L} \sum_{d=1}^{L} S_d RI(n, u, d) = - \frac{s_u}{TP(s_u)} \frac{\partial TP(s_u)}{\partial s_u}, \quad \text{w.p.1.} \quad (57)$$

and

$$\lim_{L \to \infty} \frac{1}{T_L} \sum_{d=1}^{L} S_d RI_L(n, u, d) = - \frac{s_u}{TP(s_u)} \frac{\partial TP(s_u)}{\partial s_u}, \quad \text{w.p.1.} \quad (58)$$

As we have discussed, the above equations are equivalent to

$$\lim_{L \to \infty} \frac{1}{T_L} \sum_{k\,:\, t_{u,k} \le T_L}^{L} s_{u,k}\, RI(n(t_{u,k}), u, d_k)$$

$$= - \frac{s_u}{TP(s_u)} \frac{\partial TP(s_u)}{\partial s_u}, \qquad\qquad \text{w.p.1}, \qquad (59)$$

and

$$\lim_{L \to \infty} \frac{1}{T_L} \sum_{k\,:\, t_{u,k} \le T_L}^{L} s_{u,k}\, RI_L(n(t_{u,k}), u, d_k)$$

$$= - \frac{s_u}{TP(s_u)} \frac{\partial TP(s_u)}{\partial s_u}, \qquad\qquad \text{w.p.1}, \qquad (60)$$

These two equations explain the practical implementation of perturbation analysis: At each service completion time a perturbation of a size equal to the service time $s_{u,k}$ is generated. All the perturbations generated are propagated according to the IPA propagation rules along a sample path and accumulated at T_L. The sum of all the perturbations propagated to T_L divided by T_L yields an estimate of the elasticity of the steady-state throughput. The estimate is based on only a single sample path and converges to the elasticity of the steady-state throughput with probability one. These two equations also show that for computation purpose it is not necessary to generate a small perturbation proportional to the service time. Since the constant λ in $\Delta s_{u,k} = \lambda s_{u,k}$ cancels out in the final expression of the derivative, we can simply use $s_{u,k}$ as a perturbation. This simplifies the calculation and improve the numerical variance property of the estimate.

We have achieved the goal set up at the beginning of this chapter for proving the convergence properties of the IPA estimates for closed Jackson networks. We have proved that the IPA estimate of the sensitivity of $R_L(\theta, \xi)$ in a finite period is unbiased and the IPA estimate of the throughput in a finite period (expressed in Eq. (24)) is strongly consistent. Furthermore, we have proved that the IPA estimate of the elasticity of the throughput converges both in mean and with probability one to the elasticity of the steady-state throughput as the length of the sample path goes to infinity. We have also proved that this elasticity equals the expected value of realization probability.

Before ending this section, we give two examples which verify the results about the realization probability.

Example 4.2 Consider the same system as in Example 4.1, i.e., a closed Jackson network with $\mu_i \, q_{i,j} = \mu_j \, q_{j,i}$. We have proved $f(n, u) = n_u / N$. Substituting this into Eq.(56), we get

$$\frac{s_u}{TP(s_u)} \frac{\partial TP(s_u)}{\partial s_u} = - \sum_{\text{all } n} \frac{n_u}{N} P(n) = - \sum_{n_u} \{ \frac{n_u}{N} \sum_{n_j \; j \neq u} P(n) \}$$

$$= - \sum_{n=1}^{N} \frac{n}{N} P(n_u = n) = - \frac{1}{N} Q_N(u),$$

where $Q_N(u)$ is the average queue length of the uth server in a network with a total of N customers. From Buzen's algorithm [Buzen 1973],

$$P(n) = \frac{1}{G(N)} \prod_{i=1}^{M} (y_i)^{n_i};$$

where

$$G(N) = \sum_{n_1 + \dots n_M = N} \prod_{i=1}^{M} (y_i)^{n_i};$$

and $(y_1, y_2, \dots, y_M)$ is a solution to the equation

$$\mu_j y_j = \sum_{i=1}^{M} \mu_i y_i \, q_{i,j}.$$

In our example, $\mu_i \, q_{i,j} = \mu_j \, q_{j,i}$. From this we find that $y_1 = y_2 = \dots = y_M = 1$ is a solution to the equation. Thus,

$$G(N) = \binom{M+N+1}{N}.$$

The probabilities $P(n)$ for all n have the same value $1/G(N)$. After some manipulations, we get

$$P(n_u = n) = \frac{\binom{M+N-n-2}{M-2}}{\binom{M+N+1}{N}}.$$

From this we have

$$Q_N(u) = \sum_{n=1}^{N} n\, P(n_u = n) = \frac{N}{M},$$

and

$$\frac{s_u}{TP(s_u)} \frac{\partial TP(s_u)}{\partial s_u} = -\frac{1}{M}.$$

This agrees exactly with the results derived from the well-known Jackson formulas (see Williams and Bhandiwad [1976]),

$$\frac{s_u}{TP(s_u)} \frac{\partial TP(s_u)}{\partial s_u} = Q_{N-1}(u) - Q_N(u). \tag{61}$$

Example 4.3 Consider a closed network in which $M=3$, $N=8$, $s_1=5$, $s_2=10$, and $s_3=12$. The routing probabilities are $q_{i,i}=0$ for $i=1,2,3$; $q_{1,2}=q_{1,3}=0.5$, $q_{2,1}=0.7$, $q_{2,3}=0.3$, $q_{3,1}=0.4$, and $q_{3,2}=0.6$. Eqs. (36) - (40) are solved numerically. The elasticities are then calculated using Eq.(56). They are 0.0365, 0.5133, and 0.4502. These values are exactly the same as those given by Eq.(61).

4.3.4 Calculation of Other Sensitivities

The realization probability $f(n, i)$ characterizes the sensitivity of the system throughput with respect to the service time change when the system is in state n. Similar to Eq. (48), we have

$$\lim_{L \to \infty} \frac{1}{T_L} \sum_{d=1}^{L} S_d RI(n, i, d) \chi_n(X_d) = P(n)f(n, u) \qquad \text{w.p.1.}$$

It is more reasonable to use service rate rather than service time as a system parameter in systems where the service provided by a server depends on system states. Thus, we let $\mu_{i,n}$ be the service rate of server i when the system state is **n**. By the same analysis as in Section 4.3.3, we can prove

$$\frac{\mu_{u,n}}{TP(\mu_{u,n})} \frac{\partial TP(\mu_{u,n})}{\partial \mu_{u,n}} = P(n)f(n, u).$$

The partial derivative $\partial[TP(\mu_{u,n})] / \partial \mu_{u,n}$ measures the rate of change in the steady-state throughput when the service rate $\mu_{u,n}$ changes and all other service rates stay the same. Note that the perturbed system is no longer a state independent Jackson network. Hence this sensitivity can not be obtained by using the existing formulas.

Similarly, let $\mu_{i,j}$ be the service rate of server i when there are $n_i = j$ customers in it. Then

$$\frac{\mu_{u,j}}{TP(\mu_{u,j})} \frac{\partial TP(\mu_{u,j})}{\partial \mu_{u,j}} = \sum_{n_u = j} P(n)f(n, u).$$

The partial derivative $\partial[TP(\mu_{u,j})] / \partial \mu_{u,j}$ measures the rate of change in the steady-state throughput when the service rate $\mu_{u,j}$ changes and all other service rates stay the same.

Now let us further consider the following sensitivity problem: What is the sensitivity of the system throughput in a closed Jackson network if the service time of every customer at server u changes by the same amount Δ? That is, what is the sensitivity if we let $s'_{u,k} = s_{u,k} + \Delta$ for all k (see the Exercise in Section 3.2)? In this case, $\Delta s_u = \Delta$. By perturbation analysis rules, we generate pertur-bations at every service completion time at server u. All these perturbations have the same size Δ. The essential difference between this problem and that discussed in the previous sections is that the perturbation Δ

is not proportional to the service time any more. We cannot replace the perturbation of the service time Δ by a series of perturbations of the sojourn times that the system stays in the states that it traverses during the service time. Thus, the perturbations have to be generated at the service completion times and then be propagated. Define $\Lambda_i = \{t_{i,k},\ k=1,2,...\ \}$. Then $\{T_d,\ d=1,2,...\ \} = \cup_{i \in \Gamma}\ \Lambda_i$. Let

$$\chi_{\Lambda_u}(d) = \begin{cases} 1 & \text{if } T_d \in \Lambda_u, \\ 0 & \text{if } T_d \notin \Lambda_u. \end{cases}$$

By an analysis similar to Eq. (58), we can obtain the sensitivity in the sense that every service time of server u changes by the same amount Δ:

$$\frac{s_u}{TP(s_u)} \frac{\partial TP(s_u)}{\partial s_u}$$

$$= -\lim_{L \to \infty} \{\frac{1}{T_L} \sum_{k:\ t_{u,k} \le T_L} \Delta\, RI(n(t_{u,k}),\, u,\, d_k)\} \frac{s_u}{\Delta} \qquad \text{w.p.1}$$

$$= -\lim_{L \to \infty} \{\frac{1}{T_L} \sum_{d=1}^{L} s_u\, RI(n(T_d),\, u,\, d)\chi_{\Lambda_u}(d)\} \qquad \text{w.p.1}$$

$$= -\lim_{L \to \infty} \sum_{\text{all } n} \{\frac{1}{T_L} \sum_{d=1}^{L} s_u\, RI(n(T_d),\, u,\, d)\ \chi_{\Lambda_u}(d)\ \chi_n(X_d)\}, \quad \text{w.p.1.}$$

$$\tag{62}$$

By the law of large numbers, we have

$$\lim_{L \to \infty} \{\frac{1}{L} \sum_{d=1}^{L} RI(n(T_d),\, u,\, d)\ \chi_{\Lambda_u}(d)\ \chi_n(X_d)\}$$

$$= E\{RI(n(T_d),\, u,\, d)\chi_{\Lambda_u}(d)\ \chi_n(X_d)\} \qquad \text{w.p.1.}$$

$$= E\{RI(n(T_d),\, u,\, d)\ |\ n(T_d){=}n,\ T_d{\in}\Lambda_u\}E\{\chi_{\Lambda_u}(d)\ \chi_n(X_d)\}.$$

Let $g(n, u) = E\{RI(n(T_d), u, d) \mid n(T_d) = n, T_d \in \Lambda_u\}$ be the realization proba-bility of a perturbation at server u's completion time and when the system is in state n before this time instant. Note that $g(n,u)$ is different from $f(n, u)$ because at the service completion time of server u, a customer will definitely leave the server and will enter server j with a probability $q_{u,j}$. Also,

$$E\{\chi_{\Lambda_u}(d)\, \chi_n(X_d)\} = P\,(T_d \in \Lambda_u, X_d = n) = P\,(T_d \in \Lambda_u \mid X_d = n)\, \pi(n)$$

$$= \sum_{j=1}^{M} P(X_{d+1} = n_{u,j} \mid X_d = n)\, \pi(n) = \frac{\pi(n)}{\mu(n)}\, \mu_u, \qquad n_u > 0.$$

The last equation is due to

$$P(X_{d+1} = n_{i,j} \mid X_d = n) = \frac{\varepsilon(n_i)\, \mu_i}{\mu(n)}\, q_{i,j} \qquad \text{for all d.}$$

To see that this is true, just note that $\varepsilon(n_i)\mu_i / \mu(n)$ is the probability that the jump epoch T_{d+1} is caused by a customer transition from server i to some other servers. Therefore,

$$\lim_{L \to \infty} \left\{ \frac{1}{T_L} \sum_{d=1}^{L} RI(n(T_d), u, d)\, \chi_{\Lambda_u}(d)\, \chi_n(X_d) \right\}$$

$$= \left\{ \lim_{L \to \infty} \frac{L}{T_L} \right\} \times \left\{ \lim_{L \to \infty} \frac{1}{L} \sum_{d=1}^{L} RI(n(T_d), u, d)\, \chi_{\Lambda_u}(d)\, \chi_n(X_d) \right\}$$

$$= \alpha\, \mu_u \frac{\pi(n)}{\mu(n)}\, g(n, u) = \mu_u\, P(n)\, g(n, u), \qquad n_u > 0.$$

Eq. (47) is used in the above equation. If $n_u = 0$ then we have $g(n,u) = 0$. From this equation and Eq. (62), we have

$$\frac{s_u}{TP(s_u)} \frac{\partial[TP(s_u)]}{\partial s_u} = -\sum_{\text{all } n} P(n) g(n, u). \tag{63}$$

Now the only question left is how to calculate $g(n,u)$. Since at the service completion time of server i the system is transferring from state n to state $n_{i,j}$ with probability $q_{i,j}$, we have

$$g(n, i) = \sum_{j=1}^{M} q_{i,j} \, f(n_{i,j}, i) + \sum_{j=1}^{M} [1 - \varepsilon(n_j)] q_{i,j} \, f(n_{i,j}, j), \qquad n_i > 0. \quad (64)$$

The last term represents the effect of perturbation propagation through an idle period. Using Eqs. (63) and (64) we can calculate the sensitivity of the through-put with respect to the mean service time in the sense that each customer's service time changes by the same amount.

Example 4.4 Consider a system with the same routing probabilities as the system in Example 4.3, and $N=5$, $s_1=8$, $s_2=10$, and $s_3=15$. The elasticities in the normal sense are calculated by Eq.(56). They are: 0.1655, 0.2910, and 0.5435 for $u=1,2,3$ respectively. The elasticities in the sense of this section are calculated using Eqs.(63) and (64). They are 0.1104, 0.2062, and 0.4224 for $u=1,2,3$ respectively.

Note that the elasticity when the service time changes are equal is usually less than the elasticity in the normal sense. This can be explained intuitively. If the perturbation is proportional to the length of the service time, then the longer service times acquire larger perturbations. A larger perturbation has more oppor-tunity to be realized since at the completion time of the service there tends to be more customers present at that server and fewer customers in other servers; thus other servers have more chances to become idle and the perturbation has more opportunity to be propagated to other servers.

4.4 Sensitivity Analysis of Networks with General Service Time Distributions

In this section, we discuss the extension of the results in Section 4.3 to the networks with general service time distributions. The network is the same as that discussed in Section 4.3 except that the service time distribution functions of

server i, $F_i(s, \theta)$, i=1,2,...,M, may not be exponential, where θ denotes a parameter of the distribution. Let $f_i(s,\theta) = dF_i(s,\theta)/ds$ be the probability density
function. For notational simplicity, θ sometimes does not appear explicitly. The
state process of such a network can be denoted by $\mathbf{x} = (\mathbf{n}, \mathbf{r})$, where $\mathbf{n} = (n_1, n_2,$
..., $n_M)$ with n_i being the number of customers in server i, and $\mathbf{r} = (r_1, r_2, ...,$
$r_M)$ with r_i being the elapsed service time of the customer being served at
server i. $\mathbf{n}$ is called the discrete part of the state, or the discrete state, and $\mathbf{r}$ the
continuous part. Again, let ξ be a random vector representing all the random-
ness in the network. The state process is then denoted by $X(t, \xi) = [N(t, \xi), R(t,$
$\xi)]$. Let $\pi(\mathbf{x}) = \pi(\mathbf{n}, \mathbf{r})$ be the steady-state probability density function of $X(t,\xi)$,
i.e., $\pi(\mathbf{n}, \mathbf{r})dr_1...dr_M$ is the probability that the discrete state is $\mathbf{n}$ and that the
elapsed service time of server i is in $[r_i, r_i + dr_i]$, i=1,2,...,M.

The realization probability $f(\mathbf{x}, V)$, where V is a subset of $\{1,2,...,M\}$, can
be defined in a similar manner as that for Jackson networks except that now
$f(\mathbf{x}, V)$ also depends on the continuous part of the state, $\mathbf{r}$. The dependency of
the realization probability on $\mathbf{r}$ implies that a perturbation generated at the
same discrete state $\mathbf{n}$ but with different continuous part $\mathbf{r}$ will have a different
probability to be realized by the network. Therefore, perturbations have to be
generated in each small interval $[r_i, r_i + dr_i]$ at every server; this is different
from the case of a Jackson network where a lump perturbation is generated at
each state. The perturbation generated in $[r_i, r_i + dr_i]$ can be determined by
using the perturbation generation function defined below. First, as discussed in
Chapter 2, the perturbation generated in a service period s is

$$\Delta s = \frac{\partial F^{-1}(\xi, \theta)}{\partial \theta}\Big|_{\xi = F(s,\theta)}\Delta\theta,$$

where $F^{-1}(\xi, \theta)$ is the inverse of the service time distribution function $F(s, \theta)$.
Now let r, $0 \le r \le s$, be the elapsed time of the customer and $\rho = F(r, \theta)$. As r
increases from 0 to s, ρ increases from 0 to $\xi = F(s, \theta)$. If θ changes to $\theta + \Delta\theta$,
the elapsed time r changes to $r + \Delta r$ (corresponding to the same ρ) with

$$\Delta r = \frac{\partial F^{-1}(\rho, \theta)}{\partial \theta} \Big|_{\rho = F(r,\theta)} \Delta \theta.$$

Δr can be viewed as the perturbation generated in the elapsed service time $[0, r)$. Therefore, the perturbation generated in the service interval $[r, r+dr)$ is

$$d(\Delta r) = \frac{\partial}{\partial r} \left\{ \frac{\partial F^{-1}(\rho, \theta)}{\partial \theta} \Big|_{\rho = F(r,\theta)} \right\} \Delta \theta dr. \tag{65}$$

Define the perturbation generation function as

$$G(r, \theta) = \frac{\partial}{\partial r} \left\{ \frac{\partial F^{-1}(\rho, \theta)}{\partial \theta} \Big|_{\rho = F(r,\theta)} \right\}. \tag{66}$$

Then the perturbation generated in the service interval $[r, r+dr)$ is

$$d(\Delta r) = G(r, \theta) \, \Delta \theta \, dr.$$

Pictorially, the generation of perturbation in $[r, r+dr]$ is shown in Fig. 5.

For exponential service distribution $F(s, \theta) = 1 - \exp(-s\theta)$ with $\theta = \mu$ being the mean service rate, we have $r = F^{-1}(\rho, \theta) = -(1/\theta)\ln(1-\rho)$ and $G(r, \theta) = 1/\theta$, which is independent of r. That is, for exponential distributions, the perturbation generated in any interval $[r, r+dr]$ is $-(\Delta\theta/\theta)dr$, which is proportional to dr. This linear property holds in many other situations. In fact, if θ is a scale parameter for some distribution with the form $F(s/\theta)$, then $r = \theta F^{-1}(\rho)$ and

$$G(r,\theta) = \frac{\partial}{\partial r}(\frac{r}{\theta}) = \frac{1}{\theta}.$$

In this case, the perturbation generated in $[r, r+dr]$ is $(\Delta\theta/\theta)dr$, proportional to dr. It is derived in [Cao 1991] that the realization probability for a network with general service distributions satisfies Eqs.(36) - (39) with $\mathbf{n}$ replaced by $\mathbf{x}$ $=(\mathbf{n}, \mathbf{r})$, and the following differential equation

$$\sum_{i=1}^{M} \frac{\partial f(\mathbf{n},\mathbf{r},k)}{\partial r_i} = f(\mathbf{n},\mathbf{r},k) \sum_{i=1}^{M} \varepsilon(n_i) g_i(r_i)$$

$$-\sum_{i=1}^{M}\sum_{j=1}^{M} \varepsilon(n_i)\varepsilon(n_j)g_i(r_i)q_{i,j}f(\mathbf{n}_{i,j},\mathbf{r}_{-i},k)$$

$$-\sum_{i=1}^{M}\sum_{j=1}^{M} \varepsilon(n_i)[1-\varepsilon(n_j)]g_i(r_i)q_{i,j}f(\mathbf{n}_{i,j},\mathbf{r}_{-i-j},k)$$

$$-\sum_{j=1}^{M}[1-\varepsilon(n_j)]g_k(r_k)q_{k,j}f(\mathbf{n}_{k,j},\mathbf{r}_{-k-j},j). \tag{67}$$

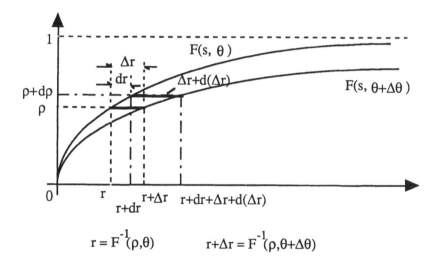

$$r = F^{-1}(\rho,\theta) \qquad r+\Delta r = F^{-1}(\rho,\theta+\Delta\theta)$$

Fig. 5 Perturbation Generated in [r, r+dr], d(Δr)

In the equation, $r_{-i} = (r_1, r_2, ..., r_i = 0,...,r_M)$, $r_{-i-j} = (r_1,r_2, ..., r_i = 0,...,r_j = 0, ...,r_M)$, and

$$g_i(r_i) = \frac{f_i(r_i)}{1 - F_i(r_i)}$$

is the hazard rate.

Finally, we can derive an equation for the sensitivity of the steady-state throughput in a network with general service time distributions; the equation is similar to Eqs. (54) and (56) for Jackson networks. Recall that $TP_L(\xi, \theta) = L/T_L(\xi, \theta)$ is the throughput measured on a sample path in $[0, T_L]$ and $TP_L(\theta)$ is the steady-state throughput. we have

$$\frac{1}{TP(\theta)} \frac{\partial TP(\theta)}{\partial \theta} = \lim_{L \to \infty} \frac{1}{TP_L(\xi,\theta)} \frac{\partial TP_L(\xi,\theta)}{\partial \theta} = - E\{G_u(r_u, \theta)f(n,r,u)\}$$

$$= - \sum_{all\ n} \int_{R^M} \{G_u(r_u, \theta)f(n,r,u)\pi(n,r)\}dr, \qquad \text{w.p.1}, \qquad (68)$$

where θ is a parameter of the service distribution function of server u, $R^M = [0,\infty)^M$ is the first quadrant in the M-dimensional real space, and the expectation "E" is with respect to the steady-state probability measure $\pi(n, r)$. In words, Eq. (68) says that in a closed network with general service time distributions the *normalized derivative* of the throughput based on a sample path in $[0, T_L]$, $[1/TP_L(\xi, \theta)][\partial TP_L(\xi, \theta)/\partial \theta]$, converges with probability one to that of the steady-state throughput, as the length of the sample path goes to infinity, and the *normalized derivative* of the steady-state throughput equals the steady-state expected value of the product of the performance generation function and the realization probability. We refer the readers to [Cao 1991] for a detailed explanation of Eq. (68). To prove (68), we need to assume that $0 \le f_i(s) \le \infty$, $|G_u(r,\theta)| \le K < \infty$, and that the state process $X(t, \xi)$ is ergodic. Some mild conditions regarding $\pi(n, r)$ are also needed. When the service time distributions are exponential, Eq. (68) is reduced to Eqs. (54) and (56).

The development of the previous and this section reinforce the belief that perturbation analysis is basically a distribution free idea. Additional results concerning the concept of realization probability and the sensitivity calculations using realization probability can be extended to general performance measures (see [Glasserman 1990a] and [Cao and Ma 1991]), networks with finite buffers in which no server may block more than one server simultaneously [Cao 1989c], and multiclass networks [Cao 1988b].

4.5 IPA Estimates of the M/G/1 and GI/G/1 Queues

For single queue-server combination, the validity of IPA can be established more directly. In this section we illustrate two such approaches.

4.5.1 M/G/1 Queue: A Direct Approach

In this section, we discuss the IPA estimates of the M/G/1 queue. We shall prove that the IPA estimate of the derivative of the mean response time in an M/G/1 queue with respect to the mean interarrival rate or a parameter of the service distribution converges with probability one to the right value as the length of the sample path goes to infinity. The approach used to prove the result is different from that discussed in the previous sections. Instead of proving the interchangeability, we directly compute the IPA estimate and compare it with the derivative obtained by taking the derivative of the existing formulas for the M/G/1 queue. The computation is based on a standard technique in queueing theory called the sub-busy period method. The result was first reported in [Suri and Zazanis 1988].

Let λ be the mean interarrival rate of the Poisson arrival process of the M/G/1 queue, s and m_2 be the mean and the second moment of the service time with a distribution $F(s, \theta)$. Then the mean response time of the M/G/1 queue is (see Appendix A or [Kleinrock 1975])

$$T = s + \frac{\lambda m_2}{2(1 - \lambda s)}. \tag{69}$$

Let θ be a parameter of the service distribution. The sensitivity of the mean response time T with respect to θ is

$$\frac{dT}{d\theta} = \frac{ds}{d\theta} + \frac{\lambda}{2(1-\lambda s)}\frac{dm_2}{d\theta} + \frac{\lambda^2 m_2}{2(1-\lambda s)^2}\frac{ds}{d\theta}. \tag{70}$$

Similarly, the sensitivity with respect to λ is

$$\frac{dT}{d\lambda} = \frac{m_2}{2(1-\lambda s)^2}. \tag{71}$$

As we showed in Section 3.5, the IPA estimate of $dT/d\theta$ based on M busy periods is

$$\frac{\Delta T}{\Delta\theta} = \frac{1}{N}\sum_{m=1}^{M}\sum_{i=1}^{n_m}\sum_{j=1}^{i}\frac{ds_{k_m+j}}{d\theta}, \tag{72}$$

where M is the number of busy periods observed, n_m is the number of customers served in the mth busy period, $k_m = n_1 + \ldots + n_{m-1}$ is the number of customers served in the first m-1 busy periods, $k_0 = 0$, $N = k_{M+1}$ is the number of customers served in M busy periods, $s_k = F^{-1}(\xi_k, \theta)$ is the service time of the kth customer.

We shall prove that as $N \to \infty$, $\Delta T/\Delta\theta$ converges with probability one to $dT/d\theta$ in (70). To prove this, we assume that the service time distribution function F is such that the following two equations hold

$$E(\frac{ds_k}{d\theta}) = \frac{d}{d\theta}E(s_k) = \frac{ds}{d\theta}, \tag{73}$$

$$E(\frac{ds_k^2}{d\theta}) = \frac{d}{d\theta}E(s_k^2) = \frac{dm_2}{d\theta}. \tag{74}$$

The validity of the interchange of the expectation and differentiation in these two equations can be verified for many distributions, e.g., exponential, uniform, deterministic distributions (see Chapter 3).

To study $\Delta T/\Delta\theta$, we first focus on one busy period. For ease of notation, we shall drop the subscript m. Thus, there are n customers served in the busy period, and s_j is the service time of the jth customer. The contribution of this busy period to (72) is

$$h = \sum_{i=1}^{n} \sum_{j=1}^{i} \frac{ds_j}{d\theta} = \sum_{i=1}^{n} \sum_{j=1}^{i} s'_j, \tag{75}$$

where $s'_j = ds_j /d\theta$. By the law of large numbers and (72), we have

$$\lim_{M \to \infty} \frac{\Delta T}{\Delta \theta} = \lim_{M \to \infty} \frac{\dfrac{1}{M}\displaystyle\sum_{m=1}^{M} h_m}{\dfrac{1}{M}\displaystyle\sum_{m=1}^{M} n_m} = \frac{E(h)}{E(n)} = (1-\lambda s)E(h), \tag{76}$$

where h_m is a quantity similar to the h in (75) for the mth busy period, $E(n) = 1/(1-\lambda s)$ is the mean of the number of customers served in a busy period [Kleinrock 1975].

To compute $E(h)$, we use the sub-busy period approach [Kleinrock 1975]. Let k be the number of customers that arrive during the service time of the first customer. As explained in [Kleinrock 1975], each of the customers arriving during the first customer's service time initiates a sub-busy period which is statistically identical to the "parent" busy period and is independent to the other sub-busy periods. The idea of sub-busy periods is based on an observation, i.e., since the customers are identical, the statistical behavior of the M/G/1 queue is independent of the order in which customers are served. Therefore, we can permute the order to follow the last come first served (LCFS) discipline. Suppose that a customer, called a parent, arrives at an idle queue and initiates a busy period; then in this parent-busy period the server serves all the customers, called the first generation siblings, that arrive at the queue during the parent's service period, and all the customers, called the second generation siblings, that arrive during the first generation siblings' service periods, and so on. Now, let k be the number of the first generation siblings. The kth customer is the last to arrive at the queue during the parent's

service period. Thus, according to the LCFS discipline, after serving the parent
customer the server starts to serve the kth customer. Note that before the server
starts to serve the (k-1)th customer, the server has to complete services to all
the customers, called the first generation siblings of the kth customer, that
arrive at the queue during the kth customer's service period, and the all the
customers, called the second generation siblings of the kth customer, that arrive
during the kth customer's first generation siblings' service periods, and so on.
It is easy to see that the period between the service starting time of the kth
customer and that of the (k-1)th customer is statistically identical to the parent-
busy period and is hence called a sub-busy period. Similarly, the (k-1)th
customer also initiates a sub-busy period, and so on. T here are a total of k sub-
busy periods in the parent-busy period. [Kleinrock 1975] contains a figure
clearly illustrating the idea explained above.

Let u_r, r=1,2,...,k, be the number of customers served in the rth sub-busy
period, and define $v_r =1+ u_1+...+ u_r$ with $u_0=v_0=1$. Then the rth sub-busy
period contains the $(v_{r-1} +1)$th to the v_rth customer. Let

$$g = \sum_{i=1}^{n} s'_i \tag{77}$$

and

$$g_r = \sum_{i=1}^{u_r} s'_{u_{r-1}+i} \tag{78}$$

be the quantity similar to g for the rth sub-busy period with $g_0=0$. Also, for the
rth sub-busy period, define the quantity similar to h,

$$h_r = \sum_{i=1}^{u_r}\sum_{j=1}^{i} s'_{u_{r-1}+j}.$$

Since sub-busy periods are statistically identical to the "parent" busy period, g_r
and h_r, have the same distributions as g and h, respectively; especially, $E(g_r)$
$=E(g)$, and $E(h_r) =E(h)$.

Next, we derive some relations between the above quantities. From (77) and (78), We have

$$g = \sum_{i=1}^{n} s'_i = s'_1 + \sum_{r=1}^{k} \sum_{i=1}^{u_r} s'_{u_{r-1}+i} = s'_1 + \sum_{r=1}^{k} g_r. \tag{79}$$

For h and h_r, we first rewrite them as

$$h = \sum_{i=1}^{n} (n-i+1) s'_i$$

and

$$h_r = \sum_{i=1}^{u_r} (u_r-i+1) s'_{u_{r-1}+i}.$$

From these two equations, after some ordering, we get

$$h = s'_1 + \sum_{r=1}^{k} [h_r + u_r \sum_{j=0}^{r-1} (g_j + s'_1)]. \tag{80}$$

Since k and g_r are independent, $E(g_1 + \ldots + g_k) = E(k)E(g)$. Taking expectation of the both sides of Eq. (79), we have

$$E(g) = E(s'_1) + E(k)E(g).$$

Note that $E(k) = \lambda s$ is the expected number of the customers arriving in the first customer's service time. Thus,

$$E(g) = \frac{E(s'_1)}{1-\lambda s}. \tag{81}$$

Now, taking the conditional expectation given s_1 of the both sides of (80), we have

$$E(h \mid s_1) = s'_1 + E\{\sum_{r=1}^{k} [h_r + u_r s'_1] \mid s_1\} + E\{\sum_{r=1}^{k} u_r \sum_{j=0}^{r-1} g_j \mid s_1\}$$

$$= s'_1 + E(k \mid s_1)[E(h) + E(u_r)s'_1] + E[u_2 + 2u_3 + \ldots + (k-1)u_k \mid s_1]E(g)$$

$$= s'_1 + E(k|s_1)[E(h)+E(u_r)s'_1] + E[(k^2-k)/2|s_1]E(u_r)E(g).$$

In the equation, $E(u_r) = E(n) = 1/(1 - \lambda s)$; $E(k \mid s_1) = \lambda s_1$ is the average number of the Poisson arrivals in an interval of length s_1. Similarly, $E(k^2 \mid s_1)=\lambda s_1+ (\lambda s_1)^2$. Using these and Eq. (81), we get

$$E(h \mid s_1) = s'_1 + \lambda s_1 \left[E(h) + \frac{s'_1}{1 - \lambda s} \right] + \frac{(\lambda s_1)^2}{2(1 - \lambda s)^2} E(s'_1).$$

Taking expectation with respect to s_1 yields

$$E(h) = \frac{E(\frac{ds_1}{d\theta})}{1-\lambda s} + \frac{\lambda}{(1-\lambda s)^2}E(s_1\frac{ds_1}{d\theta}) + \frac{\lambda^2 m_2}{2(1-\lambda s)^3}E(\frac{ds_1}{d\theta}).$$

Substituting this into Eq. (76), we have

$$\lim_{M\to\infty} \frac{\Delta T}{\Delta \theta} = E(\frac{ds_1}{d\theta}) + \frac{\lambda}{(1-\lambda s)}E(s_1\frac{ds_1}{d\theta}) + \frac{\lambda^2 m_2}{2(1-\lambda s)^2}E(\frac{ds_1}{d\theta}). \tag{82}$$

With Eqs.(73) and (74), we see that the right-hand side of (82) is just the same as the right-hand side of (70). Thus, we have proved that the IPA estimate converges with probability one to the derivative of the mean response time with respect to a service parameter in an M/G/1 queue.

Next, let us discuss the IPA estimate for $dT/d\lambda$. Let a_k be the interarrival time of the kth customer. Then, similarly to (72), the IPA estimate for $dT/d\lambda$ is

$$\frac{\Delta T}{\Delta \lambda} = -\frac{1}{N}\sum_{m=1}^{M} \sum_{i=1}^{n_m} \sum_{j=2}^{i} \frac{da_{k_m+j}}{d\lambda}, \tag{83}$$

where the minus sign indicates that a customer who arrives a little later will have to wait less. In (83), the index j runs from 2 to i because the change in the first customer's arrival time does not affect the response time. Let

$$\frac{\Delta T_{k_m, i}}{\Delta\lambda} = -\sum_{j=2}^{i} \frac{da_{k_m+j}}{d\lambda}$$

be the contribution of the perturbation of the response time of the ith customer in the k_mth busy period to the IPA estimate. According to the perturbation gene-ration rule, for exponential distributed interarrival time, we have (see Chapter 3 or Section 4.2)

$$\frac{da_{k_m+j}}{d\lambda} = -\frac{a_{k_m+j}}{\lambda}.$$

Therefore, the contribution of the ith customer of the k_mth busy period is

$$\frac{\Delta T_{k_m, i}}{\Delta\lambda} = \frac{1}{\lambda}\sum_{j=2}^{i} a_{k_m+j}.$$

Substituting the above equations into (83), and letting M→∞, we get

$$\lim_{M\to\infty}\frac{\Delta T}{\Delta\lambda} = E(\frac{\Delta T_{k_m, i}}{\Delta\lambda}) = \frac{1}{\lambda} E(\sum_{j=2}^{i} a_{k_m+j}). \tag{84}$$

Note that the sum on the right-hand side is the total time that has elapsed from the beginning of the busy period up to the arrival of a customer. Now, Poisson arrivals take a random look at the system and it is known [Kleinrock 1975] that the conditional mean of this quantity, given that the system is found to be busy, is $m_{b2}/2m_{b1}$, with m_{b1} and m_{b2} being the first (mean) and second moment of the length of the busy period. Also, $m_{b1}=s/(1-\lambda s)$ and $m_{b2}=m_2/(1-\lambda s)^3$. Thus, $m_{b2}/2m_{b1} = m_2/2s(1-\lambda s)^2$. Multiplying the conditional mean by the probabi-lity that the system is busy, λs, we get

$$E\{\sum_{j=2}^{i} a_{k_m+j}\} = \frac{m_2\lambda}{(1-\lambda s)^2}$$

and

$$\lim_{M\to\infty}\frac{\Delta T}{\Delta\lambda} = \frac{m_2}{2(1-\lambda s)^2}.$$

This is the same as Eq. (71). That is, the IPA estimate of the derivative of the mean response time with respect to the mean interarrival time in an M/G/1 queue converges with probability one to the right value.

Finally, the consistency of the IPA estimate also holds for the GI/G/1 [Suri and Zazanis 1987] and GI/G/m queue [Fu and Hu 1991a], which gives the algorithms of the IPA estimates for the derivatives of the mean response time with respect to service time and interarrival time distribution parameters for the GI/G/m queue as well as a direct proof for the consistency of the IPA estimate for the M/M/m. The main difference between a GI/G/1 queue and a GI/G/m queue can be explained by the concept of *"local"* busy periods. In a GI/G/m queue, there are m servers in the service station. A local busy period is a busy period of any server in the station; while a global busy period is a busy period of the station. (The station is in a global busy period if any one of the servers is busy.) Thus, a global busy period may contain any number (including 0) of local busy periods of a particular server. The complexity of the IPA estimate comes from the fact that a perturbation of a customer's arrival or service completion time in one server can only be propagated to the customers in the same local busy period of that server. We will not discuss the details of the proof, which involve calculating the average length of local busy periods, etc. Of course, another way to verify the consistency of the IPA estimate is by viewing the GI/G/m queue as a GI/G/1 queue with load dependent service rates (see Appendix A).

Exercise 1.   Consider the problem of developing an IPA algorithm for the M/M/2 queue.   Sketch out the principal difference from that of the M/M/1 queue.

Exercise 2 An M/M/2 queue can be equivalently thought of as an M/M*/1 queue where the notation (*) indicates that the server is load dependent with

$$\mu(i) = \begin{cases} i\mu & \text{if } i \leq 2 \\ 2\mu & \text{if } i > 2 \end{cases}$$

Suppose you apply the load dependent IPA algorithm developed in Section 3.5 to the M/M*/1 queue. What difference, if any, will you observe when comparing this to the IPA algorithm developed in Exercise 1 for the M/M/2 queue?

4.5.2 GI/G/1 Queue: The Approach Based on Stochastic Convexity

Another method of proving the strong consistency of the IPA estimate due to [Hu 1991] is based on the stochastic convexity of the random variable to be estimated. The fundamental of the method is the following result from [Rockafellar 1970]. Let $f_n(\theta)$, $n=1,2,...$, be a sequence of functions with the property $\lim_{n\to\infty} f_n(\theta) = f(\theta)$. If every $f_n(\theta)$ is convex and $f(\theta)$ is differentiable at θ_0, then

$$\lim_{n\to\infty} \frac{d^- f_n(\theta_0)}{d\theta} = \lim_{n\to\infty} \frac{d^+ f_n(\theta_0)}{d\theta} = \frac{df(\theta_0)}{d\theta}, \tag{85}$$

where "+ " and " - " denotes the right- and left-side derivatives, respectively. Eq.(85) also implies that the two limits on the right-hand side of (85) exist. Intuitively speaking, $\lim_{n\to\infty} f_n(\theta) = f(\theta)$ means that $f_n(\theta)$ approaches $f(\theta)$ as $n\to\infty$; the convexity assures that $f_n(\theta)$ does not oscillate around $f(\theta)$, i.e., convexity rules out the pathological cases illustrated in Fig. 6.

As an example of the application of the above result to the IPA estimates, we study the waiting time in a GI/G/1 queue. The nth customer's waiting time, W_n, satisfies the Lindley equation [Lindley 1952]:

$$W_{n+1} = \begin{cases} W_n + s_n - a_{n+1} & \text{if } W_n + s_n - a_{n+1} \geq 0, \\ 0 & \text{otherwise.} \end{cases}$$

where s_n is the service time of the nth customer and a_n is the interarrival time of the nth customer; the equation is equivalent to

$$W_{n+1} = \max(0, W_n + s_n - a_{n+1}), \tag{86}$$

with initial condition $W_0=0$. Let

$$\overline{W}_n = \frac{1}{n}\sum_{i=1}^{n} W_i$$

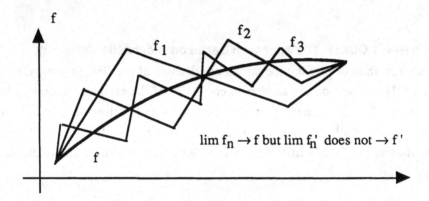

lim $f_n \to f$ but lim f_n' does not $\to f'$

Fig. 6 Convergent f_n But Nonconvergent f_n'

be the average of the waiting times of the first n customers. By the law of large numbers, $\lim_{n\to\infty}\overline{W}_n = E(W)$, where $E(W)$ is the mean waiting time. Now if $\overline{W}_n$ is convex for all n, then we have $\lim_{n\to\infty} d(\overline{W}_n)/d\theta = d[E(W)]/d\theta$, i.e., $d(\overline{W}_n)/d\theta$ is a strongly consistent estimate of the derivative of the mean waiting time. Note that

$$\frac{d\overline{W}_n}{d\theta} = \frac{1}{n}\sum_{i=1}^{n} \frac{dW_i}{d\theta}$$

is the IPA estimate which can be obtained on a single sample path of the GI/G/1 queue. In other words, the IPA estimate is strongly consistent if $\overline{W}_n$ is convex. This is true if all W_n's are convex. Now assume that all s_n's and a_n's are convex. Then $W_n + s_n - a_{n+1}$ is convex. Because the maximum of two convex functions are convex, W_{n+1} is convex. By induction, all W_n's are convex. Therefore, if all s_n's and a_n's are convex then the IPA estimate of the derivative of the waiting time of a GI/G/1 queue is strongly consistent. Many random variables in the form of s_n (or a_n) $= F^{-1}(\theta, \xi)$ can be shown to be convex for any realization of ξ. Such random variables are called strong stochastic convex (SSCX) in [Shanthikumar and Yao 1989].

Note that this approach does not require the assumption that s_n's and a_n's are independent and identically distributed. The approach may be applied to other networks such as open tandem networks if the convexity can be established. In general, however, this is not an easy task .

4.6 Some Technical Proofs

This section provides explanations for some technical results used in this chapter.

A. Ergodicity of the Imbedded Markov Chain in a Closed Jackson Network
As we explained in Section 4.2.3, for some special networks such as a cyclic network the imbedded Markov chain **X** may be aperiodic and non-ergodic. However, if in a network $q_{i,i} \neq 0$, for some i, then **X** is aperiodic. The reason is: Any state **n** of this network with $n_i > 0$ is not a periodic state since the probability that the system jumps from state **n** to the same state is positive. Thus, all the states of **X** are also aperiodic since **X** is irreducible ([Cinlar 1975]). Next, suppose that $q_{i,i} = 0$ for all i and the Markov chain **X** is periodic. We can construct an auxiliary network by replacing any server, say server k, by an equivalent server. The equivalent server has a mean service time $s'_k = s_k/2$, and routing probabilities $q'_{k,j} = q_{k,j}/2$, for $j \neq k$. After the completion of the service in the equivalent server, a customer has a probability of 1/2 of being fed back to the same server. According to FCFS discipline, the fed back customer would be put at the end of the queue in the server. However, since all customers are identical, we can exchange the position of the fed back customer with that of the first customer in the queue; thus, the fed back customer can be considered as receiving successive services from the server. The average number of the services a customer receives from this equivalent server is 2. It is easy to check that the total service time a customer receives

from the equivalent server (i.e., the sum of all service times the customer receives from the server) before the customer leaves the server is exponentially distributed with a mean $2s'_k = s_k$. Furthermore, the conditional routing probabilities given that the customer leaves the equivalent server is $q'_{k,j} /(1/2)$ $= q_{k,j}$ for $j \neq k$. Thus, a server with a mean s_k' and routing probabilities $q'_{k,j}$ is equivalent to a server with a mean s_k and routing probabilities $q_{k,j}$. The imbedded Markov chain of the auxiliary network is aperiodic. The Markov processes of the original and the auxiliary networks are the same. Therefore, by constructing an auxiliary network if necessary, we can always assume that X is aperiodic. Since X is also irreducible and has a finite number of states, X is ergodic (see [Cinlar 1975] or Appendix A).

B. The proof of

$$TP(\theta) = \frac{1}{R(\theta)}.$$

The balance equation of P(n) is

$$\mu(n)P(n) = \sum_{i=1}^{M}\sum_{j=1}^{M}\epsilon(n_j)\mu_i q_{i,j}P(n_{j.\,i}), \qquad (87)$$

where $n_{j,i} = (n_1, ..., n_j -1,..., n_i +1,..., n_M)$ is a neighboring state of n. On the other hand, $\pi(n)$ satisfies

$$\pi(n) = \sum_{i=1}^{M} \sum_{j=1}^{M} \frac{\epsilon(n_j)\mu_i q_{i,j}}{\mu(n_{j.\,i})}\pi(n_{j.\,i}). \qquad (88)$$

Note that Eq. (87) can be written as

$$\mu(n)P(n) = \sum_{i=1}^{M} \sum_{j=1}^{M} \frac{\epsilon(n_j)\mu_i q_{i,j}}{\mu(n_{j.\,i})} \{\mu(n_{j.\,i})\, P(n_{j.\,i})\}. \qquad (89)$$

Eq. (89) is in the same form for $\mu(n)P(n)$ as Eq. (88) for $\pi(n)$. Since the Markov chain X is irreducible and aperiodic, we have

$$\alpha \, \pi(n) = P(n)\mu(n)$$

where α is a normalizing constant (cf. Theorem 2.1 in Chapter 6 of [Cinlar 1975]). From this equation, we have

$$TP(\theta) = \sum_{\text{all } n} P(n)\mu(n) = \sum_{\text{all } n} \alpha \, \pi(n) = \alpha$$

and

$$\alpha \sum_{\text{all } n} \frac{\pi(n)}{\mu(n)} = \sum_{\text{all } n} P(n) = 1.$$

Therefore, $\alpha R(\theta) = 1$ and $TP(\theta) = 1/R(\theta)$.

C. If f_n, n=1,2, ... are uniformly integrable and $|g_n| \le |f_n|$ for all n, then g_n, n=1,2,... , are also uniformly integrable.

This can be shown as follows. By the definition of uniform integrability, we have

$$\lim_{\alpha \to \infty} \sup_n \int_{\{|f_n| \ge \alpha\}} |f_n| d\mu = 0.$$

From $|g_n| \le |f_n|$, we get $\{|g_n| \ge \alpha\} \subseteq \{|f_n| \ge \alpha\}$. Thus, for all n, we have

$$0 \le \int_{\{|g_n| \ge \alpha\}} |g_n| \, d\mu \le \int_{\{|f_n| \ge \alpha\}} |g_n| \, d\mu \le \int_{\{|f_n| \ge \alpha\}} |f_n| \, d\mu.$$

Therefore,

$$0 \le \sup_n \int_{\{|g_n| \ge \alpha\}} |g_n| \, d\mu \le \sup_n \int_{\{|f_n| \ge \alpha\}} |f_n| \, d\mu.$$

Taking the limit of $\alpha \to \infty$ in the above inequality, we obtain

$$\lim_{\alpha \to \infty} \sup_n \int_{\{|g_n| \ge \alpha\}} |g_n| d\mu = 0.$$

This means that g_n, n=1,2,..., are uniformly integrable .

D. The proof of Eq. (49).

By (48), Eq. (49) is equivalent to

$$\lim_{L \to \infty} \frac{1}{T_L} \sum_{d=1}^{L} S_d \{RI_L(n, i, d) - RI(n, i, d)\} = 0 \qquad \text{w.p.1.}$$

Let $n_1 = (N, 0,...,0)$, $L_0 = 0$, and $L_r = \min \{l > L_{r-1}, X_l = n_1 \}$, r=1,2,.... The sequence $\{T_{L_r}\}$ is a delayed renewal process with T_{L_r} being the regenerative times. Since any perturbation propagated to state n_1 is already realized by the network, we have $RI_L(n, i, d) = RI(n, i, d)$ for all $d \leq L_r < L$. Therefore,

$$\lim_{L \to \infty} \frac{1}{L} \sum_{d=1}^{L} S_d \{RI_L(n, i, d) - RI(n, i, d)\}$$

$$= \lim_{L \to \infty} \frac{1}{L} \sum_{d=L_r+1}^{L} S_d \{RI_L(n, i, d) - RI(n, i, d)\}, \qquad \text{w.p.1.}$$

Now,

$$\sum_{d=L_r+1}^{L} S_d |RI_L(n, i, d) - RI(n, i, d)| \leq 2 \sum_{d=L_r+1}^{L} S_d \leq 2(T_{L_{r+1}} - T_{L_r}).$$

By the law of large numbers,

$$\lim_{r \to \infty} \frac{1}{r} \sum_{k=1}^{r} \{T_{L_{k+1}} - T_{L_k} \} = E\{T_{L_{r+1}} - T_{L_r}\} \qquad \text{w.p.1.}$$

Thus, we have

$$\lim_{r \to \infty} \frac{1}{r} \{T_{L_{r+1}} - T_{L_r}\}$$

$$= \lim_{r \to \infty} \frac{1}{r} \sum_{k=1}^{r} \{T_{L_{k+1}} - T_{L_k} \} - \lim_{r \to \infty} \frac{1}{r} \sum_{k=1}^{r-1} \{T_{L_{k+1}} - T_{L_k} \} = 0 \qquad \text{w.p.1.}$$

The result follows directly from the above equation and the fact that r/L<1.

CHAPTER FIVE

Extensions of Infinitesimal Perturbation Analysis

The IPA algorithm described in Chapters 3 and 4 is very efficient whenever applicable. However, it is also clear that one can easily construct examples of DEDS which will not satisfy the conditions for interchange discussed in Chapter 4. Generally, such cases involve event order changes on the nominal and perturbed sample paths leading to different state sequences of these two sample paths and discontinuities in the sample performance measure. In such circumstances, one can always adopt directly the general techniques to be described in Chapters 6 and 7 which overcome these difficulties at some cost. In this chapter, however, we would like to present a series of examples or classes of examples which on the surface violate the requirements of the simple IPA yet can still be solved by extensions of the IPA approach. Two main ideas are involved. First (in Section 5.1), we show that the model representation of a DEDS can be important. The same system can be represented in more than one way, some of which are hostile to the simple IPA algorithm while others are not. Secondly, discontinuities in sample performance can often be smoothed out. By replacing infrequent but finite perturbations with equivalent frequent infinitesimal perturbations, we can often extend the applicability of IPA (Sections 5.3 - 4).

5.1 System Representation and PA

An example of discontinuous $L(\theta, \xi)$ occurs when we consider a simple finite stochastic birth-death process using a Markov chain model [Glasserman 1988b, Heidelberger et al. 1988]. In state (with population) n, we have the state dependent birth (death) rate λ_n (μ_n). Let $E[L(\theta,\xi)] = p_0$ be the stationary probability of state 0 and consider the following method of simulating the birth-death process. For each state, say n, we generate two exponentially distributed lifetimes B_n and D_n according to the rates λ_n and μ_n, respectively. The next state will be n+1 if $B_n < D_n$, and n-1 otherwise. Now if we perturb the birth rate from λ_n to $\lambda_n + \Delta\lambda_n$, then it is entirely possible that on some occurrences of state n, $B_n > D_n$ on the nominal but $B_n + \Delta B_n < D_n$ on the perturbed sample path. A disconti-nuity results since the next state for the nominal will be n-1 while for the perturbed, n+1. Continuing the two trajectories from this point on could lead to two completely different sample paths. It is shown in [Heidelberger et al. 1988] that if we ignore such changes in the state sequence of the nominal and the perturbed paths then IPA will incorrectly estimate the value of $dp_0/d\lambda_n$ (in fact with the wrong sign!). This can be seen intuitively. Increasing the birth rate in state n will increase population and hence decrease the value of p_0, i.e., $dp_0/d\lambda_n < 0$. But by the perturbation generation rule with the simple IPA, increasing λ_n results in $\Delta B_n < 0$. Under deterministic similarity, or the state sequence invariance, the state holding times at all the birth states (i.e., the states followed by a birth event) will increase proportionally, and that at all the death states (i.e., the states followed by a death event) will not change. Thus, since state 0 is a birth state, the simple IPA algorithm applied to the above model yields $dp_0/d\lambda_n > 0$, which is wrong. However, as pointed out by [Glasserman 1988b], this inconsistency is only a result of the particular model of birth-death process we have adopted for constructing the simulation and the IPA algorithm. Suppose we consider an alternative model (or representation): A cyclic queueing network in which N customers alternate service at two exponential servers with load dependent rates. Server 1 works at rate μ_n when its queue contains n customers and server 2 works at rate λ_{N-n} when its queue contains n customers. Then the number of customers at server 1 is exactly the same as the population in the above described finite birth-death process in the sense that the trajectories generated by these two different representations are statistically indistinguishable (see Fig. 1).

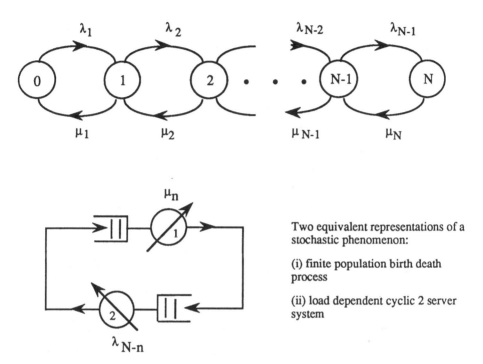

Two equivalent representations of a stochastic phenomenon:

(i) finite population birth death process

(ii) load dependent cyclic 2 server system

Fig. 1 Two Equivalent Representations of a Stochastic Phenomenon

However, if we follow the natural way of constructing a sample path for this network, we get a structurally different (and much smoother) sample path. Suppose at time t a customer just arrives at server 1 and finds $n-1$ customers waiting there. This means server 2 has $N-n$ customers in the queue and is beginning to serve a new customer at the rate λ_n. Let the service time generated be B_n. This B_n now will compete with the residual service time at server 1 which we shall denote as D_n'. Note D_n' was scheduled the last time a customer departed from server 1. Now if $B_n > D_n'$ then server 1 will complete service first and send a customer to server 2. The state (the number of customers) will become $n-1$. If $B_n < D_n'$, then the state will become $n+1$ with server 1 again receiving a customer from server 2. The cycle is repeated with the initiation and generation of new service for either server 1 $(B_n > D_n')$ or server 2 $(B_n < D_n')$. The important difference here is that at each event only one random duration instead of two is generated, and the comparison is between the new lifetime and the old residual lifetime rather than between the two new lifetimes as in the Markov chain case. From the viewpoint of the memoryless property of the exponential distribution, the two different cases are probabilistically equivalent[1]. However, this difference produces a very different perturbed behavior when we consider changing from λ_n to $\lambda_n+\Delta\lambda_n$. Suppose we have $B_n>D_n'$ on the nominal path but $B_n + \Delta B_n<D_n'$ on the perturbed path. Once again the state sequences will diverge to n-1 and n+1, respectively. However, the respective remaining lifetimes at server 2 and server 1 for the next comparison will be B_n-D_n' and $D_n'-B_n-\Delta B_n$. These are likely to be very small numbers compared to the newly generated lifetime D_{n-1} or B_{n+1} (Since $B_n>D_n'$ and $B_n + \Delta B_n<D_n'$, we have $0<B_n-D_n'<-\Delta B_n$ and $0<D_n'-B_n-\Delta B_n<-\Delta B_n$.) Consequently the state sequences will return to n in both the nominal and the perturbed cases after the next state. Thus, the result of such an occurrence ($B_n>D_n'$ on the nominal but $B_n + \Delta B_n<D_n'$ on the perturbed) for this network is a temporary deviation of the two paths (in state n+1 and n-1 respectively) over an interval in the same order as "$\Delta\lambda_n$." Since the probability of the occurrence of such a deviation is also of order "$\Delta\lambda_n$," then according to the development of Chapter 4, we can ignore such deviations or state sequence changes in the computation of first derivative estimates. We illustrate this graphically in Fig. 2. [Glasserman 1988b] performs a detailed analysis of this case and demonstrates that under the network (vs. Markov chain) model of the birth-death process, simple IPA estimates correctly converges to the value of $dp_0/d\lambda_n$. The consistency of the IPA estimate for birth-death processes can also be proved through the use of a load-dependent M/M/1 queue (see [Hu 1991]).

---

[1] In the sense that we could equally well in the tandem queue case discard the remaining service time of server 1, D_n', and re-schedule it by generating a new sample using the exponential distribution with mean μ_n.

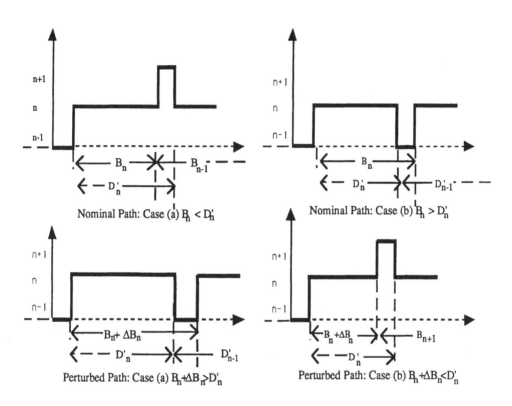

Fig. 2 The Nominal and Perturbed Paths for the Cyclic Two-Server Queue

Exercise Construct an example of two different representations of a random variable, x, that have different smoothness properties with respect to a parameter, θ. (Hint: let the value of x be dependent on whether or not certain random outcome is greater or less than θ.)

An even more direct example of different model representations can be given in terms of a simple random variable as follows: Consider the problem of generating samples of a random variable with a given distribution $F(x)$ via the method of inverse transform (see Appendix B.2) as shown in Fig. 3.

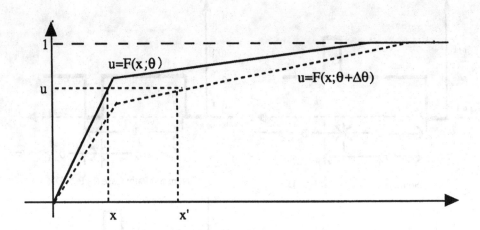

Fig.3 Model of Mapping From u to x

The function F maps u.i.i.d. u's into F(x)-distributed x's. However this is only the natural and the usual way of doing so. The same mapping can be achieved by rearranging the function F(x) as illustrated in Fig. 4.

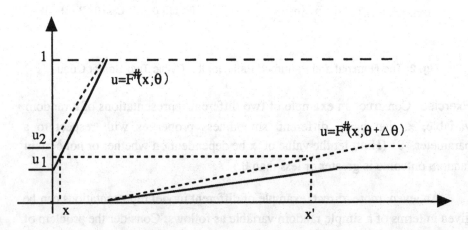

Fig. 4 Alternative Model of Mapping from u to x

But in this case, it is clear that the mapping from u to x is not continuous. In other words for the **same** value of θ, the distribution of x is the same under Fig.

3 or 4. The behavior of the mapping under perturbation of θ is, however, not the same! As u crossed either u_1 or u_2, the mapping $F^{\#}(x)$ is not continuous[2].

The above discussion illustrates the point that just as there are different ways to simulate a random variable, there are different ways to simulate a sample path. Each choice of simulation implicitly defines a construction of the process under study. On a common probability space, we can define paths x(t; θ,ξ) and y(t; θ,ξ) such that for each θ, x and y are the same process statistically though x(t; $\cdot,\xi$) and y(t; $\cdot$, ξ) are different functions of θ. If x(t; θ_0,ξ) = y(t; θ_0, ξ) but x(t; $\theta_0+\epsilon,\xi$) $\neq$ y(t; $\theta_0+\epsilon$, ξ) then sample derivatives obtained from the same sample path under the two different representations will be generally different. In the case under discussion the queueing network model produces a much smoother and tolerant sample path than the birth-death model. Thus, while we ignore state sequence changes in both IPA calculations, one set of calcu-lations leads to correct results and the other doesn't. It also cannot be over emphasized that if we observe an actual birth-death process it is perfectly legal and correct to calculate the $dp_0/d\lambda_n$ using the queueing network model even if the process is actually constructed using the Markov chain model. Although the choice of the representation is important in the derivation of the IPA algorithm used, for the actual implementation it is usually the case that any simulation based on some valid representation will do—provided the IPA algorithm is incorporated into the simulation as though the "right" represen-tation were being used. IPA works only on the data of the actual sample path; no parameter change is ever really introduced.

More generally, this idea of using a representation of a stochastic process which under parametric perturbation only induces temporary deviations from the nominal sample path has been generalized. [Glasserman 1989] shows that so long as a continuous time Markov chain (discrete state Markov process) satisfies the condition CM described below, then there exists a representation for which the simple IPA algorithm will yield unbiased estimates of the gradient for any finite time performance measure.

---

[2] We are indebted to Prof. S. Strickland for this example.

Condition CM (Common successor property): if a pair of states, x and y, have a common predecessor, then they have a common successor.

Because of the memoryless property of the Markov lifetimes, for any systems satisfying common successor condition (CM), it is always possible to assign the shortest remaining lifetime at the occurrence of an event to the successor state which will bring together the nominal and the perturbed states. This is, in fact, the natural assignment for the cyclic two-server example above. Let us examine this in some detail. At time t, let us consider the case where both servers 1 and 2 are busy with queueing lengths n and N-n and remaining service times RT_1 and RT_2 where $RT_2=RT_1+\Delta$, $\Delta>0$ ($RT_1\equiv D_n'$ and $RT_2\equiv B_n$ in Fig. 2 case (b)). Thus, on the nominal path, server 1 will finish service first at time $t+RT_1$ and send a customer to server 2 which by now has remaining service time Δ. The system is now in state n-1 and server 1 has acquired a new service time ST_1 ($=D_{n-1}'$)generated by an exponential distribution with rate μ_{n-1}; server 2's remaining service time, Δ, will be modified to Δ' by the new rate λ_{N-n+1} according to the perturbation propagation rules for load-dependent servers. But for small Δ or Δ' (which is of the same order as Δ), it is highly likely (with probability $\to 1$ as $\Delta\to 0$) that $\Delta<ST_1$ and server 2 will now finish service first and send a customer to server 1. Thus, the next state will be "n." Now assume that because of perturbation, we have, on the perturbed path at time t, two remaining service times RT_1' and RT_2' with $RT_1'=RT_2'+\delta$ and $\delta>0$. Server 2 will finish first at time $t+RT_2'$ and the state jumps to n+1. In this case, server 1 will have a remaining service time δ, and server 2 a newly generated service time ST_2' with rate λ_{N-n-1}. We shall now be comparing δ vs. ST_2' with the same highly probably result that server 1 will finish first causing the state to transition to "n." Therefore, the state sequence of the nominal path is n, n-1, n, and that of the perturbed path is n, n+1, n. Moreover, the times that the nominal and the perturbed paths stay in states n-1 and n+1, respectively, are only Δ and δ, respectively. As $\Delta\to 0$ and $\delta\to 0$, the effect of the difference of the two sample paths on any performance is negligible.

The scenario of the four states n, n-1, n+1, and n in the state sequence of the above two-queue system can be generalized to what is shown in Fig. 5 for a Markov chain. The figure illustrates only the transitions between four states in the Markov chain; states "b" and "c" correspond to states n-1 and n+1 and have a common predecessor "p," which corresponds to the first state n in the state sequence, and a common successor "s," which corresponds to the second state n in the state sequence.

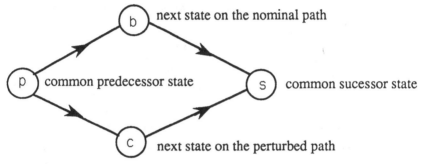

Fig. 5 Common Predecessor and Successor of a Pair of States

Let θ be the parameter being perturbed. For simplicity of discussion, we assume that the mean lifetimes of the four transitions from state "p" to "b," from state "p" to "c," from state "b" to "s," and from state "c" to "s" are the same. If the nominal path and the perturbed path corresponding to $\Delta\theta$ start to differ by going to different states "b" and "c" from the same state "p," then the difference between the lifetimes t_{pb} and t_{pc} must be so small such that $|t_{pb}-t_{pc}| = O(\Delta\theta)$. Consequently, by **assigning** the remaining lifetime to the alternative "b $\rightarrow$ s" in state "b" on the nominal path and to the alternative "c $\rightarrow$ s" in state "c," we guarantee the two paths shall converge once again to state "s." This is possible because of the memoryless property of the Markov chain. Furthermore, using the same reasoning as above we see that such an assignment will imply that both the temporary deviation between the nominal and the perturbed paths and the probability of its occurrence are of order $O(\Delta\theta)$ which in turn insures the satisfaction of condition (IC), i.e., Eq.(4.1), which validates IPA. For the cyclic two-server example above, the service times RT_1

and RT_2 are identified as t_{pb} and t_{pc}. The natural assignment of Δ to server 2 on the nominal and δ to server 1 on the perturbed insures that the deviation of the two sample paths is only temporary. It cannot be emphasized too strongly that from a probabilistic viewpoint, we could have just as easily thrown away the remaining service time Δ or δ, and used instead a regenerated lifetime from the appropriate exponential distribution. However, in this case we can no longer guarantee the immediate returning of both the nominal path and the perturbed path to the the same state. The common successor condition and the assignment scheme simply generalize this idea.

Note that the assumption of the equal mean transition lifetimes (t_{pb} and t_{pc}) implies that the transition probabilities from state "p" to "b" (i.e., the probability of $t_{pb} < t_{pc}$) is the same as that from "p" to "c" (i.e., the probability of $t_{pb} > t_{pc}$). Now we consider the general case where the transition probabilities are different and denoted as $q(p,b)$, $q(p,c)$, $q(b,s)$, and $q(c,s)$, respectively. The Markov process can be constructed as follows. Let r_1, r_2 be two i.i.d. exponential distributed random variables with mean one. Let us take $r_1/q(p,b)$ and $r_2/q(p,c)$ as the lifetimes, t_{pb} and t_{pc}, of the transitions from state "p" to "b" and from state "p" to "c," respectively. We determine the state transition as follows: if $t_{pb} < t_{pc}$ then the state transition is from "p" to "b;" otherwise, it is "p" to "c." In this setting, the state transition probability from "p" to "b" is $q(p,b)$ and that from "p" to "c" is $q(p,c)$. The means of t_{pb} and t_{pc} are $1/q(p,b)$ and $1/q(p,c)$, respectively, and the mean sojourn time at state "p" is one. (This can be adjusted by multiplying r_1 and r_2 by a constant.) If $t_{pb} < t_{pc}$ then the remaining lifetime of the transition from "p" to "c" is $t_{pc} - t_{pb} = r_2/q(p,c) - r_1/q(p,b)$, which has a mean $1/q(p,c)$. Thus, given $t_{pb} < t_{pc}$

$$r_3 = r_2 - \frac{q(p,c)}{q(p,b)} r_1$$

is exponentially distributed with mean one. We use $r_3/q(b,s)$ as the lifetime of the transition from state "b" to "s." Similarly, if $t_{pc} < t_{pb}$ then the next state is "c," and the remaining lifetime of the transition from "p" to "b" is $t_{pb} - t_{pc} = r_1/q(p,b) - r_2/q(p,c)$, which has a mean $1/q(p,b)$. Also, given $t_{pc} < t_{pb}$

$$r_4 = r_1 - \frac{q(p,b)}{q(p,c)} r_2$$

is exponentially distributed with mean one. We use $r_4/q(c,s)$ as the lifetime of the transition from state "c" to "s."

From the above construction, we can explain why IPA works. Suppose in the nominal path the state transition is from "p" to "b" (i.e., $t_{pb} < t_{pc}$), and because of some change in a parameter the state transition becomes from "p" to "c" (i.e., $t_{pc} < t_{pb}$) in the perturbed path. Then as the parameter change goes to zero, $|t_{pb} < t_{pc}|$ goes to zero. Hence t_{bs} or t_{cs} goes to zero. This means that the lifetimes of the transition from "b" to "s" in the nominal path and that from "c" to "s" in the perturbed path also converge to zero. That is, the nominal path converges to "p-b-s," and the perturbed path to "p-c-s," with the sojourn times in states "b" and "c" converge to zero. As discussed above, this assures the continuity of the performance function.

The limiting case $q(b,s) \to 0$ seems paradoxical. Let $q(b,s) = \varepsilon$. Then the above approach provides an unbiased IPA estimate whenever $\varepsilon > 0$; the approach, however, does not work if $\varepsilon = 0$ because condition CM no longer holds. Thus, it seems that the intrinsic property of the approach has a dramatic change at $\varepsilon = 0$. One may think that by adding a sufficiently small probability $\varepsilon > 0$ to a transition with zero probability one can make a Markov process that violates the condition (CM) satisfy (CM) without substantial change of performance. This paradox can be explained by looking at the variance of the IPA estimate. Indeed, for any $\varepsilon > 0$ we can construct an unbiased IPA estimate by using the approach described above, but the variance of such an estimate depends on ε. If $q(b,s) = \varepsilon$ is very small then the variance of the lifetime $r_3/q(b,s) = r_3/\varepsilon$ is very big. This naturally leads to a big variance of the IPA estimate. As $\varepsilon \to 0$, the variance of r_3/ε as well as that of the IPA estimate goes to infinity. Thus, the fact that IPA does not work at $\varepsilon = 0$ can be explained as that the estimate has an infinite variance.

It should be emphasized that in order for IPA to give unbiased estimates a special construction of the Markov processes is required, namely, the assignment of the shortest remaining lifetime to the common successor. In a simulation environment, this means extra work. In a real world experiment where clock readings cannot be observed, extra memories are required to implement an IPA algorithm based on condition (CM). For an arbitrary Markov chain with a large state space this may be a practical problem. More technical details and the scope of this condition (CM), see [Glasserman [1989]].

Exercise Show that reversible Markov chains (i.e., the transition probability $q(x,y)>0 => q(y,x)>0$) and single-class Jackson networks, open or closed, satisfy the common successor condition.

In sum, the remaining lifetime of a Markov event can be replaced by a newly generated exponential lifetime without changing the probability law of the process. This idea can be applied in either direction. The discussion in this sub-section emphasizes the use of this idea to create smoother models for IPA. We shall show in Section 5.4 another application where this statistical equivalence is used to preserve deterministic similarity but where we apply this equivalence in the reverse direction. Finally, we cannot emphasize too strongly that the subject of model representation is barely explored. This is primarily because the issue did not arise until derivative estimation became an interesting subject for research. The main point is that while two representations may be statistically equivalent with respect to various performance measures, there is no a priori reason that their sample derivatives are also equivalent. The fact that while under one representation, say the GSMP model, a DEDS sample path may not satisfy the conditions (CM) or (CU) to be described below, does not preclude the possi-bility that there may exist a more benign representation for it to which IPA may be applicable. One purpose of this chapter is to display the rich array of research topics in this subject area.

5.2 Another Sufficient Condition for Interchangeability
The previous section discussed a condition (CM)—the common successor condition—under which we can adopt a particular construction for general

Markov chains such that IPA can be made to work. In this section we shall discuss a related condition, applicable more generally, known as the commuting condition (CU) [Glasserman 1990bc, Li-Ho 1989] on the underlying DEDS model which will guarantee the applicability of IPA. We should emphasize that, despite its similarity to condition (CM), condition (CU) does not call for any special construction involving IPA. It is merely a sufficient condition which insures the interchangeability of expectation and differentiation as required in Chapter 4. Furthermore, condition (CU) does not require that the inter-transition time be exponentially distributed. It is applied to the sample paths of generalized semi-Markov processes and leads to more explicit and directly verifiable conditions than the (IC) condition in Chapter 4. The reason that we postponed its discussion until after the presentation of property (CM) is for comparison and adding insight.

5.2.1 The Basic Idea

As discussed in Chapter 4, infinitesimal perturbation analysis (IPA) gives an unbiased estimate if and only if the expectation and the differentiation are interchangeable for the sample performance function. We have established the interchangeability for closed Jackson networks. We see that the main reason causing a violation of the interchangeability is the discontinuity of the sample performance function. Once the continuity is proved, the interchangeability can be obtained via Lebesgue's bounded convergence theorem using some mild conditions on the slopes of the sample performance function. Based on this understanding, Glasserman [1990bc] proposed a sufficient condition for the interchangeability which guarantees the unbiasedness of the IPA estimates. The basic idea is as follows. Recall that for a closed Jackson network, we showed in Chapter 4 that the sample performance function $T_L(\theta, \xi)$, the time of the occurrence of event τ_L, is both left and right continuous at $\theta_2(\xi)$, where $\theta_2(\xi)$ is a value of the parameter θ at which two transitions happen at the same time on the particular sample path ξ. This implies that as θ increases from $\theta_2(\xi)-0$ to $\theta_2(\xi)+0$, two events on the sample path (θ, ξ) change their order of occurrence. The state sequences for the two sample paths $(\theta_2(\xi)\text{-}0, \xi)$ and $(\theta_2(\xi)\text{+}0,\xi)$ are different.   However, the essential fact is that, as θ approaches $\theta_2(\xi)$ from the both sides $\theta_2(\xi)\text{-}0$ and $\theta_2(\xi)\text{+}0$, the two state sequences of these two sample paths become the same. The state between the two events that change their order disappears as θ approaches $\theta_2(\xi)$. The state-independen trouting mechanism of a closed Jackson network guarantees that after these two events change the order, the system reaches the same state. In other words, the state of the system after these two events does not depend on the order of these two events. This property assures the continuity of $T_L(\theta, \xi)$. By the same argument, the continuity holds for any other sample performance functions that are defined as functionals of the sample path.

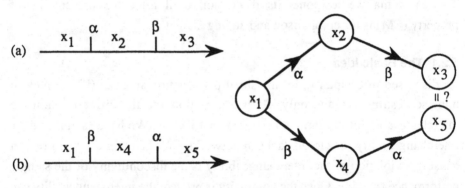

Fig. 6 The Effect of an Order-Changing of Two Events

In general cases, the above property may not hold. Fig. 6 depicts the problem. Fig. 6.a illustrates the nominal path in which event α occurs before event β. The system stays in state x_1 before event α occurs and transfers to state x_2 after event α and then transfers to state x_3 after event β. Suppose that because of a change in a parameter θ, the occurrence of α is delayed so that β happens before α. Then in the perturbed path (illustrated in Fig. 6.b) the system first transfers to state x_4 after event β and then transfers to state x_5 after event α. x_4 and x_5 may not be the same as x_2 and x_3. (To help fix ideas, consider the example of a CPU under a multitask priority nonpreemptive service discipline where α and β represent the arrivals of low and higher priority tasks respectively which may entail different consequences.) Therefore, as θ increases to $\theta_2(\xi)$, at which the two events α and β happen at the same time, the occurrence time of β approaches that of event α from the right side and the state sequence approaches x_1, x_2 (for an infinitesimal time), x_3,...., while as θ decreases to $\theta_2(\xi)$, the occurrence time of β approaches that of event α from the left side and the state sequence approaches x_1, x_4 (for an infinitesimal time), x_5,.... This indicates that, since x_2 and x_4 disappear in the limiting cases, the fact that they differ is not important to the continuity of the sample perfor- mance function. More importantly, if $x_3 = x_5$, then as θ approaches $\theta_2(\xi)$ from both sides $\theta_2(\xi)$-0 and $\theta_2(\xi)$+0, the corresponding state sequences converge to the same sequence x_1, x_3 ($=x_5$), ..., this implies that the sample performance function converges to the same value from both sides of $\theta_2(\xi)$. Thus, the sample performance function is continuous at $\theta_2(\xi)$. Loosely speaking , the condition for the continuity of the sample performance function can be stated as follows: Let x_3 be the state reached by the system after two events α and β with α occurring before β, and x_5 be the state reached by the system after two events β and α with β occurring before α. Then the sample performance function is continuous if $x_3 = x_5$, i.e., the state reached by the system does not depend on the order of these two events and the contributions of the different inter-mediate states, x_2 and x_4, are infinitesimal.

To state the condition more precisely, we have to study the probabilistic nature of the state transition mechanism in a GSMP model. We shall do this in the next subsection. But first, conceptually we note that the above statement of order independence should hold w.p.1 for all sample paths (θ, ξ)'s. Let Ξ_0 be the set of ξ representing the sample paths on which the state reached from x_1 via the consecutive occurrences of α and β is the same as that reached from x_1 via the consecutive occurrences of β and α for different values of θ in a neighborhood of $\theta_2(\xi)$. Choosing $\theta = \theta_2(\xi)$ -0, we can see that the probability of Ξ_0, $P(\Xi_0)$, equals the probability that the system transfers from x_1 via α and

β. On the other hand, letting $\theta = \theta_2(\xi) + 0$, we see that $P(\Xi_0)$ equals the probability that the system transfers from x_1 via β and α. Thus, in order for the state reached by the system from state x_1 via two events α and β does not depend on the order of the occurrences of these two events on almost every sample path we require that the probability that the system transfers from x_1 to x_3 via α and β is the same as that via β and α[3]. This statement is a sufficient condition for the continuity of the sample performance functions; it is almost necessary for the continuity, except for that in some cases the system may have some particular properties such as that the contributions of x_5 and x_3 to the performance function $L(\theta, \xi)$ are the same and that the probability characterization of the process after x_3 and x_5 are the same.

5.2.2 A Sufficient Condition Based on the Generalized Semi-Markov Process Model

We first describe a GSMP formally following the notations of Appendix B and Chapter 3 Section 3. The set X represents the possible states, the set Γ (usually denoted as a set of integers $\{1,2,...,I\}$) represents the possible events. For any $x \in X$, an event list of x, $\Gamma(x)$, is a subset of Γ and represents the set of events that can occur in state x. The evolution of the process is determined by the occurrence of events. If event $\alpha \in \Gamma(x)$ occurs in state x, then the system will transfer to the next state x' with probability $p(x'; x, \alpha)$. We assume that for all x and α, there are finitely many x' such that $p(x';x,\alpha) > 0$.

For any state x in the process, each event in $\Gamma(x)$ is associated with a clock. A clock reading is a vector $c = (c(1), ..., c(I))$ of nonnegative real numbers. $c(\alpha) > 0$ if and only if $\alpha \in \Gamma(x)$. $c(\alpha)$ represents the remaining time to when event α occurs, assuming no other events occur before α. The state transition happens whenever an event occurs. Let τ_n be the nth state transition time and a_n the nth event. With a fixed initial state x_0, the sequence $\{\tau_n\}$ can be

---

[3] Note despite the similarity of this condition with that of the condition (CM), this condition calls for the equality of certain probabilities while the condition (CM) calls for special constrcutions and the remaining life time assignments. See the example at the end of Section 5.2.2 for further clarifications.

determined as follows: Suppose that at τ_n the system enters state x with clock $c_n = (c_n(1), ..., c_n(I))$. Then (see Eqs.(3.8) - (3.9))

$$\tau_{n+1} = \tau_n + \min \{c_n(\alpha): \alpha \in \Gamma(x_n)\}$$

$$a_{n+1} = \min\{\alpha \in \Gamma(x_n): c_n(\alpha) = \min \{c_n(\alpha'): \alpha' \in \Gamma(x_n)\}\} \equiv \alpha^*$$

and the state transition is triggered by event α^*. At τ_{n+1}, a new state x' is selected according to p(x'; x, α^*). If $\beta \in \Gamma(x')$ and $\beta \notin \Gamma(x) - \{\alpha^*\}$, then a new clock value $c_{n+1}(\beta)$ is sampled for event β from a c.d.f. $\phi_\beta(x)$. If $\beta \in \Gamma(x')$ and $\beta \in \Gamma(x) - \{\alpha^*\}$, then $c_{n+1}(\beta)$ is set to $c_n(\beta) - (\tau_{n+1} - \tau_n)$, i.e., it is reduced by the time elapsed since the last transition. See Eqs.(3.12-13). The state transition in a GSMP with transition probability p(x'; x, α) is implemented as follows. A natural way is using a doubly indexed sequence of random numbers U(α,n) uniformly distributed on [0, 1) to determine the next state after the nth transition triggered by event α according to the following rule: First we partition the interval [0, 1) into disjointed intervals, each of length p(x'; x, α) for each x' such that p(x'; x, α) >0. If U(α,n) falls in an interval corresponding to x', then event α triggers a transition from state x to state x'. This way the probabilistic state transition can be represented as a functional mapping x'$\equiv$ ψ(x,a,U(a,n)). See also Chapter 3 Eq.(3.11). With this GSMP state transition construction, it is not hard to see that we can insure that the state reached via the occurrence of events α and β to be independent of the order of occurrence, i.e.,

$$\psi(\psi(x,\alpha, U(\alpha,n)),\beta,U(\beta,n+1)) = \psi(\psi(x,\beta,U(\beta,n+1)),\alpha, U(\alpha,n)) \qquad (1)$$

by requiring that the following conditions hold.

Conditions CU (Commuting Conditions)

i) If $\alpha, \beta \in \Gamma(x)$ and p(x';x, α)>0, then $\beta \in \Gamma(x')$,

> ii) If α, $\beta \in \Gamma(x_1)$ and x_2 and x_4 satisfy $p(x_2; x_1, \alpha) \, p(x_4; x_2, \beta) > 0$, then
> there is an x_3 for which $p(x_3; x_1, \beta) = p(x_4; x_2, \beta)$ and $p(x_2; x_1, \alpha) =$
> $p(x_4; x_3, \alpha)$,

In fact, if the CU conditions hold, then the sample performance function $L(\theta, \xi)$ defined as

$$L(\theta, \xi) = \int_0^T f[x(t)]dt \tag{2}$$

where T is a stopping time[4] and f is a bounded real function on X with $|f| < f^*$ $< \infty$, is a continuous function of θ. See Fig. 7.

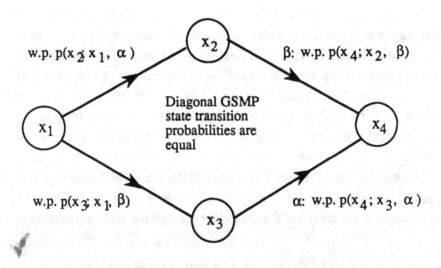

Fig. 7 Commuting Condition (CU)

This is a formal statement of the basic idea described in Section 5.2.1. Having established the continuity of the sample performance function $L(\theta, \xi)$, we can discuss the sufficient conditions for the interchangeability of the expectation and differentiation on $L(\theta, \xi)$ very much similarly to what was

---

[4] A random time that has a finite expectation and is dependent only on past history.

done in Chapter 4. Rather than reproducing detailed developments which can be found in [Glasserman 1990c], we illustrate instead with a simple example comparing condition (CU) with condition (CM) of Section 5.1. Consider the example of a simple cyclic two-server system with feedback as shown in Fig. 8. The feedback routing probability p_n may depend on the state, n, the population at the

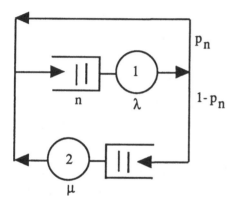

Fig. 8 A Cyclic Queue with Feedback

queue of server 1. So long as all $1 > p_n > 0$, then we submit property (CM) is satisfied since the two-server network is equivalent to a simple birth-death process with rates μ and $\lambda(1 - p_n)$ respectively. Similarly, from the GSMP viewpoint, if the p_n's are independent of n, then we can directly verify that condition (CU) is satisfied. Otherwise, IPA is not applicable

Exercise Convince yourself of the assertions associated with the example of Fig. 8 with respect to conditions (CU) and (CM).

5.2.3 Queueing Network Examples

Although the interchangeability condition was proved for several cases such as the response time of a GI/G/1 queue, the throughput and response time of a closed Jackson network, and the throughput and response time of a queueing network in which there is no simultaneous blocking, the sufficient condition above provides a uniform way of establishing the interchangeability based on easily verifiable structural conditions of the system under question. Glasserman [1990bc] gives examples of networks for which the interchangeability holds and for which it does not hold. We shall briefly state the results.

Networks for which the interchangeability hold:
1. Simple queueing networks: A simple network consists entirely of FCFS, infinite buffer, single-server nodes and a single class of customers with a state-independent Markovian routing mechanism. A GI/G/1 queue and Jackson networks are examples of simple queueing networks.

2. Simple networks with multiclass customers in which every node visited by more than one class of customers is fed by only a single source. In such a network, a customer cannot change the order of arriving at a server with any customer of a different class.

3. Networks with blocking in which every node with a finite buffer is fed by a single source. In such a network, no server can directly block more than one server simultaneously. A cyclic queue with finite buffers is an example of such networks.

4. Some networks with state-dependent routing mechanism described as follows: For any (arrival or departure) event α, let $\{X_1(\alpha), ..., X_N(\alpha)\}$ be the partition of the states such that x and x' are in the same subset $X_i(\alpha)$ if and only if the transition probabilities for the customer moving upon the occurrence of α are the same in x and x'. Then the interchangeability holds if that event α is the only event that can trigger a state transition between two states that belong to two different subsets $X_i(\alpha)$ and $X_j(\alpha)$, $i \neq j$.

Networks for which the interchangeability does not hold:
1. Simple networks with multiclass customers in which there exists at least one node that is visited by more than one class of customers and is fed by more than one single source. An example of such a network is the given in [Cao 1987c].

2. Networks with simultaneous blocking in which there is at least one server which may directly block more than one server at the same time. The discontinuity of the sample performance function of such a network is first observed in [Cao and Ho 1987b].

3. Many networks with state-dependent routing mechanism. An example of such a network is the GI/G/1/K finite capacity queue in which arrivals that find the buffer full are simply lost.

It is clear by now that the basic requirement for IPA applicability is the continuity of the sample path with respect to parameter perturbation. Roughly speaking, under either condition CM or CU we can construct a sample path in such a way that a change in θ can only make the perturbed path deviate temporarily from the nominal one. More precisely, a small change in θ can only change the sample state sequence $(x_1, ..., x_k, x_{k+1}, ..., x_n)$ to $(x_1, ..., x_k^*, x_{k+1}, ..., x_n)$ and the holding times at states x_k and x_k^*, denoted as w_k and w_k^*, go to zero as $\Delta\theta \to 0$. This guarantees the continuity of the performance function L. If the condition CM or CU does not hold, the process may not come back to the same state x_{k+1} after reaching x_k^*. The succeeding sections and chapters will deal with such non-IPA-applicable cases. However, before we do this it is instructive to examine one direct approach to deal with this problem of sample path discontinuity. Taking a cue from the condition CM and CU, we shall attempt a transformation which will make any discontinuities due to parameter perturbation insignificant. In other words, suppose the nominal and the perturbed trajecotries start to differ at states x_k and x_k^* and we cannot guarantee the quick return of the trajectories to the same state, x_{k+1}. Then, let us make sure that the trajectories after x_k and x_k^* do not amount to anything that has substantial effect on the performance. One way to accomplish this is to introduce a nonlinear transformation of time scale which will render the trajectory length after x_k and x_k^* to be of the order of $\Delta\theta$. This way even if discontinuities occur, they will have no first order effect on the performance and IPA can be applied to the transformed problem. To help fix ideas, consider a Markov chain that does not satisfy condition CM, i.e., there are no common successors to x_k and x_k^*. Motivated by the above observation, we propose an algorithm which imbeds the Markov chain in a non-Markov process, X. The non-Markov process is constructed as follows. For any state i, the interval $[0,1)$ is partitioned into subintervals $[a_{i,j}, a_{i,j+1})$ with $a_{i,m} = 0$ and

$$a_{i,j} = \sum_{m=1}^{j-1} Q_\theta(i,m).$$

where $Q_\theta(i,m)$ is the transition rate from state i to state m. Let $\xi_1, \xi_2, ...$ be a sequence of independent and uniformly distributed random variables in $[0, 1)$. If $x_k = i$, and $\xi_k \in [a_{i,j}, a_{i,j+1})$ then we choose $x_{k+1} = j$. The initial state x_0 is given. This mechanism shows that the imbedded chain X is Markov. Let the holding time w_k of X be determined by w_{k-1}, ξ_k, and $[a_{i,j}, a_{i,j+1})$ as follows (see illustration below)

$$w_k = r_k(\xi_k)w_{k-1}, \text{ with } w_0 = 1, \text{ and}$$

$$r_i(\xi_k) = \frac{4 \min(\xi_k - a_{i,j}, a_{i,j+1} - \xi_k)}{Q_\theta(i,j)}.$$

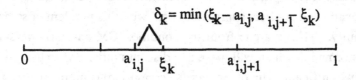

It is easy to check that $E[r_i(\xi_k)] = 1$ and $E(w_k) = 1$ for all k. The process thus constructed is not a Markov process because w_k depends on the previous state holding times. From the construction, for a given sequence $\xi_1, \xi_2, ...$ suppose that x_k changes to x_k^* because of a small change in θ. This can only happen when ξ_k is close to either $a_{i,j}$ or $a_{i,j+1}$. Thus, as $\Delta\theta \to 0$, $r_i(\xi_k) \to 0$, and $w_k \to 0$. We also have $w_m \to 0$ for all $m > k$, i.e., the sample path length after x_k shrinks to zero as $\Delta\theta \to 0$. Thus, the performance function will be continuous with respect to θ, despite the fact that all the states after x_k^* in the perturbed path are different from those in the nominal path. Once the continuity is verified, the unbiasedness of the IPA estimate can be established similarly as before by simply finding a bound for the sample derivative. We shall not pursue this study here. What is interesting is that the algorithm based on the non-Markov process, which depends on the state holding times, can be converted to a discrete time version, which depends solely on the state of the Markov chain X. This can be done by taking the expectation of the state holding times. Surprisingly, [Glasserman 1990] shows that this discrete version is identical to the algorithm given by the likelihood ratio method to be discussed in Chapter 7. In this sense, the likelihood ratio method can be considered as an estimate yielded by IPA with a special representation. As discussed in Section 7.3, we expect a big variance for this method. Intuitively, we can also see that such a nonlinear transformation is numerically unsound. Thus, we do not expect this approach to the problem of sample path discontinuity to be practical useful. It is presented here primarily because of the way it illuminates the basic issue of IPA validity and sample path continuity.

5.3 Routing Probability Sensitivity

One of the important parameters in a queueing network is the routing proba-bility, q_{ij}, which denotes the probability of a customer's going to station j after completing service at station i. In practice, q_{ij} can characterize the percentage of parts following different operations in a manufacturing context or the traffic density on different links in a communication network. When using IPA to estimate the sensitivity of a performance measure L, say throughput, with respect to this parameter, $\theta = q_{ij}$, one immediately encounters the difficulty that $L(\theta,\xi)$ is discontinuous in θ for some set of ξ. As explained in Chapter 4, $L(\theta,\xi)$ is in fact a piecewise constant function of θ. In a finite length sample path, Δq_{ij} can always be made so small as never to cause any perturbation in the nominal path, i.e., $\Delta L(q_{ij},\xi)/\Delta q_{ij} = 0$ for any ξ. On the other hand, regardless of how small Δq_{ij} is, $\Delta L(q_{ij},\xi)$ will be finite for any ξ as the length of the path goes to infinity since sooner or later some customers will change direction because of a nonzero q_{ij}[5]. Such perturbations cannot be ignored since they constitute the very reason for a nonzero $dE[L(q_{ij},\xi)]/dq_{ij}$. It

---

[5] If we fix the length of the path, then we can always find some ξ's such that on these paths customers will switch direction because of a nonzero q_{ij}.

would appear that IPA cannot be applied to such problems.   This section will illustrate one more approach for the resolution of this problem.

Intuitively, we realize that changing q_{ij} simply means changing the rate of the customer stream along different paths in the steady state.   In queueing network terminology, this means changing the steady-state visit ratios to a station. Specifically, we have for closed networks the traffic equations (Appendix A)

$$v_j = \sum_{i=1}^{M} q_{ij} v_i \tag{3a}$$

and the probability constraints,

$$\sum_{j=1}^{M} q_{ij} = 1 \tag{3b}$$

where v_i is the visit ratio to station i. For any feasible dq_{ij} satisfying Eq.(4b) below, we can calculate the induced dv_i's via Eqs.(3a).   Thus, taking differentials of both sides of (3a) and (3b) we get

$$dv_j = \sum_{i=1}^{M} [\, dq_{ij} v_i + q_{ij} dv_i \,], \qquad j=1,2,...,M \tag{4a}$$

$$\sum_{j=1}^{M} dq_{ij} = 0, \qquad i=1,2,...,M. \tag{4b}$$

However, for given dq_{ij}, Eq.(4a) is not linearly independent. But assuming the additional condition $dv_M = 0$, we can then solve for the remaining dv_i. Assume for the moment that we are dealing with a Jackson network. We know from Appendix A that the throughput can be expressed as a function of the products of the visit ratio and the mean service time, $v_i s_i$, i.e.,

$$E[L] = f(v_1 s_1, v_2 s_2,..., v_M s_M) \tag{5}$$

Thus the throughput is only affected by v_i through the product $v_i s_i$. In this case the effect of any change in v_i on the throughput can be equivalently achieved by some change in s_i through

$$s_i dv_i = v_i ds_i, \quad \text{or} \quad ds_i = \frac{s_i}{v_i} dv_i \tag{6}$$

Thus, at least for Jackson networks, we have shown that we can convert any dq_{ij} into equivalent ds_i's. Consequently, the sensitivity of L with respect to q_{ij} can be evaluated in terms of the sensitivities of L with respect to s_i, which can be calculated using the IPA methods of Chapters 3 and 4. Intuitively, this is entirely reasonable: Changes in routing probability certainly amount to changes in the arrival rates at the stations involved and changes in arrival rates can be viewed equivalently as changes in the service rates on a relative basis. Of course for Jackson networks the above demonstration is superfluous since analytical formulas exist for the various sensitivities involved and there is no need for the use of IPA. *However, the more important and basically distribution-independent idea here is that an infrequent but finite perturbation (in customer routing) can be replaced by a series of frequent and infinitesimal perturbations (in sample service time).* We will be using this idea many times in different guises in succeeding sections.

More generally, the product-form networks (e.g., the BCMP network) have their throughput dependent on q_{ij} through the product $v_i s_i$ thus making this approach applicable. If in a nonproduct form network the performance depends on q_{ij} only through the visiting ratio v_i, then we can make the following conjecture based on dimensional analysis in physics:

$$E[L] = f(v_1 m_{11}, v_1^2 m_{12}, .., v_1^k m_{1k}, ...; ...;$$
$$v_M m_{M1}, v_M^2 m_{M2}, ..., v_M^k m_{Mk}, ...) \tag{7}$$

where m_{ik} is the kth moment of the service distribution of station i and has the dimension of $(t)^k$. From (7) if the elements of the routing probability matrix Q = $[q_{ij}]$ change such as to induce changes dv_i in v_i, then the change in E[L] is just the same as if Q does not change but all moments m_{ik} change according to the following analog of Eq.(6)

$$(v_i + \Delta v_i)m_{i1} = v_i(m_{i1} + \Delta m_{i1}) \qquad \Rightarrow \Delta m_{i1} = \frac{\Delta v_i}{v_i} m_{i1}$$

$$(v_i + \Delta v_i)^2 m_{i2} = v_i^2(m_{i2} + \Delta m_{i2}) \qquad \Rightarrow \Delta m_{i2} = 2\frac{\Delta v_i}{v_i} m_{i2}$$

.

.

.

$$(v_i + \Delta v_i)^k m_{ik} = v_i^k(m_{ik} + \Delta m_{ik}) \qquad \Rightarrow \Delta m_{ik} = k\frac{\Delta v_i}{v_i} m_{ik} \qquad (8)$$

Thus if Eq.(7) holds then the sensitivity of E[L] with respect to Q can be calculated through the sensitivity of E[L] with respect to all these changes Δm_{ik}. At first glance, this may appear to be even more work than calculating the original sensitivity dE[L]/dQ. Fortunately, we have the following useful fact. Suppose m_k is the kth moment of the distribution function F(s), then the kth moment of the perturbed distribution $H(s) \equiv F(s(1-\Delta v/v))$ is $m_k + \Delta m_k$ with $\Delta m_k = k(\Delta v/v)m_k$. This can be seen by directly computing the kth moment of H(s)

$$\int_0^\infty s^k\, dH(s) = \int_0^\infty s^k dF[s(1-\frac{\Delta v}{v})]$$

$$= \int_0^\infty \left(\frac{s}{1-\frac{\Delta v}{v}}\right)^k dF(s) = \int_0^\infty s^k(1+k\frac{\Delta v}{v})\, dF(s)$$

$$= (1+k\frac{\Delta v}{v})m_k = m_k + \Delta m_k \qquad (9)$$

This fact provides an easy way of implementing IPA to obtain the sensitivity of $E[L]$ with respect to all changes of Δm_{ik}. As we discussed in Chapter 3, the basic perturbation generation rule is to compute the perturbation in a service time duration due to a change in the corresponding service time distribution. In the current case, if the perturbed service time is generated according to the distribution $H_i(s) \equiv F_i(s(1-\Delta v/v))$ then the moments will be perturbed as shown in Eq.(8). Moreover, in this case

$$\Delta s_i = H_i^{-1}[\ F_i\ (s_i)] - s_i = (1 + \frac{\Delta v_i}{v_i})\ s_i - s_i = \frac{\Delta v_i}{v_i}\ s_i, \tag{10}$$

that is, the time scale of server i is multiplied by a factor of $(1+\Delta v_i/v_i)$. Therefore, we can generate the perturbed service time durations satisfying Eq.(8) by generating a perturbation according to (10). In other words, (8) is automatically implemented by (10). Eqs.(7) and (8) can be viewed as a power series generalization of Eqs.(5) and (6). If (7) is true, then the algorithm based on (8) is exact[6]. If not, then (8) can be regarded as a power series approximation to the calculation of $E[L]$. The point to be noted is that in this method of calculating $\Delta E[L]/\Delta Q$ via $\Delta E[L]/\Delta m_{ik}$ we need not know the explicit form of the function f in Eq.(7). Experimental results validating this approach can be found in [Ho and Cao 1985].

The above development also illustrates an important "modeling idea" for IPA. The difficulty of the routing probability sensitivity problem for IPA is due to the fact that perturbation of routing probability, however small, will cause finite perturbations in arrival sequence between different servers (and hence the state sequence), however infrequently. Yet it is precisely such occurrences that contribute to sensitivity calculations in this case. The idea here is to convert or smooth such infrequent but finite perturbations to frequent but infinitesimal perturbations for which IPA is applicable. In the above example, small perturbations in routing probabilities are first converted to small perturbations in the

---

[6] Note the conjecture of Eq.(7) implicitly assumes that the structure of the functional dependence of L is not an explicit function of the parameter θ. The dependence on θ only comes through m_{ik}'s.

visit ratios to the servers. Perturbations in the visit ratios are then converted to a series of equivalent perturbations in the service times through the product term $v_i s_i$. The following example uses the same idea but illustrates this idea even more explicitly [Vakili and Ho 1987; 1989].

Example Small perturbations in routing probabilities always produce infrequent but finite changes in a sample path. Consider the case of routing customers to one of two possible servers with different mean service times s_1 or s_2 according to the proportion d:1-d as follows:

$$s = \begin{cases} -s_1 \ln (v) & 0 \leq u \leq d, \\ -s_2 \ln (v) & d < u \leq 1, \end{cases} \tag{11}$$

where u and v are both uniform random variables over [0, 1), and "d" plays the role of routing probability. Notice that under the condition that $0 \leq u \leq d$ (resp. $d < u \leq 1$), u/d (resp. (1-u)/(1-d)) is also a uniform random variable over [0, 1). So instead of using (11) to determine service time we can use

$$s = \begin{cases} -s_1 \ln \left(\dfrac{u}{d}\right) & 0 \leq u \leq d, \\ -s_2 \ln \left(\dfrac{1-u}{1-d}\right) & d < u \leq 1. \end{cases} \tag{12}$$

It can be observed that if we generate a service time from Eq. (12), the service time of *every* customer changes when d is changed. The important advantage of using (12) is that the service time changes of customers who switch between servers because of the change of "d" are no longer finite. These changes are now infinitesimal since switching takes place only at values of "u" near "d", i.e., when s is near $\ln(1) = 0$. It is clear that the service time changes are now of the order Δd.   Thus, when switching takes place, the discontinuities will be of the same order. On the other hand, the probability that a customer in the nominal path is to be switched is also in the order of Δd. Therefore, the total change in performance,   caused by switched customers,   is of order $(\Delta d)^2$ and can be ignored in calculating the first derivative. This is equivalent to saying that event order changes are ignorable and IPA will give us strongly consistent estimates for the sensitivities with respect to routing probability "d."

Exercise   Eq. (11) calls for the use of two u.i.i.d. random numbers while Eq. (12) needs only one. Are we getting something for nothing here?

5.4 The Multi-Class M/G/1 Queue and Networks

5.4.1 The Two-class M/M/1 Queue

Sections 5.1 and 5.3 gave examples of the two main ideas of extending the applicability of IPA to cases involving discontinuous $L(\theta,\xi)$. In this section we show an example which involves the use of both ideas together, i.e., (i) using an equivalent but more benign representation, (ii) smoothing out infrequent and finite perturbations into frequent and infinitesimal perturbations. Let us consider an M/M/1 queue with two arriving classes of customers with arrival rates λ_1 and λ_2, respectively, and mean service times s_1 and s_2, respectively. If we take the performance E[L] to be the average waiting time of all customers and the system parameter to be, say, λ_1, then simple IPA algorithm will in general produce a biased estimate of the sensitivity $dE[L]/d\lambda_1$. This can be seen intuitively. A perturbation in λ_1, no matter how small, will sooner or later cause two successive arriving customers of different classes to change the order of their arrivals. As a result, the first and the second customers (in terms of arrival at the queue) on the nominal path change places on the perturbed path. Thus, the first customer in the nominal path,   being the second one in the perturbed path, has to wait an extra sample service time in the perturbed path. This difference in waiting time due to the difference in mean service time for the two classes is in general finite even though the cause may be due to an infinitesimal change in arrival time of the two customers. Even in the case of equal mean service time for the two classes, the ordinary IPA estimate can still be biased for some performance measures. Under deterministic similarity, the event order of the arrivals of the two classes are assumed to be invariant, but in reality class 1 arrivals will continuously and indefinitely slide ahead of those of class 2 (assume $d\lambda_1 > 0$). These difficulties can be overcome via the following development.

Consider the following problem. Let T be the mean system time of a customer (for both classes of customers), and we want to estimate $\partial T/\partial \lambda_1$. Notice that the total arrival process can be regarded as a Poisson process with rate $\lambda_1 + \lambda_2$, and the ratio between class 1 customers and class 2 customers is λ_1/λ_2. Now instead of considering the original multiclass M/M/1 queue, we set up an "equivalent" single-class M/G/1 queue. The arrival rate of this M/G/1 queue is $\lambda \equiv \lambda_1 + \lambda_2$. The service time at the server is hyperexponentially distributed with two parallel stages having means s_1 and s_2, and routing probabilities d and 1-d, respectively, where $d \equiv \lambda_1/(\lambda_1+\lambda_2)$. It is now a single-class queue since we no longer distinguish the class identity of the customers once they are in the queue. When λ_1 is changed, *both* the arrival rate and the service times of customers in this equivalent queue will be changed (because of the change in "d"). If we denote T_N as the steady-state system time of a customer in this new queue, then it is clear that $T_N = T$. Furthermore, we have

$$\frac{\partial T_N}{\partial \lambda_1} = \frac{\partial T_N}{\partial \lambda} \frac{\partial \lambda}{\partial \lambda_1} + \frac{\partial T_N}{\partial d} \frac{\partial d}{\partial \lambda_1} \tag{13}$$

From definitions, $\partial \lambda/\partial \lambda_1$ and $\partial d/\partial \lambda_1$ can be evaluated via simple calculus. We also know that $\partial T_N/\partial \lambda$ can be obtained by using IPA since we are now dealing with a single-class queue. But it is not obvious that we can use IPA to estimate $\partial T_N/\partial d$. Actually, applying IPA directly to calculating $\partial T_N/\partial d$ will give us the wrong answer. This is again due to the fact that when d is changed, sooner or later, some customers will change the service stages in the hyperexponential distribution (i.e., their means are changed either from s_1 to s_2 or from s_2 to s_1). These will cause finite changes in the service time. So we seem to be back to square one again. However, this discontinuity can be easily smoothed out. In fact, the example on routing probability in Section 5.3 solves this problem[7].If we identify the "d" in Eqs.(11 - 12) with $\lambda_1/(\lambda_1+\lambda_2)$ then by using the "trick" in Eq.(12) we can get a strongly consistent estimate of $\partial T_N/\partial d$ via IPA and by Eq.(13) a strongly consistent estimate of $\partial T/\partial \lambda_1$. To show this, consider the equivalent single-class M/G/1 queue. It is very easy for us to get the following formulas for T and T_N (see [Kleinrock 1975] or Appendix A)

$$T = T_N = \bar{x} + \frac{\lambda m_2}{2(1-\lambda\bar{x})}, \tag{14}$$

where

$$\bar{x} = ds_1 + (1-d)s_2 = \frac{\lambda_1 s_1 + \lambda_2 s_2}{\lambda_1 + \lambda_2}$$

is the mean service time of the equivalent M/G/1 queue, and

$$m_2 = 2[ds_1^2 + (1-d)s_2^2] = \frac{2(\lambda_1 s_1^2 + \lambda_2 s_2^2)}{\lambda_1 + \lambda_2}$$

is the second moment of the service time. In Chapter 4 and, more explicitly, in [Suri and Zazanis 1988], it has been proved that IPA gives a strongly consistent estimate of the first derivative of the mean system time of a customer with respect to the arrival rate for a single class M/G/1 queue. Thus direct differentiation of (14) gives

$$\left[\frac{\partial T_N}{\partial \lambda}\right]_{IPA} = \left[\frac{\partial T_N}{\partial \lambda}\right]_{EXACT} = \frac{m_2}{2(1-\lambda\bar{x})^2}. \tag{15}$$

Since we can ignore the changes caused by switched customers, then according to (12) the change of service time with respect to "d" is given by

---

[7] The example of Eqs.(11-12) deals with routing customers to two different exponential servers according to "d;" this is equivalent to the case here of a single server with an hyperexponential service time distribution characterized by the two different mean service s_1 and s_2 and probability of "d" for switching between them.

$$\frac{\partial s}{\partial d} = \begin{cases} \dfrac{s_1}{d} & 0 \le u \le d, \\[2mm] -\dfrac{s_2}{1-d} & d < u \le 1. \end{cases} \tag{16}$$

From (16), we have directly

$$E\left[\frac{\partial s}{\partial d}\right] = s_1 - s_2; \quad E\left[s\frac{\partial s}{\partial d}\right] = \frac{1}{2}(m_{12} - m_{22}), \tag{17}$$

where $m_{12} = 2s_1^2$ and $m_{22} = 2s_2^2$ are the second moments of the two exponentially distributed service stages in the hyperexponential server, respectively. It also has been shown in Chapter 4 Section 5 (Eq.(4.82)) and by [Suri and Zazanis 1988] that by using IPA we can get

$$\left(\frac{\partial T_N}{\partial d}\right)_{IPA} = E\left(\frac{\partial s}{\partial d}\right) + \frac{\lambda E\left(s\frac{\partial s}{\partial d}\right)}{1 - \lambda\overline{x}} + \frac{\lambda^2 m_2 E\left(\frac{\partial s}{\partial d}\right)}{2(1 - \lambda\overline{x})^2}. \tag{18}$$

Direct differentiation of Eq.(14) and the definition of "d" give us

$$\left[\frac{\partial T}{\partial \lambda_1}\right]_{exact} = \frac{\lambda_2(s_1 - s_2)}{(\lambda_1 + \lambda_2)^2} + \frac{s_1^2 + \lambda_2 s_1 s_2(s_1 - s_2)}{(1 - \lambda_1 s_1 - \lambda_2 s_2)^2} \tag{19}$$

and

$$\frac{\partial d}{\partial \lambda_1} = \frac{\lambda_2}{(\lambda_1 + \lambda_2)^2}. \tag{20}$$

Putting Eqs. (15) and (17 - 20) together, it is possible to verify after laborious algebra the identity

$$\left[\frac{\partial T}{\partial \lambda_1}\right]_{EXACT} = \left[\frac{\partial T_N}{\partial \lambda}\right]_{IPA}\frac{\partial \lambda}{\partial \lambda_1} + \left[\frac{\partial T_N}{\partial d}\right]_{IPA}\frac{\partial d}{\partial \lambda_1}.$$

So this revised IPA gives a strongly consistent estimate.

Exercise Extend the above proof to the case of a two class M/G/1 queue. What is the analog of Eqs.(7) and (8)?

Again the above illustration is used primarily for ease of demonstration. Experimentally, there is no inherent limitation to extending this idea to a multi-class GI/G/1 queue. In what follows, some numerical results on such GI/G/1 cases are provided to test our algorithm. In all these experiments, we focus on estimating $dT/d\lambda_1$, with T being the mean system time. Since there does not exist a close form formula to compute $dT/d\lambda_1$ for general multiclass GI/G/1 queues, the brute force (BF) estimates are employed to compare with our IPA estimates from (13). We use symmetric difference estimates in the brute force method. That is,

$$\left(\frac{\partial T}{\partial \lambda_1}\right)_{BF} = \frac{T(\lambda_1+\Delta\lambda/2)-T(\lambda_1-\Delta\lambda/2)}{\Delta\lambda} . \tag{21}$$

The length of every sample path is 1,000,000 customers, and we run 100 replications for every experiment to get the confidence interval at a 95% level by using normal distribution approximation. The results are presented for three traffic intensities $\rho = 0.2, 0.5$, and 0.8 ($\rho=\lambda_1 s_1 + \lambda_2 s_2$). It is interesting to notice that IPA estimates have tighter confidence intervals than BF estimates. This is an illustration of the superior variance property of IPA often observed in experiments (see also a discussion in Section 7.3)

Experiment 1 Let the arrival distributions of the two classes, $A_i(x)$, be the uniform distribution over $[0, 2/\lambda_i]$; the service distributions $G_i(s)$ are exponential with mean $\bar{s}_i$ (i=1, 2). We choose $\Delta\lambda=0.1$ for $\lambda_1=2$, $\Delta\lambda=0.05$ for $\lambda_1=0.6$ and 0.2. The results are given in Table 1.

λ_1	λ_2	$\bar{s}_1$	$\bar{s}_2$	ρ	BF est.	Revised IPA est.
2	2.5	0.3	0.08	0.8	1.3030 ± 0.1189	1.2525 ± 0.0258
0.6	4	0.5	0.05	0.5	0.6563 ± 0.1175	0.6423 ± 0.0112
0.2	0.5	0.6	0.16	0.2	0.8124 ± 0.0249	0.8840 ± 0.0053

Table 1 Experimental Comparison of IPA of (13) With Brute Force Estimates

Experiment 2 $A_1(x)$ is an exponential distribution with λ_1; $A_2(x)$ is a uniform distribution over $[0, 2/\lambda_2]$; $G_1(s)$ is a uniform distribution over $[\lambda_1, 2\bar{s}_1 - \lambda_1]$; $G_2(s) = \alpha(s - \lambda_2)^3$, $s \in [\lambda_2, 1 + \lambda_2$, where α is a constant. We choose $\Delta\lambda = 0.1$ for $\lambda_1 = 1$, $\Delta\lambda = 0.05$ for $\lambda_1 = 0.5$ and 0.3. The results are given in Table 2.

λ_1	λ_2	$\bar{s}_1$	$\bar{s}_2$	l_1, l_2	ρ	BF est.	Revised IPA est.
1	0.5	0.35	0.9	0.15	0.8	1.5394 ± 0.1281	1.7274 ± 0.0723
0.5	0.2	0.8	0.5	0.1	0.5	1.6043 ± 0.0642	1.6615 ± 0.0239
0.3	0.1	0.6	0.2	0	0.2	0.6212 ± 0.0151	0.6359 ± 0.0062

Table 2 Experimental Comparison of IPA of (13) With Brute Force Estimates

5.4.2 The Rescheduling Approach

In this section, we explore one more alternative way of solving the two-class M/M/1 example of Section 5.4.1 due to [Gong, Cassandras, and Pan 1990]. This involves another application of the representation idea discussed in Section 5.1. For simplicity, we shall assume that the mean service rates for the two classes are the same. But instead of mixing the two arrival classes into one equivalent class, we shall keep the distinction between the classes of the arrival customers. To avoid the problem of the class 1 arriving customer continuously sliding ahead of class 2 arrivals as $\Delta\lambda_1 > 0$, the idea of "re-scheduling" is introduced. At the instant of every class 1 arrival that is followed by a class 2 arrival, there is a remaining time to the arrival for the class 2 customer.   Since arrivals are memoryless, we can in fact regard this remaining-time-to-arrive for the next class 2 customer as a newly generated lifetime from an exponential distribution with mean $1/\lambda_2$.   Under this interpretation, a simulation of the arrivals is carried out as in Fig. 9a as opposed to that of Fig. 9b in a more traditional multiclass simulation.

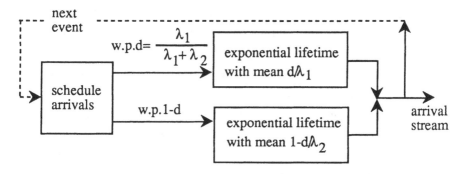

Fig. 9a Re-Scheduling Approach to Generate Two-Class Arrivals

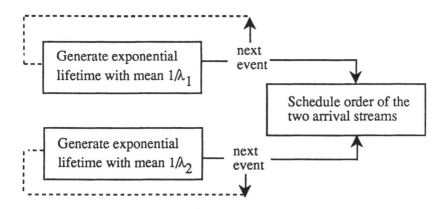

Fig. 9b Traditional Simulation Event Scheduling for Two-Class Arrivals

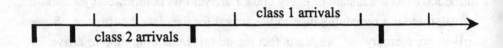

Fig. 9c The Arrivals of the Two Class Customers

Note that either model could have generated the sample two-class arrival stream of Fig. 9c. The main difference being that in Fig.9b the two independent streams of random lifetimes are used while in Fig. 9a only one stream is used. As a result, the behavior of the sample arrival streams under perturbation of λ_1 are very different. In the case of Fig. 9b, $\Delta\lambda_1$ has no effect on the class 2 sample arrivals. Perturbations are generated, accumulated, and propagated only among class 1 arrivals within a busy period. On the other hand, in Fig. 9a, every class 2 arrival inherits the perturbation of its immediate predecessor regardless of its class. In other words, perturbations generated by class 1 arrivals are propagated to class 2 arrivals. The only difference is that no perturbation generation takes place at class 2 arrivals. Because of this interclass propagation, class 1 arrivals never slide ahead of those of class 2 and deterministic similarity is preserved.

Exercise 1 Show that the model of Fig. 9a is equivalent to that of Fig. 9b.

Exercise 2 Work out the IPA algorithm for Fig. 9a. See [Gong, Cassandras, and Pan 1990].

Exercise 3 In Fig.9a when we change λ_1 we will also change "d" which in turn will cause switching of the output of the event scheduler. In the description of the IPA for Fig. 9a, we ignore such switchings. What is the basis for this?

5.4.3. Approximate Multiclass Network Analysis and IPA
With such a methodology for dealing with multiclass GI/G/1 queue, we can visualize an approximation method for solving sensitivity problems involving

multiclass queueing networks. The idea is based on the QNA (Queueing Network Analyzer) approach [Whitt 1983] and is illustrated in Fig. 10.

Fig. 10 Relationship Between System Parameter and Performance

We are interested in the sensitivity of some performance measure associated with a particular server in a complex network with respect to an arbitrary system parameter, θ, in another part of the network. The viewpoint of QNA is that the performance of the server in question (whether it is queue length, system time, or utilization) can only be affected by θ through the arrival distribution, F. The sensitivity of the performance of the server with respect to a parameter of F for the multiclass arrival is a local matter. The problem has just been dealt with in the previous section. Therefore we need only determine the sensitivity of the arrival distribution F with respect to θ. This global sensitivity can be approximately captured via a set of algebraic equations known as the **QNA traffic and variance equations,** which relate the mean λ and the variance σ of the arrival rate in all parts of the network to parameters of the network. We can directly differentiate these traffic equations to get an approximation of the sensitivity of F with respect to θ by computing dλ /dθ and dσ/dθ. Chain rule of calculus then let us compute the performance sensitivity dJ(θ)/dθ, i.e.,

$$\frac{dJ(\theta)}{d\theta} = \frac{dJ}{d\lambda}\frac{d\lambda}{d\theta} + \frac{dJ}{d\sigma}\frac{d\sigma}{d\theta} + \cdots \tag{22}$$

where λ and σ are respectively the mean and variance of the arrival distribution. Eq.(22) is approximate for two reasons: (i) in general the arrivals at a particular server in a complex network are neither independent nor completely characterized by the first two moments, i.e., we have neglected additional terms in Eq.(22) which may not be small, (ii) although the QNA traffic equation for the mean λ is exact, the one for the variance is approximate. There is no reason to believe that $d\sigma/d\theta$ based on this QNA formula will be any more accurate. Nevertheless, this is in the same spirit of the QNA calculation where the performance $J(\theta)$ is calculated via this two step approach, namely, using the traffic and variance equations to capture global relationships and GI/G/1 analysis for local considerations. The only difference is that while QNA handles the GI/G/1 part by an approximate analytical formula we propose to use perturbation analysis and on-line arrival data directly to calculate $dJ/d\lambda$ and $dJ/d\sigma$ via the results of the previous section and chapters. Single-queue IPA results are well understood and promise to be more accurate than analytical approximations and more applicable to situations involving additional complications, e.g., GI/G/m server or finite queue size subject to loss. We call this idea of capturing global sensitivity via QNA equations and local sensitivity via PA calculation Local Perturbation Analysis, or LPA. However, the application of IPA using on line data requires one more assumption. As shown in Chapter 3, IPA calculation for the G/G/1 queue requires the evaluation of the derivative of the interarrival time with respect to changes in λ and σ. Since in general we do not know the exact inter-arrival distribution, some assumption must be made to enable this calcu-lation. The following exercise indicates one solution.

Exercise Suppose the inter-arrival time distribution can be characterized by a generalized scale-location parametrization

$$A(\lambda,\sigma) = \alpha(\lambda,\sigma) + \beta(\lambda,\sigma) \ y, \qquad\qquad (23)$$

where y is a random variable independent of λ and σ, calculate what are $dA/d\lambda$ and $dA/d\sigma$ for this case. Show that they can be represented exclusively as functions of A, E[A], Var[A], $dE[A]/d\theta$, $dVar[A]/d\theta$ which are quantities that can be measured on line without knowing the form of α and β. Show also that Eq. (23) includes most well-known distributions.

Note this problem of modeling the arrival distribution at the server in question is not unique with PA. The QNA approach also must make similar assumption when deriving analytical approximation for the performance at the server.

5.5 Smoothed Infinitesimal Perturbation Analysis (SPA)

By now it is clear that the basic difficulty in the application of infinitesimal perturbation rules is the effect of possible discontinuities in the sample performance with respect to parameter perturbations. For any sample path or an ensemble of sample paths, event chain invariance, or deterministic similarity between the nominal path and the perturbed path, will be violated with probability one if the sample path is long enough or the ensemble is large enough regardless of the size of $\Delta\theta$. Once violated, the nominal path and the perturbed path may look entirely different from that time on. In other words, an arbitrarily small perturbation in θ will always lead to discontinuities in sample path and hence in $L(\theta,\xi)$. Sample path discontinuity is the colloquial equivalent of the mathe-matical question of interchanging expectation with differentiation in Eq. (4.1). Elementary analysis tells us that such discontinuities may rule out any possibility for the above mentioned interchange to be valid. Hence, we have the mathe-matical objection that PA may not provide a consistent estimate of the derivative of expected L by computing the expectation of the derivatives. The previous sections illustrated the simple idea of smoothing out infrequent but finite perturbations into frequent but infinitesimal perturbations by some transfor-mation of the perturbing variables.

We now present another more formal approach to the smoothing idea. Mathematically, let us rewrite Eq. (4.1) as

$$dE[L(\theta,\xi)]/d\theta = d\{E_z E_{/z}[L(\theta,\xi)]\}/d\theta$$

$$=?= E_z \{d[E_{/z}[\Delta L(\theta,\xi)]/d\theta]\} \tag{24}$$

In other words, we decompose the "expectation" into a conditional expectation on L first, followed by an expectation on the conditioning variable z. We can expect $E_{/z}[L(\theta,\xi)] \equiv L(\theta,z)$ to be smoother than $L(\theta,\xi)$ because of the smoothing property of conditional expectations; and hence may make the interchange between E_z and $d/d\theta$ possible. Now it may appear counter-intuitive that averaging over ξ can smooth out the discontinuities in L due to θ. We answer this first intuitively via the illustration of Fig. 4.3 where a series of staircase functions average to a differentiable function with well-defined slopes. In fact, assuming that $J(\theta)=E[L(\theta,\xi)]$ is differentiable, the usual brute force way of computing sensitivity via

$$\lim_{\substack{n\to\infty \ \Delta\theta\to 0}} \left\{ \frac{1}{n}\sum_{i=1}^{n}L(\theta+\Delta\theta;\xi_i) - \frac{1}{n}\sum_{i=1}^{n}L(\theta;\xi_i) \right\}$$

is a simple statement of the above fact of averaging in the extreme where the conditional expectation becomes the unconditional expectation. In other words, even though $L(\theta,\xi)$ may be discontinuous in θ, a sufficient amount of averaging in ξ can make $E[L(\theta,\xi)]$ differentiable. Now the trick with Eq.(24) is to average just enough to avoid discontinuities but not to require the duplication of another experiment as in the brute force case. Between the extremes of differentiating the expectation and taking the expectation of the differentiation, a whole spectrum of partial expectation and smoothing possibilities exists. This is workable provided $L(\theta,z)$ is computable, i.e., z must be based on data available on the sample path.

Let us illustrate this by considering the example of estimating the sensitivity of the average number of customers served in a busy period, E[n], with respect to the mean service time, s, for a G/G/1 queue. Fig. 11 illustrates the typical situation.

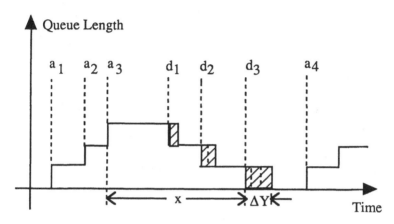

Fig. 11 A Busy Period of a GI/G/1 Queue

It is clear that for a finite time interval we can always choose a perturbation in the mean service time so small that no two busy periods coalesce into one. Consequently, the sample derivative of n with respect to s, based on any finite length sample path, is always zero. Yet for any fixed Δs, no matter how small, soon or later we shall have occasion to meet the situation where the perturbation accumulated in a busy period, ΔY, is greater than the length of the idle period between two busy periods. To analyze this case consider time d_3 illustrated in Fig. 9. At d_3 we have accumulated perturbation $\Delta Y = \Delta s_1 + \Delta s_2 + \Delta s_3$ representing the total change in d_3 due to the service time changes of customers C_1, C_2, and C_3. Also, $x = d_3 - a_3$ units of time have elapsed since the last arrival at time a_3. Whether or not we will encounter a discontinuity in the sample performance (or the busy period will coalesce with the next) depends on the size of ΔY and the duration of x. For this reason, we choose $z = \{\Delta Y, x\}$ as the conditioning random variable. The conditional expectation of the change in the number of customers served in a busy period is given by the conditional probability of coalescing and the expected number of customers added to a busy period because of each coalescing. The latter is the expected value of the number of customers served in a busy period. Thus,

$$\Delta\{E[n]/\Delta Y, x\} = \text{Prob[coalescing takes place} / \Delta Y, x]* E[n]$$
$$= \{[g(x)/1-G(x)]*\Delta Y *E[n]\}.$$

Then

$$\Delta E[n] = E_{\Delta Y, x}\{[g(x)/1-G(x)]*\Delta Y *E[n]\} \tag{25}$$

where $g(x)$ and $G(x)$ are respectively the density and the distribution function of x, and $g(x)/[1-G(x)]$ is the well-known hazard rate in reliability theory, which is the rate of failure that may occur in dx given that it did not occur in x units. If we further specialize to the case of an M/M/1 queue with arrival rate λ and service rate $1/s$, then Eq. (25) can be explicitly evaluated to give

$$E[\Delta\{E[n]/\Delta Y, x\}/\Delta s] = [E(n)]^{2}*\lambda \tag{26}$$

which agrees with the well-known formula [Kleinrock 1975, p.213]. Incidentally, since

$$E[n] = \lim_{N\to\infty} \frac{N \ (= \text{total no. of customers served})}{N_{BP}(= \text{no. of busy periods or no. of arrivals to an idle server in } N)}$$

$$= 1/r_0 \ (= \text{probability of finding the server idle})$$

E[busy period length] = E[n]*s,

the sensitivities of r_0 and that of the mean busy period length can be easily derived from (26). Finally, from (25) and (26), we have

$$\frac{dE[n]}{ds} = \frac{d\{E_{\Delta Y, x}E_{|\Delta Y, x}[n]\}}{ds} = E_{\Delta Y, x}\left\{\frac{dE_{|\Delta Y, x}[n]}{ds}\right\}.$$

This shows that we can exchange the order of d/ds and $E_{\Delta Y, x}$ and use $\{dE_{/\Delta Y, x}[n]/ds\}$ as an unbiased estimate of $dE[n]/ds$. This estimate is called the SPA (smoothed PA) estimate. For more details and further examples of this approach, see [Glasserman and Gong 1989b, Gong and Ho 1987].

It is instructive to examine further the ideas of SPA behind this example. Let us consider the performance measure

$$J(\theta) = E[f(x_T(\theta))] \tag{27}$$

where T is some terminal time and x is the discrete state of the system, e.g., the number of jobs in the system. Since x is discrete, the sample performance will be inherently discontinuous with respect to $d\theta$. PA is interested in the calculation of

$$\lim_{\Delta\theta\to0}\frac{1}{\Delta\theta}E[f(x_T(\theta+\Delta\theta)) - f(x_T(\theta))]$$

which can be expanded to

$$= \lim_{\Delta\theta\to0}\frac{1}{\Delta\theta}E[E[f(x_T(\theta+\Delta\theta)) - f(x_T(\theta))]/z]$$

$$= \lim_{\Delta\theta\to0}\frac{1}{\Delta\theta}E[\sum_s P(x_T(\theta+\Delta\theta)=s/z)(f(s)-f(x_T(\theta)))] \tag{28}$$

$$= E[\lim_{\Delta\theta\to0}\frac{1}{\Delta\theta}\sum_s P(x_T(\theta+\Delta\theta)=s/z)(f(s)-f(x_T(\theta)))]$$

provided that the expectation and the limit are interchangeable. In (28), z may be any data observable along the nominal path $x_t(\theta)$. The term

$$P(x_T(\theta+\Delta\theta)=s/z)$$

is recognized as the jump rate of the state x_T and $(f(s)-f(x_T(\theta)))$ the size of the jump. Roughly speaking, we have traded the differentiation of $f(x_T)$ with the differentiation of the probability of a state change (P' is the hazard rate in the example above). We expect the latter to be a more continuous function. Eq.(28) is also in the same spirit of the development of the interchangeability condition (IC) of Chapter 4 where the discontinuity problem is studied via the two components of the probability of discontinuity and the size of the discontinuity.

Glasserman and Gong [1989b] have studied this formal approach to SPA methodology under a GSMP framework and showed that the crucial structural condition to permit the application of the SPA idea is in fact the commuting condition (CU) discussed earlier in Section 5.2, i.e., Eq.(1)

$$\psi(\psi(x,\alpha,\ U(\alpha,n)),\beta,U(\beta,n+1))=\psi(\psi(x,\beta,U(\beta,n+1)),\alpha,\ U(\alpha,n))$$

In words, we require that the timings of events be continuous functions of θ, and when events change order, the times of their occurrences change continuously. At the points when they change order, the state may be discontinuous but they come back to the same state, i.e., the order of occurrence of the two events does not change the final state. The discontinuity is infinitesimal and does not contribute to discontinuity in the sample path. However, sample path continuity does not necessarily insure sample performance continuity since the performance measure can be inherently discontinuous as in (27). Conditional expectation in (28) then smooths out the performance discontinuities. In short, general SPA is based on two ideas: (i) use the condition (CU) to rule out sample path discontinuity, (ii) use conditional expectation to smooth out performance discontinuity. The example of this section clearly utilizes both. The exercises below give two further examples.

Exercise Consider the simple Jackson queueing network of Fig. 12,

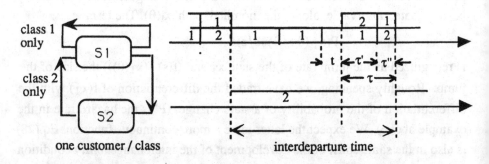

Fig. 12 A Simple Two-class Network and a Typical Regenerative Cycle

where the mean service time for both classes in S_1 is $1/\lambda$ and that of S_2 is θ. All distributions are exponential. Use SPA to determine the sensitivity of mean interdeparture time from S_1 with respect to θ. Check this against a direct derivation of the sensitivity via differentiating the mean interdeparture time.

Exercise In an M/M/1/K queue, the maximum number of customers in the queue (including the one in service) is K. If an arrival customer finds the queue full, it leaves the system immediately and is considered lost. Let λ and μ be the mean arrival rate and the mean service rate, respectively. We have shown in the exercise in Chapter 4 that this problem is equivalent to a two server cyclic network with K customers and service rates λ and μ, respectively. Simple IPA can be applied to the equivalent network directly. Now, attempt to solve the problem directly by using SPA. (Hint: Solution for both exercises can be found in [Gong and Ho 1987]).

At this point, it is clear that the SPA estimate depends on the conditioning variable z. A particular z may provide unbiased SPA estimates for some systems, but not for other systems. On the other hand, if the SPA estimate based on a random variable is biased, the SPA estimates based on other variables may be unbiased. For example, let us consider the same problem of Fig. 11 again but with a D/M/1 queue instead of the GI/G/1 queue. The SPA estimate based on $z = \{\Delta Y, x\}$ shown in Fig. 11 will now be biased. In fact, the SPA estimate is $\lim_{\Delta s \to 0}\{\Delta E_{/\Delta Y, x}[n]\}/\Delta s = 0$, where s is the mean service time. This can be shown as follows. Let d=constant be the interarrival time. For any two busy periods, we can choose Δs small enough such that $d > x + \Delta Y$ since by definition d>x. This means that the probability that these two busy periods coalesce is zero. Because of the deterministic feature of the arrival process, there is no randomness once z is fixed. Therefore, $\Delta E_{|\Delta Y, x}[n] = 0$ for this small Δs. However, we may choose other random variables as the conditioning variables and obtain an unbiased SPA estimate for the derivative of the average number of customers served in a busy period with respect to s. Let $z' = \{\Delta Y', x'\}$ shown in Fig. 13, with x' being the elapsed interarrival time at the beginning of the last customer's service period and $\Delta Y'$ being the perturbation at that time. Given z', the event that the two busy periods coalesce is again a random event. We leave the proof of the unbiasedness of the SPA estimate using z' as an exercise. A similar discussion about the second derivative is in [Fu and Hu 1991b].

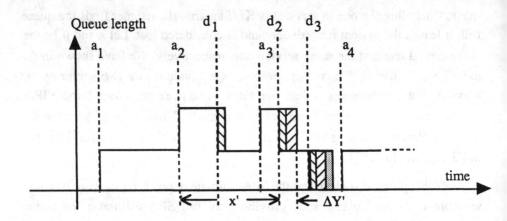

Fig. 13 Conditioning Variable z' for a D/M/1 Queue

Exercise Prove that the SPA estimate of the derivative of the average number of customers served in a busy period with respect to the mean service time in a D/M/1 queue, based on the conditioning variable z' shown in Fig. 13, is unbiased.

In traditional simulation, statistics about the occurrence of an event is typically collected only when the event occurs. This is very natural. However, in the case of rare events, e.g., estimating the tail probability of a distribution, this natural approach to collecting statistics can be very inefficient. In Appendix B, we discuss the method of importance sampling which is the inverse of the likelihood ratio method of Chapter 7 as a way to improve efficiency. SPA can be viewed as another approach to improve efficiency. The point is this: the nonoccurrence of an event also carries information about the probability of an event. For example, if a customer arrives at a queue and finds it nonempty, the fact also conveys information about r_0 is the probability of finding the server idle upon arrival. Thus, instead of collecting statistics based on the occurrence or nonoccurrence of an event, a zero-one situation, and then average the results, we accumulate data at every event occurrence but use analysis to determine the continuous variable amount of weighting to be used in the accumulation. Conceptually, this is quantitatively implementing the idea of converting "infrequent by finite perturbations into frequent but infinitesimal perturbations."

Finally, the purpose of this chapter is to demonstrate the mind-set of perturbation analysis. In the stochastic performance evaluation and the DEDS environment, there are numerous ways to take advantage of the notions of "statistical equivalence," "conditioning," and other analytical devices. In succeeding chapters, we shall further illustrate this mind set via the technique of "aggregation," "cut-and-paste," and "likelihood ratio estimations."

Finite Perturbation Analysis

One can view the basic goal of PA for DEDS as the reconstruction of an arbitrarily perturbed sample path from a nominal path. Under deterministic similarity, the simple infinitesimal perturbation analysis (IPA) rules described in Chapters 3 and 4 and extended in Chapter 5 compute a perturbed path infinitesimally different from the nominal for the purpose of gradient calculation. The computation of the perturbation propagations can be efficiently done since the "critical timing path" or the "future event schedule" between the nominal and perturbed paths remains the same. However, as pointed out before in the limit of a path of a very long duration or an experiment with very large ensembles of runs, deterministic similarity will always be violated[1]. In such cases, the IPA rules which ignore the order changes of events have been proved to give unbiased and consistent estimates for performance gradients of only certain classes of DEDS (see Chapters 4 and 5). Although the domain of application of IPA is constantly being expanded, it is nevertheless important to devise methodologies which can overcome the basic problem of "event sequence order change," or "discontinuous performance measure." In other words, the key question is *"how can one reconstruct the perturbed path from the nominal path short of essentially running a separate new simulation / experiment?"* In

---

[1] For any **given** sample path of a finite duration or a **given** finite ensemble of sample paths of a finite duration, it is always possible to find a perturbation small enough such that determi-nistic similarity will hold. This only requires the assumption that no two events occur simultaneously, which is basic in all DEDS work. See Chapter 4 Section 1 and Appendix B.3. However, the flip side of this argument is that, for any **given** perturbation no matter how small, one can always (with probability one) find a sample path long enough or an ensemble of paths large enough such that deterministic similarity will be violated.

this and the next chapter we address this question directly and suggest an alternative which we believe is more efficient than brute force reconstruction. To initiate the discussion, let us first dispel the seemingly intuitive notion that one cannot generate a sample path x(t; θ+Δθ,ξ) from an x(t;θ,ξ) when Δθ≠0. A simple view of this general problem of perturbation analysis of DEDS and the efficient construction of multiple sample paths of a DEDS under different values of θ can be obtained by appealing to some fundamental procedures in discrete event simulation. We review this briefly below.

The transform method of generating samples of a random variable with an arbitrary distribution

It is well-known that given samples of a uniform distribution on [0, 1), $u_1, u_2, ...,$ etc., we can transform these into samples, $x_1, x_2, . . .,$ from an arbitrary distribution F(x) by using the inverse transform $F^{-1}(u)$, provided it exists (see Appendix B and Fig. 1 below).

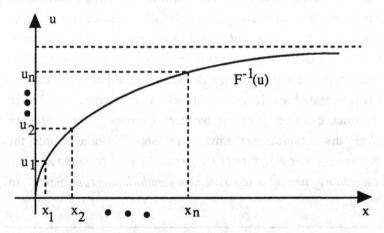

Fig. 1 Inverse Transform Method for Generating F(x)-distributed Random Variables.

This is an elementary way of using samples from one distribution (say, parameterized by θ) to get samples from, and hence information about, another distribution (governed by θ'). The so called "likelihood ratio / score function / change of measure method" of single-run sample path analysis can be viewed as based on the same principle [Aleksandrov et al. 1968, Rubinstein 1986b, Reiman and Weiss 1989b, Glynn 1987, Section 7.3 and Appendix B.4.4]. Let $f(x;\theta)$ be the density function of the stochastic process under discussion and $J(\theta) = \int L(x)f(x;\theta)dx$ be some performance measure. Let $g(x) \equiv f(x;\theta+\Delta\theta)$, then we have

$$J(\theta+\Delta\theta) = \int L(x)\, g(x)\, dx = \int [L(x)\frac{g(x)}{f(x)}]f(x)dx.$$

In other words, we can compute $J(\theta+\Delta\theta)$, which is based on $g(x)$ using samples from $f(x)$ provided we appropriately **re-scale** the sample performance measure $L(x)$ as above. This is conceptually the same as re-scaling the uniform random variables by the inverse transform to get $F(x)$-distributed random variables. More details on this method can be found in Section 7.3.

The rejection method of generating samples of a random variable with an arbitrary distribution

The other well-known method of generating random variables of arbitrary distributions is the so-called rejection method which is best illustrated using Fig. 2 below (see Appendix B.2):

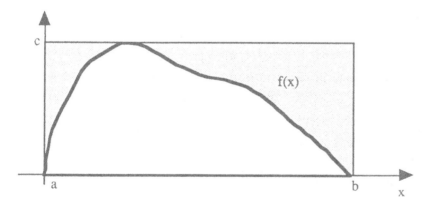

Fig. 2 The Rejection Method of Random Variable Generation

We generate two-dimensional uniformly distributed random numbers over the rectangle with sides ab and ac through the use of two one-dimensional random number generators over [a,b] and [a,c]. The samples thus generated are thrown away if they are in the shaded region or retained if in the unshaded region which are defined by the density function f(x). The accepted samples by definition and construction are distributed according to f(x). This is another way of getting f(x)-distributed samples from uniformly distributed samples. As we shall described below, the "cut-and-paste via state matching" method of Extended Perturbation Analysis (EPA) is simply the application of the "rejection" idea to the case of stochastic processes in discrete event simulations.

6.1 The Idea of "Cut-and-Paste"

Informally, consider a discrete event dynamic system modeled as a continuous time Markov process or a GSMP with exponential lifetimes. Let $x(t;\theta,\xi)$ denote the time history of the evolution of the GSMP, i.e., the state sequence and the holding times in each state. Now suppose the system parameter, θ, is changed to $\theta' \equiv \theta + \Delta\theta$ for some finite $\Delta\theta \neq 0$. We ask the question "what will be the perturbed trajectory $x(t; \theta+\Delta\theta,\xi)$?" Ordinarily, the question is answered by essentially a separate generation of the perturbed path via another experiment on or simulation of the DEDS. However, we submit that a considerable simplifi-cation is possible, provided we are willing to accept $x(t;\theta+\Delta\theta,\xi')$ with $\xi' \neq \xi$, i.e., a statistically equivalent perturbed path. The point here is that often we are only interested in $E[L(x(t;\theta+\Delta\theta,\xi))]$ or, in the ergodic cases we are interested in $\lim_{t\to\infty} L[x(t;\theta+\Delta\theta,\xi)]/t$, where L is some functional of the sample path. In either case, having the statistically equivalent $x(t;\theta+\Delta\theta,\xi')$ is perfectly accept-able. Put another way, so long as the basic structural model of the DEDS did not change except for some parameter values, the separate generation of sample paths in traditional simulation entails a great deal of duplicate effort that can be and should be leveraged to improve computational efficiency.

The basic idea, which is called "cut-and-paste," is as follows: a GSMP trajectory is characterized by the sequence of states and the holding times of each state as the system evolves in time. A nominal (corresponding to θ) and a perturbed trajectory (corresponding to θ') which start out from the same state at $t=t_0$ will begin to differ either in event occurring times and/or state sequences as the system evolves due to accumulated perturbations caused by the system parameter perturbation. If the difference is only in the state holding times, then we can simply keep track of the current incremental difference in the timing of events between $x(t;\theta+\Delta\theta,\xi)$ and $x(t;\theta,\xi)$ in a separate register (e.g., the A_i registers in Chapter 3). Using this difference together with $x(t;\theta,\xi)$ **only**, we can immediately generate $x(t;\theta+\Delta\theta,\xi)$. This is the essence of the IPA algorithm which, compared to the brute force generation of $x(t;\theta+\Delta\theta,\xi)$, represents a considerable computational simplification. However, if the two trajectories should begin to differ also in the state sequence, say at $t=t_1$ with $x(t_1;\theta+\Delta\theta,\xi)$ $\neq x(t_1;\theta,\xi)$, then simple IPA no longer applies since the evolution of the perturbed trajectory starting at t_1 may become completely different. Conventional practice calls for a separate experiment or simulation of the DEDS from this time onward in order to generate $x(t;\theta+\Delta\theta,\xi)$. However, this is not necessary. What can be done is this. Let us suspend the generation of $x(t;\theta+\Delta\theta,\xi)$ from $x(t;\theta,\xi)$ starting at t_1 until at some $t=t_2$ where $x(t_2;\theta,\xi) = x(t_1;\theta+\Delta\theta,\xi)$. At that time, by virtue of the Markov property we can re-start the generation of $x(t;\theta+\Delta\theta,\xi)$ from $x(t;\theta,\xi)$. Now we continue from t_2 until some t_3 when the state sequence differs once more and the whole process repeats. Note that from t_0 to t_1 and from t_2 to t_3, IPA applies by construction. The only additional thing we need to do is to recognize event order changes which is much easier than to generate the complete perturbed path. Thus, by selectively "cutting" out segment of $x(t;\theta,\xi)$ and "pasting" the ends together we can construct a trajectory $x(t;\theta+\Delta\theta,\xi')$ which is statistically indistinguishable from the actual perturbed trajectory $x(t;\theta+\Delta\theta,\xi)$ generated by a separate experiment / simulation. This is the basic idea behind "cut-and-paste" and the extended perturbation analysis (EPA) technique reported in [Ho and Li, 1988]. For the simplest example of "cut-and-paste," consider the case of cutting off the queue content history of an M/M/1 queue when it reaches K to get a path for an M/M/1/K queue.

Exercise Draw a typical sample path of an M/M/1 queue and convince yourself that from it a legitimate sample path of an M/M/1/K queue can be derived by cut-and-paste.

Exercise Suppose you have a biased coin with probability for heads, θ, equal to 2/3. A typical coin toss sequence may appear as [. . . H, T, H, H, H, T, T, H, H, . . .]. Consider the following two methods of generating from this sequence another sequence that could be produced from a fair coin, i.e. θ=0.5:

  (i) deterministically cut out every other "H" in the above sequence.

 (ii) use a purely random device to cut out half of the "H"'s in the above
     sequence.

Comment on these two approaches.

It should also be noted that the opposite of "cut-and-paste" is "insert-and-paste," which is equally applicable under the Markov assumption. By "insert-and-paste," when the nominal trajectory matches the perturbed one, i.e.,

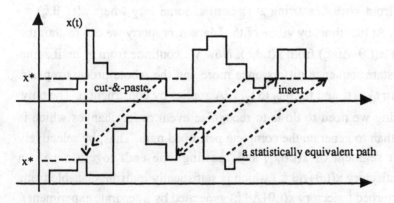

Fig. 3 Markov Trajectory under "Cut, Insert, and Paste"

$x(t_2;\theta,\xi) = x(t_1;\theta+\Delta\theta,\xi)$, instead of generating the nominal trajectory, we generate the perturbed one and insert the segment of the perturbed trajectory to the nominal. Fig. 3 graphically illustrates this. Also see the discussion in Section 6.1.2 and Section 6.2.

As a side remark, it is also interesting to note that this "cut-and-paste method" can be viewed as a dual approach in relation to the "likelihood ratio method," or the "score function method," of single-run gradient estimation mentioned earlier in this section and to be described in detail in Section 3 of Chapter 7. In the likelihood ratio approach, we ask the question: "what is the likelihood that a trajectory $x(t;\theta,\xi)$ can in fact be generated by the system operating under the parameter value $\theta+\Delta\theta$?" The answer to this question provides a quantitative measure to *re-scale* the contribution of the particular trajectory $x(t;\theta,\xi)$ in the estimation of the performance $E[L(x(t;\theta+\Delta\theta,\xi))]$. In our cut-and-paste approach, instead of re-scaling the weight attached to the trajectory, we *re-assemble* segments of the trajectory to make it statistically equivalent to $x(t;\theta+\Delta\theta,\xi)$. In the former method, the longer the duration of the trajectory, the less likely will $x(t;\theta,\xi)$ be generated by $\theta+\Delta\theta$. Consequently, we can expect numerical or variance difficulties in using long (in values of t) non-regenerative trajectories in the likelihood method, particularly for large $\Delta\theta$. the cut-and-paste method, a large $\Delta\theta$ tends to cause frequent "cuts" and subsequent waits for "paste." The efficiency issue is addressed in [Ho and Li, 1988 and Li 1988] and later on in this chapter.

6.1.1 A Simple Example of "Cut-and-Paste"

Now let us offer a simple example of the cut-and-paste idea of PA. Picture a finite state machine consisting of three states only: A, B, C and a system parameter θ which can assume values of 0 and 1. Under $\theta=0$, the trajectory of the machine is periodic, $A \rightarrow B \rightarrow C \rightarrow A \rightarrow B \rightarrow \ldots\ldots$ While under the condition $\theta=1$, the trajectory from A has an equal probability to go to the other two states, and similarly for the states B and C resulting in a completely symmetrical and stochastic system. Now consider a nominal path (NP) under $\theta=0$,

A,B,C,A,B,C,A,B,C,A,B,C,A,B,C,A,B,C,A,B,C,A,B,C,A,B,C,. . . .   **(NP)**

and one realization of a perturbed path (PP) with θ=1 which may look like

A,B,A,B,C,A,B,C,B,C,B,C,A,B,A,B,C,A,B,C,B,C,A,B,C,B,C,. . . .   **(PP)**

We claim that we can construct an equivalent perturbed path, denoted as the Constructed Perturbed Path or CPP, from the data of NP as follows: At any state, say A, we flip a fair coin to see where the perturbed path could have gone. If the outcome of the flip is to B, then we continue to follow along the NP since no deviation from the NP took place. If on the other hand, the outcome is C, then we simply skip the next state along the NP which is B and wait until the state C actually occurs on the NP. At that time we can follow the NP until the next occurrence of the coin-flip deviation and the process repeats . . . . Exactly the same "coin toss - test - accept or wait" scheme can be followed for other states along the NP. Thus, by selectively (according to the outcome of the coin toss) discarding sections of the NP, we constructed an equivalent perturbed path, CPP, as below

A,B,A,B,C,A,B,A,B,C,B,C,A,B,C,A,B,A,B,C,A,B,C,A,. . . . . .   **(CPP)**

which is obtained from NP by skipping segments (shown in outline font) of the state sequence

A,B,C,A,B,C,A,B,C,A,B,C,A,B,C,A,B,C,A,B,C,A,B,C,A, . . .   **(NP)**

Note that the CPP thus constructed need not be the same as the particular PP above. Nevertheless we submit that the CPP thus constructed is statistically indistinguishable from the PP since they were constructed out of the same probabilistic mechanism[2]. This idea, which we shall call "The Statistical

---

[2] Readers may wish to point out that in this case the construction of CPP is computationally not that different from actually implementing PP itself. Two points should be made. An actual comparison can show that even in this case there is a small saving in terms of codes required to implement CPP as a part of NP vs. to implement PP separately. This is because, for CPP one only has to mechanize the functions of "test" and "wait" which do not become any more

Invariance of Markov State Sequence Under Cut-and-Paste," depends on two things:

(i) a long enough trajectory or a large enough ensemble of trajectory segments of an irreducible finite state machine eventually produces all conceivable behavior of the system under all parameter values. The only difference in trajectories under different parameter values is in the frequency of occurrence of different states and sequences.

(ii) using short term calculations (e.g., coin toss or the next state transition function) which are given system knowledge, we can pick out selected segments of a trajectory to reconstruct a CPP under the operating condition of an arbitrary parameter value. Only state transitions calculations are needed to enable us to construct the CPP. There is no need to extrapolate the path deviation indefinitely into the future which is what construction of a PP requires. Markov property allows the construction of CPP to be repeatedly restarted.

Alternatively, we can take the viewpoint that system behavioral information under θ' can be squeezed out from that of θ in the following way, which is the analog of the inverse transform method for random variable generation and the likelihood ratio method of single-run sample path analysis to be discussed later. It is entirely possible for a path such as ABCABC to be generated under $\theta=1$ instead of under $\theta=0$. Its probability of occurrence is different under the two different parameter values. Thus, if we appropriately re-scale the contribution of any path to an ensemble average by weighting it with different probabilities (i.e., change of measure), then one can in fact

---

complex with systems of increasing complexity. On the other hand, for PP one has to perform additional tasks such as "future event list generation," "process updating," and "trajectory evolution for the perturbed path," which can be complex for practical simulation experiments. More importantly, as the dimension of the parameter vector becomes larger than 1, say n, we can recreate equivalent CPPs for different parameter perturbations from a single NP without making the length of NP n times as long. This is because segments of NP skipped as a result of one parameter may very well be used in the construction of another CPP for a different parameter. Consequently, we preserve the inherent N:1 advantage of perturbation analysis. Section 6.2 and Chapter 7 will discuss in more detail the computational issues.

compute statistical averages of system performance under the probability distribution θ' using samples from that of another distribution θ. Instead of changing the measure or probability of the path sequence, in PA under "cut-and-paste" we change the path directly. This can be thought of as a dual approach. The point is that there is nothing counter-intuitive about the fact that a given trajectory under θ can provide information about the system's behavior under θ' so long as one uses the given system knowledge to help the analysis.

Quite often we are interested in the case of $\Delta\theta \rightarrow 0$. let us consider a simple variation on the above example. Suppose now we can perturb the value of $\theta=0$ by a small amount $\Delta\theta$. This perturbation does not change the behavior (order) of the state sequence $A \rightarrow B \rightarrow C \rightarrow A \ldots$ under $\theta=0$. It merely changes the duration of the time interval between the occurrences of the states (events) A and B by an amount Δt_{AB}, which is proportional to $\Delta\theta$ and can be denoted as $\Delta t_{AB} = a * \Delta\theta$. Now, we may ask the simple question: "what is the change in the termination time of the 1,000,000th occurrence of the state (event) C with $\theta'=0+\Delta\theta$?" Since the state sequence did not change, all Δt_{AB} generated will accumulate and propagate to other events intact. We have by inspection $\Delta T_C(1,000,000) = 1,000,000a * \Delta\theta$. The crucial point here is the invariance of the state sequence between the NP and the PP. This makes the propagation of perturbation from one event to another using NP alone very easy. More generally, if the state sequence, or the event chain, or the critical timing path of a simulation / experiment remains invariant under perturbation, i.e., deterministic similarity holds, then the calculation and *propagation of perturbations* of any event in the system can be efficiently done using the NP alone. This is, of course, the basis of IPA.

The above two ideas: "Statistical Invariance of Markov State Sequence Under Cut-and-Paste" and "Perturbation Propagation under Deterministic Similarity" can now be combined to answer a non-trivial question about the system under discussion. Imagine now θ can be varied continuously from 0 to 1. As θ changes from 0 to $\Delta\theta$, not only the duration between events A and B change as described above, but the probability of state transition from A to B (similarly for B→C and C→A) changes from 1 to 1-0.5Δθ and that of A to C (similarly for B→A and C→B) from 0 to 0.5Δθ in a complementary fashion. By combining the above two ideas, we can answer the question in the following not-so-trivial exercise.

Exercise For the above assumed behavior of the system under perturbation of θ, and the NP trajectory with θ=0, what is the expected termination time of the 1,000,000th occurrence of the state C under, say, θ=0+Δθ = 0.01? (For simplicity let us assume that the time durations from A→B, B→C, C→A are always the same and equal to 1 unit except for A→B when perturbed.)

(Answer Using the first idea we can identify the segments of the NP that should be skipped to construct a CPP. Using the second idea we only generate and propagate perturbations along the CPP. In other words, all calculations are frozen during the skipped segments. We leave as an exercise for the reader to prove that for any Δθ the expected change in the termination time is 1,000,000a(Δθ-Δθ^2/2)/3.)

Generalizing the idea of this exercise, we can extend IPA to situations where deterministic similarity does not hold. Basically, we use IPA on the nominal path until a state sequence change is encountered. At that point, we cut the PP and freeze any PA calculation which is only resumed when a paste operation is achieved. In other words, small perturbations are tracked by IPA rules while discontinuous perturbations in the state sequence are realized via cut-and-paste. This is then the essence of Extended PA [Ho and Li 1988].

6.1.2 State vs. Event Matching

For the purpose of discussion here, we shall also denote the idea outlined above as "**cut and paste via state matching.**" The reason for this is our wish to demonstrate the generality and flexibility of the basic approach of "cut-and-paste" via another variation on the same theme. We shall call this variation "**cut and paste via event (list) matching.**" Note that in the discussion and the simple example above, we cut and paste based on the criterion of matching the perturbed state $x(t_i;\theta+\Delta\theta,\xi)$ with the nominal state $x(t_{i+1};\theta,\xi)$ with $t_{i+1} > t_i$. This is only a sufficient condition[3]. Many times, two different states will have the same event list and the same future lifetimes associated with the events. For example, consider the case of routing a stream of Poisson arrivals to servers S_1 and S_2 using probabilistic routing. Suppose that because of a perturbation in the routing probability, a customer instead of going to server S_1 goes to S_2. This is shown in Fig. 4.

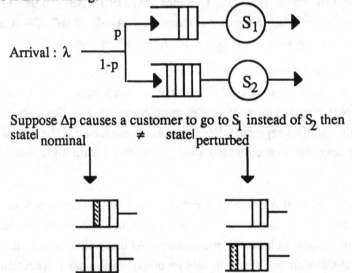

Suppose Δp causes a customer to go to S_1 instead of S_2 then state|nominal $\neq$ state|perturbed

Fig. 4 Different State Sequences With the Same Event List

The nominal and the perturbed states are now different. However, unless server S_2 happens to be idle on the nominal trajectory, the fact that the states of the two trajectories are now different does not change immediately the events and the lifetimes associated with the two "next" states. The customers being served in servers S_1 and S_2 are not different. The difference is only in the customers

---

[3] This is not a very practical requirement except for small problems. Since the number of states of a DEDS explodes combinatorially, the mean length between the recurrence of a particular state may be uncomfortably long. This is the same problem encountered by regenerative simulation and is amenable to the same approximations. See also Sections 6.2 and 8.4.

waiting in the queues at servers S_1 and S_2. Such a difference may cause **future** differences in the evolution of the trajectories. As such there is no reason to wait until the occurrence of exact state matching to re-start the generation of the perturbed trajectory so long as we simultaneously also keep track of the perturbed state in another set of registers. The only time one actually has to "cut" is when in the future an event occurrence causes the nominal and the perturbed states to become "incompatible," or to have different event lists. For example, if server S_2 were "idle" on the nominal but became busy as a result of routing perturbation or if the queue in server S_1 on the perturbed trajectory became depleted before that of the nominal because of earlier routing changes, then the two states no longer give rise to the same event lists and two different sets of lifetimes will be involved. At that point in time we must cut and wait until the two set of states become compatible again. In this sense, event (list) matching is a form of aggregated "State Matching" in which all states with the same event lists are aggregated together. In order to make the evolution of the trajectories exact for both the nominal and the perturbed, we use separate registers to keep track of the actual nominal and perturbed states; however, **only one** event list is used to generate the events and lifetimes. This is often possible since event lists deal only with the active elements (e.g., customers being served) of the DEDS while state represents all the elements (e.g., including waiting customers). For queueing systems where the state is the number of customers in the queues, so long as the event lists of the nominal and the perturbed trajectories match, we can say that the nominal and the perturbed trajectories are *stochastically similar in increments or incrementally similar in law* [Vakili 1989]. This is in the sense that the incremental changes in state (e.g., addition or deletion of waiting custo-mers at each server) for both the nominal and the perturbed processes obey the same probabilistic laws. In such a case they may as well be represented by the same incremental sample path. Once the event lists stop being the same, we have to freeze the evolution of one of the trajectories and let the other go on. At some future time, t^*, the event lists become the same again. At that time, simultaneous evolution can begin again since the incremental processes from t^* onwards between the nominal and the perturbed are stochastically similar once again. Note also that when event lists cease to match, there is a complete symmetry in designating which trajectory is called the nominal and which the perturbed. In other words, there is complete freedom in choosing which trajectory to freeze and which is allowed to evolve. The choice can be made on the basis of efficiency depending on which evolution of the two trajectories will lead to a match sooner. For example, if the DEDS is operating under heavy traffic condition and mismatch occurs because of an idle server on the perturbed path and if we freeze the perturbed path and wait for the nominal server to become idle, then we may have to wait a long time. On the other hand if we freeze the nominal and permit the perturbed path to evolve, the idle server most probably will soon be busy and we have matched event lists once again. Conceptually we can further distinguish the former approach as "cut-and-paste," and the latter, as "insert-and-paste." Generically, we shall, however, use "cut-and-paste" as a general terminology for both "cut" and "insert." Viewed in this light, the distinction between the "nominal" and the "perturbed" becomes irrelevant. The underlying purpose is simply to minimize the computational duplication when one is attempting to generate several trajectories from the same system under different operating parameters.

We believe that letting "cut-and-paste" be governed by **event matching** will result in further computational savings when compared to **state matching**. The additional cost is the necessity of keeping track of the perturbed state sequence in a set of registers which represents minimal added burden. The next section demonstrates this analytically.

6.2 The State Matching vs. Event Matching Algorithms

In this section, we use the terminology and model of Generalized Semi-Markov Processes (GSMP) to present, via a simple example, the idea of event matching algorithms, which will be shown to be an aggregated version of state matching used in the Extended Perturbation Analysis (EPA) algorithms. We shall see that the basic building blocks of a GSMP, i.e., the *state* and the *event*, give rise to two broad classes of algorithms which we will refer to later as *state matching* and *event matching* algorithms.

6.2.1 Analytical Comparison of State Matching Algorithms vs. Event Matching Algorithms for a Simple System

In this section, we consider a simple queueing system. The example serves as an illustration of formalizing a DEDS in the framework of GSMP, as well as an intuitive example to compare the idea of state and event matching algorithms.

Example Consider a symmetric closed queueing network with two single-server service stations. Let the two servers be exponential servers with equal service rates μ. Customers served at one server are routed to the other server with probability p and are routed back to the same server with probability 1-p. There are N customers circulating in the system. Let the buffer sizes at both queues be greater than N and the service discipline at both servers be FCFS. The system is shown in Fig. 5. Our goal is to construct a perturbed sample path for the perturbed system with $p'=p-\Delta p$ ($\Delta p>0$).

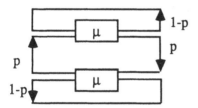

number of customers in the system= N

Fig. 5 Example for Comparison of State vs. Event Matching

We can model this queueing system as a GSMP as follows: Let the state space be $X=\{0,1,...,N\}$ representing the number of customers in the first (top) service station. Let the set of different types of events be $\Gamma=\{e_1,e_2\}$, where e_i denotes a service completion at server i, for i=1,2. According to the above definition, $\Gamma(0)=\{e_2\}$, $\Gamma(N)=\{e_1\}$, and $\Gamma(i)=\{e_1,e_2\}$ for $1\leq i<N$. Let the

distribution function of event lifetimes or clock readings be the exponential distribution: $\phi_i(s)=1- e^{-\mu s}$, $s\geq0$, for $i=1,2$. Let $p(x+1;x,e_i)=p$, if $i=2$, and $p(x-1;x,e_i)=p$, if $x>0$, $i=1$.

Now let us take a look at the effects of the perturbation $p\rightarrow p-\Delta p$. Suppose on the nominal path each time a customer completes service we decide the custo-mer's next service station by generating an independent uniformly distributed random variable u on $[0, 1)$. If $0\leq u<p$, then the customer goes to the other service station; otherwise, the customer joins the end of the queue at the same station. When p is changed to $p-\Delta p$ $(\Delta p>0)$, it may happen that $p-\Delta p<u<p$ at some completion time of a customer. As a result, the customer on the nominal path goes to the other service station, while the same customer on the perturbed path goes to the same service station. Clearly, then the nominal

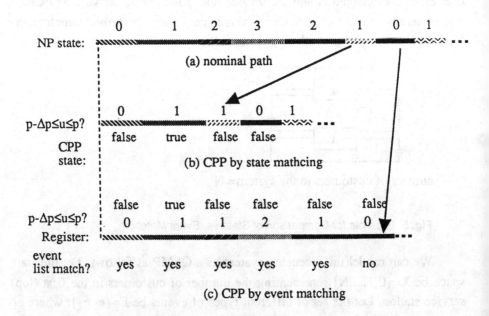

Fig. 6 a,b,c Sample Paths Produced by EPA and EMPA.

state and the perturbed state will be different. A possible sample path of the process is drawn in Fig. 6(a). For example, before a service completion, the state is x=n>0 on both NP and PP, but because of a customer finishing service at station 1, and p-Δp<u<p, the next state on NP and PP will be n-1 and n, respectively. If this happens, we stop the construction of PP until at a later completion time we find that the next state on NP is the one on PP at the stopping point, i.e., state n. Then we resume the construction of PP. The "cut-and-paste" or extended PA algorithm applied on the sample path of Fig. 6(a) is shown in Fig. 6(b). As we have argued above, this constructed perturbed path for a Markov GSMP is a legitimate sample path of the process $x_{\theta'}(t,\xi)$. In the EPA algorithm, each time the states on NP and PP become different, we stop PP until later we find on NP the state we want. Therefore, we refer to this class of algorithms as *state matching* algorithms. One computational disadvantage of EPA is that the cut portion on NP, when we wait to find the required state, may be very large if the state space is big. EPA matches a particular state among all the states in the state space X. The event matching method of perturbation analysis, which we shall denote for short as EMPA, tries to aggregate the state space when we do "cut-and-paste." It can be viewed as an aggregated approach of state matching algorithms. The idea is to match event lists instead of states. Thus we refer to this class of algorithms as *event matching* algorithms. A trajectory of a GSMP is determined by the sequence of states and the associated events with their life-times. For each fixed state, there is a uniquely defined event list. On the other hand, there may be a set of different states associated with the same event list. Thus, when we do state matching, the associated set of events (i.e., the event list) is uniquely determined. But when we do event matching, we need to maintain some additional information to distinguish different states. The non-uniqueness of the mapping from event lists to states very naturally aggregates the state space. For example, by this aggregation, the state space of our two-server example aggregates into three sets i.e., {0}, {N}, and {1,...,N-1}, corresponding to the three kinds of event lists.

Let us now show how to do event matching for our two-server queueing system. Suppose that we start NP and PP from the same state $x_\theta(0,\xi) = x_{\theta'}(0,\xi)$ $= x_0$. Suppose that NP and PP are the same until at some completion time t, we find $p-\Delta p < u < p$; then $x_\theta(t^+,\xi) \neq x_{\theta'}(t^+,\xi)$. However, in the event matching algorithm, if $\Gamma(x_\theta(t^+,\xi)) = \Gamma(x_{\theta'}(t^+,\xi))$, we still do not need to cut. What we need is to remember the particular state $x_{\theta'}(t^+,\xi)$ on PP associated with this event list $\Gamma(x_{\theta'}(t^+,\xi))$. In other words, we maintain a separate register to remember the state on the perturbed path, while sharing the event list on the nominal path with the perturbed path whenever they are the same. If on the other hand, the next event lists are not the same, i.e., $\Gamma(x_\theta(t^+,\xi)) \neq \Gamma(x_{\theta'}(t^+,\xi))$, then we need to stop the generation of PP until at a later completion time s on NP, we find $\Gamma(x_\theta(s^+,\xi)) = \Gamma(x_{\theta'}(t^+,\xi))$. Then we start generating PP again. The event matching procedure is illustrated in Fig. 6(c) for the sample path in Fig. 6(a). It can be readily seen that the only cases where we need to do "cut-and-paste" are when NP and PP are on different aggregated states of {0}, {N}, and {1,...,N-1}. Furthermore, each cut portion using event matching is shorter than that of state matching since we only require matching the event list of a future state rather than the state itself.

It is clear how this idea can be generalized. We first partition our state space X according to an equivalence relation. Two states, x and x', are said to belong to the same equivalence class if $\Gamma(x) = \Gamma(x')$ where $\Gamma(x)$ is the event list for the state x. Let there be D different kinds of event lists for the GSMP (e.g., in the above example D=3). Let z(x) be the set of states having one of these D event lists, i.e., $z(x) = \{x' \in X, \Gamma(x') = \Gamma(x)\}$. Let Z(X) be the set of all such D z(x)'s. Thus we have aggregated the state space X into a set Z(X) with only D members. The number of states in X is usually much bigger than D. For instance, in the above example, D=3, but the number of states in X is N+1. For another example, it has been shown [Li, 1988] that for a multiclass closed queueing network with FCFS discipline, M=5 exponential servers, K=3 classes of customers, $N_1=3$, $N_2=3$, and $N_3=4$ for class 1, 2 and 3 customers respectively, the cardinality of Z(X) is D=1024, while the total number of states in X is approximately 1522050. Hence, the degree of state reduction of this aggregation is considerable. It has been observed that the length of a cut portion by matching a state or an event list depends not only on the number of states (or aggregated states), but also on whether the state or the event list is a rare one on the sample path. It should be pointed out that even though a great state reduction is achieved by aggregation, the cut portion may still be very long in some cases. Thus, variations in using cut, paste, and insert (as discussed in the introduction and will be discussed in the next section) should be practiced. The point to be emphasized here is the flexibility that comes with this viewpoint. To further compare the state matching and event matching algorithms, we proceed to analyze the specific example in more detail.

As mentioned before, our goal in using cut-and-paste algorithms is to simultaneously construct sample paths of the nominal and the perturbed systems. Therefore, any time we suspend the construction of one of the paths ("cut" if we suspend the construction of the perturbed and "insert" if we suspend the construction of the nominal), the algorithm is inefficient. Hence, as a measure of the inefficiency of the algorithms, we consider the proportion of the time that cut or insert is being used compared to the total length of the construction. To compare the efficiency of the state matching and the event matching algorithms, we consider the symmetric closed network above. We use both algorithms to construct sample paths of the perturbed system along a sample path of the nomi-nal system. For each algorithm, we calculate analytically the proportion of the cut and insert part of the construction and use the result as a basis for comparing the efficiency of the algorithms. We shall show that for large N, the event match-ing algorithm is significantly more efficient than the state matching algorithm.

Let $x(t) = x(t; p, \xi)$ be the state of the nominal system at time t, and $x'(t) = x'(t; p-\Delta p, \xi)$ and $x''(t) = x''(t; p-\Delta p, \xi)$ be the states of the perturbed system at time t for the state matching and event matching algorithms, respectively. We provide a new interpretation of "cut" and "insert" as follows (this interpretation utilizes the augmented chain idea of Cassandras and Strickland [1989b,c], which is also discussed in Section 7.2): Let us consider one of the algorithms, say the state matching. We consider the two processes $x(t)$ and $x'(t)$ together as one augmented process $\mathbf{x}(t) = (x(t), x'(t))$. The augmented process is well defined from t_0, the initial time, to t_1 when one of the operations, cut or insert, becomes necessary. Assume from t_1 to t_2 we suspend the perturbed sample path while the nominal path is active (a cut operation). We define $\mathbf{x}(t) = (x(t), x'(t))$ on $t_1 < t < t_2$ as follows: $x'(t) = x'(t_1)$ for $t_1 < t < t_2$; $x(t)$ is already defined on $t_1 < t < t_2$. In other words, we freeze the $x'(t)$ component of $\mathbf{x}(t)$ while $x(t)$ evolves. It is easy to see that the operation insert is equivalent to freezing $x(t)$ while $x'(t)$ evolves. Therefore $\mathbf{x}(t) = (x(t), x'(t))$ is well defined during the cut and insert operations and hence for all $t \geq t_0$. Our analytic calculation is based on identifying those states of the process $\mathbf{x}(t)$ where being in those states means one of the operations, cut or insert, is being performed. The steady-state probability of being in these states, therefore, can be used as a measure of the cut and insert part of the construction.

(I) State matching algorithm For simplicity we assume $p = \frac{1}{2}$. We start the nominal and the perturbed systems from the same initial point, say $x(t_0) = x'(t_0) = n$. These two processes remain equal until at some instant in time, e.g., t_1, when because of different routing decisions at the nominal and the perturbed systems $x(t_1) = i$ and $x'(t_1) = i \pm 1$ (the augmented process leaves the diagonal, Fig.7). At this point "cut" or "insert" is necessary. Our criterion for choosing between cut and insert operations is: which one provides the more efficient route (smaller expected first passage time) to return to the diagonal (where the states of the two processes match again). It is not difficult to observe that: if $x(t_1) = i \leq \frac{N}{2}$ and $x'(t_1) = i+1$, then a "cut" is more efficient and the augmented process moves vertically (for example, the expected first passage time from (1, 2) to (2, 2) is less than the expected first passage time from (1, 2) to (1, 1). The PP can go from state 2 to states 3, 4, . . indefinitely before returning to state 1 while the NP can at most wander to state 0 (See Fig. 7). On the other hand, if $x(t_1) = i \leq \frac{N}{2}$ and $x'(t_1) = i - 1$ then an "insert" is more efficient and the augmented process should move horizontally for the same reason. Because of the symmetry of the system, the choices of cut and insert for $x(t_1) = i > \dfrac{N}{2}$ follow from the choices for $x(t_1) = i \leq \dfrac{N}{2}$. Given the above cut and insert operations, the process $x(t)$ is a Markov process and the state transition rate diagram of the process is as shown below in Fig. 7.

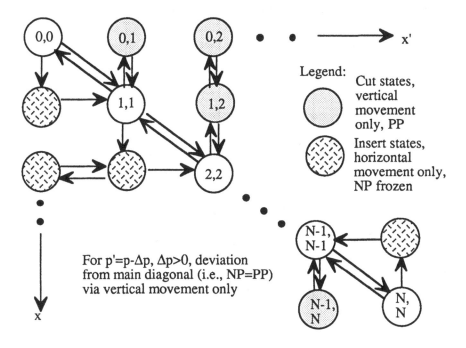

Fig. 7 State Transition Rate Diagram (State Matching Algorithm)

In Fig. 7, The transition rate for all horizontal and diagonal moves is $(\dfrac{1}{2} - \Delta p)\mu$, that for all vertical moves out of the diagonal is $\Delta p\mu$, and that for all other vertical moves is $\dfrac{1}{2}\mu$.

Let $\pi(i,j)$ be the steady-state probability of being in state (i,j). We assume that N is even and N=2M (the same approach can be used when N is odd). From the upper-half part ($i \leq M$) of the diagram under the diagonal we derive the following equations by balancing the transition rates across appropriate boundaries. Defining a boundary around all insert states on row i+1 and balancing the transition rates, we get:

$$\pi(i+1, i)(\tfrac{1}{2} - \Delta p)\mu = \pi(i, i)\Delta p\mu, \qquad\qquad 0 \le i \le M\text{-}1. \qquad (1)$$

Defining a boundary around all states above and including the (i+1)th row and to the left of the (i+1)th column and balancing the transition rates, we get:

$$\pi(i+1, i)(\tfrac{1}{2} - \Delta p)\mu + \pi(i, i)(\tfrac{1}{2} - \Delta p)\mu = \pi(i+1, i+1)(\tfrac{1}{2} - \Delta p)\mu,$$

$$0 \le i \le M\text{-}1. \qquad\qquad (2)$$

Balancing the rates between insert states on row i+1, we get:

$$\pi(i+1, j) = \pi(i+1, i), \qquad 0 \le j \le i\text{-}1, \qquad\qquad 0 \le i \le M\text{-}1. \qquad (3)$$

It follows from Eqs.(1), (2), and (3) that:

$$\pi(i, i) = (1 - 2\Delta p)^{M\text{-}i}\,\pi(M,M), \quad 0 \le i \le M\text{-}1, \qquad\qquad (4)$$

$$\pi(i+1, i) = 2\Delta p(1 - 2\Delta p)^{M\text{-}(i+1)}\pi(M,M), \qquad 0 \le i \le M\text{-}1. \qquad (5)$$

A similar procedure can be used for the part of the diagram above the diagonal. Defining a boundary around all cut states on column i and balancing the transition rates, we get:

$$\pi(i, i)\Delta p\mu = \pi(i\text{-}1, i)\tfrac{1}{2}\mu, \qquad\qquad 1 \le i \le M. \qquad (6)$$

Balancing the rates between cut states on column i, we get:

$$\pi(j, i) = \pi(i\text{-}1, i), \qquad 0 \le j \le i\text{-}1, \qquad\qquad 1 \le i \le M, \qquad (7)$$

and it follows from these equations that:

$$\pi(i\text{-}1, i) = 2\Delta p(1 - 2\Delta p)^{M\text{-}i}\pi(M,M), \qquad 1 \le i \le M. \qquad (8)$$

Focusing on the states vertically above (M, M) and using Eqs. (7) and (8), we get:

$$\pi(j, M) = \pi(M-1, M) = 2\Delta p \, \pi(M,M), \qquad 0 \le j \le M-1. \qquad (9)$$

Therefore,

$$\sum_{j=0}^{M-1} \pi(j, M)) + \pi(M,M) = (2\Delta\pi \, M + 1)\pi(M,M) \le 1 \qquad (10)$$

and

$$\pi(M,M) \le \frac{1}{2\Delta p \, M + 1}. \qquad (11)$$

Let A be the steady-state probability of the augmented process being in one of the states (i, i), $0 \le i \le 2M$, i.e., the steady-state probability of being on the diagonal. Then A can be used as a measure of the duration of time when the sample paths are being constructed simultaneously (no cut or insert is being used). Then

$$A = \pi(M,M) + 2 \sum_{i=0}^{M-1} \pi(i, i) =$$

$$\pi(M,M) + 2\pi(M,M)[\sum_{i=0}^{M-1} (1 - 2\Delta p)^{M-i}], \qquad (12)$$

$$A = \pi(M,M) \frac{1 - \Delta p - (1 - 2\Delta p)^{M+1}}{\Delta p}. \qquad (13)$$

Therefore,

$$A \le \frac{1 - \Delta p - (1 - 2\Delta p)^{M+1}}{\Delta p} \frac{1}{2\Delta p \, M + 1}. \qquad (14)$$

Notice that when M (or equivalently N) approaches infinity we have

$$\lim_{N \to \infty} A \le 0 \implies \lim_{N \to \infty} A = 0. \qquad (15)$$

In other words, for a large N the state matching algorithm becomes very inefficient and almost always one of the cut or insert operations is being used.

(ii) **Event matching algorithm** We calculate the same quantity (i.e., proportion of time when no cut or insert is being used) for the event matching algorithm. We start the nominal and the perturbed sample paths from the same initial value. As was discussed above, we partition the state space, $\{0, ..., N\}$, into three equivalence classes $\{0\}, \{N\}, \{1, . . . , N-1\}$. As long as the two processes are in the same equivalence class we can continue the simultaneous construction of both sample paths. The choice of cut or paste in this case is again quite clear: if $x(t_1) \in \{0\}$ or $\{N\}$ and $x''(t_1) \in \{1, . . . , N-1\}$, then cut is used; if $x''(t_1) \in \{0\}$ and $x(t_1) \in \{1, . . . , N-2\}$ or $x''(t_1) \in \{N\}$ and $x(t_1) \in \{2,...., N-1\}$, then insert is used.

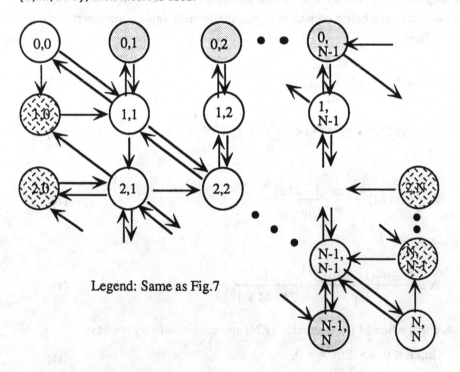

Legend: Same as Fig.7

Fig. 8 State Transition Diagram (Event Matching Algorithm)

The state transition rate diagram is shown in Fig. 8. In the figure, the transition rate of all horizontal moves and all moves in the diagonal direction is $(\frac{1}{2} - \Delta p)\mu$. The transition rate of all vertical moves from $(0, i)$ to $(1, i)$ and from (N, i) to $(N-1, i)$ is $\frac{1}{2}\mu$. The transition rate of all other vertical moves is $\Delta p\mu$.

Focusing on the first column from the left we can write

$$\pi(i, 0)(\tfrac{1}{2} - \Delta p)\mu = \pi(i+1, 1)(\tfrac{1}{2} - \Delta p)\mu, \quad 2 \le i \le N-2, \tag{16}$$

$$\pi(1, 0)(\tfrac{1}{2} - \Delta p)\mu = \pi(2, 1)(\tfrac{1}{2} - \Delta p)\mu + \pi(0, 0)\Delta p\mu, \tag{17}$$

$$\pi(0, 0)[(\tfrac{1}{2} - \Delta p)\mu + \Delta p\mu] = \pi(1, 1)(\tfrac{1}{2} - \Delta p)\mu \tag{18}$$

From Eqs.(17) and (18) it follows that

$$\pi(1, 0) = \pi(2, 1) + 2\Delta p\, \pi(1, 1) \le \pi(2, 1) + \pi(1, 1).$$
$$\text{(since } 2\Delta p \le 1\text{)} \tag{19}$$

Therefore, Eqs.(16) and (19) imply

$$\sum_{i=1}^{N-2} \pi(i,0) \le \sum_{i=1}^{N-1} \pi(i,1). \tag{20}$$

From the equality of transition rates between columns, it follows that

$$(\tfrac{1}{2} - \Delta p)\mu \sum_{i=1}^{N-1} \pi(i,j) = (\tfrac{1}{2} - \Delta p)\mu \sum_{i=1}^{N-1} \pi(i,1), \quad 2 \le j \le N-1. \tag{21}$$

From Eqs.(20) and (21) we conclude:

$$(N-1) \sum_{i=1}^{N-2} \pi(i,0) \le \sum_{i \& j=1}^{N-1} \pi(i,j). \tag{22}$$

Therefore,

$$\sum_{i=1}^{N-2} \pi(i,0) \le \frac{1}{N-1}. \tag{23}$$

Balancing the rates between first and second rows, we have

$$\frac{1}{2}\mu \sum_{i=1}^{N-1} \pi(0,i) \le [\Delta p\mu + (\frac{1}{2} - \Delta p)\mu]$$

$$\sum_{i=2}^{N-1} \pi(1,i) + \Delta p\mu\pi(1, 1). \tag{24}$$

Therefore,

$$\sum_{i=1}^{N-1} \pi(0,i) \le \sum_{i=2}^{N-1} \pi(1,i). \tag{25}$$

Balancing the rates between other columns, we have

$$\sum_{i=1}^{N-1} \pi(j,i) = \sum_{i=1}^{N-1} \pi(1,i), \quad 2 \le j \le N-1. \tag{26}$$

An argument similar to the one given for equation (22) results in

$$\sum_{i=1}^{N-2} \pi(0,i) \le \frac{1}{N-1}. \tag{27}$$

Notice that the states associated with the cut and insert operations are the states on the perimeter of the diagram, except for states (0,0) and (N,N). If B is the steady-state probability of the augmented process being in one of these states then we have

$$B= \sum_{i=1}^{N-2} \pi(0,i) + \sum_{i=1}^{N-2} \pi(N,i) + \sum_{i=1}^{N-2} \pi(i,0)$$

$$= \sum_{i=1}^{N-2} \pi(i,N) = 2 \sum_{i=1}^{N-2} \pi(0,i) + 2 \sum_{i=1}^{N-2} \pi(i,0). \tag{28}$$

From Eqs.(23) and (27), it follows that

$$B \leq \frac{4}{N-1}.$$

Let A be as defined in the state matching algorithm, then A + B = 1 and therefore

$$A \geq 1 - \frac{4}{N-1}.$$

When N approaches infinity we have

$$\lim_{N \to \infty} A = 1. \tag{29}$$

In other words, for a large N the event matching algorithm is very efficient, and compared to the length of the sample paths, the cut and insert parts are very short. Therefore it is clear that, when N is large in our example, the event matching algorithm is significantly more efficient than the state matching.

Remark Note that intuitively, the efficient states of the augmented process for the state matching algorithm, i.e., the diagonal states where the nominal and the perturbed paths are simultaneously generated, is a one-dimensional set in the two-dimensional set of all states and hence when the number of states goes to infinity the relative weight of the efficient states goes to zero. By contrast, the inefficient states for the event matching algorithm, i.e., the states where only one of the perturbed or the nominal paths is generated, are on the perimeter of the diagram and therefore constitute a one-dimensional set in the two-dimensional set of all states. Thus, when the number of states goes to infinity, the relative weight of the inefficient states goes to zero.

6.2.2 Experimental Validation

In this section, we compare the event matching and state matching algorithms by applying both algorithms to a multiclass queueing network. First we specify how a general multiclass queueing network can be modeled as a GSMP and then we introduce the specific network that is used to compare the algorithms.

Let us consider a closed queueing network with M single server stations $S_1,...,S_M$, and K classes of customers $C_1,C_2,...,C_K$ with population $N_1,N_2,...,N_K$, respectively. Assume that the queues have infinite capacity (this assumption can be dropped without much complication to EPA and EMPA). Let $F_{mk}(\cdot\,;s_{mk})$ be the distribution function of the service times for class k customers at server m, parametrized by $s_{mk} \in R$, which is the mean service time. The discipline is FCFS at all queues. The routing probability matrix for class k customers is denoted by $P^{(k)}=[p_{ij}^{(k)}]$, where $p_{ij}^{(k)}$ is the probability that C_k chooses the routing from S_i to S_j. We assume that all the service times at a server are independent of each other and are independent of the service times of other servers and the routings. In general, this network cannot be solved analytically except for the case when s_{mk} is independent of k, and F_{mk} is exponentially distributed, for all m=1,2,...,M [Baskett et al. 1975]. These general multiclass queueing networks can be modeled by a GSMP with the state space defined as:

$$X=\{x: x=(x_1,...,x_M); x_i=(x_{i1},...,x_{in(x,i)})^T, x_{ij} \in \{1,...,k\}, i=1,...,M\},$$

where M is the number of servers, x_{ij} is the class of customer at the jth position in queue i, and n(x,i) is the number of customers in queue i including the one at service. If n(x,i)=0, we let $x_i=(0)$. There is a total of M×K types of possible events, i.e., K types of service completion events of the K different classes of customers at each server i, i=1,...,M. Thus we define:

$$\Gamma=\{i_{gh}: 1 \leq g \leq M, 1 \leq h \leq K\},$$

where i_{gh} denotes the event that a customer of class h finishes service at server g; and

$\Gamma(x) =\{i_{gh}: x_{g1} = h, 1 \leq g \leq M, 1 \leq h \leq K\}$, and $\phi_{i_{gh}} = F_{gh}(., s_{gh}), \mu_{i_{gh}} = \frac{1}{s_{gh}}$.

Now let us specify the routing probability distribution $p(x';x,i)$. Clearly, we can choose:

$p(x';x,i_{gh})=p_{gk}^{(h)}$, if $x'=(x_1,...,x_{g-1},x_g',x_{g+1},...,x_{k-1},x_k',x_{k+1},...,x_M)$,

where $1 \leq k \leq K$, and $x_k'= (x_{k1},...,x_{kn(x,k)}, h=x_{g1})$ and $x_g'=(x_{g2},...,x_{gn(x,g)})$ (i.e., a customer of type h finishes at server g). In other cases, $p(x'; x, i_{gh})=0$.

Given this GSMP model, we can define the aggregated set $Z(X)$ for the event matching algorithm as follows: two states $x=(x_1,...,x_M)$ and $x'= (x_1',..., x_M')$ belong to the same equivalence class if $x_{i1}=x_{i1}'$, for all $i=1,...,M$. Thus there is at most a total of $D=(K+1)^M$ elements in $Z(X)$. The maximum number of states in X can be computed by a combinatorial formula given in [Ho and Li 1988]. Having defined a GSMP model for multiclass queueing networks, we now introduce the experimental multiclass network that we use to compare the event matching and state matching algorithms.

The network consists of three exponential servers and two classes of customers. The mean service times of the servers are $s_{11}= s_{12} = s_1 = 1.2$, $s_{21}= s_{22} = s_2 = 1.2$ and $s_{31}= s_{32} = s_3= 1.0$. The routing probabilities are defined as $p_{11}^{(1)}= 1-p$, $p_{13}^{(1)}= p$, $p_{23}^{(2)}= 1$, $p_{31}^{(1)}= 1$, $p_{32}^{(2)}= 1$. We denote the number of customers of class 1 and class 2 by N_1 and N_2, respectively (Fig. 9).

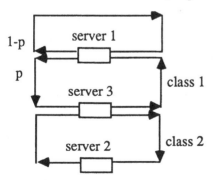

Fig. 9 Multiclass Network for Comparing Event vs. State Matching

The performance measure is the throughput of class 2 customers at server 3, TP_{32}. We use both the event matching and state matching algorithms to estimate $TP_{32}(p)$ and $TP_{32}(p + \Delta p)$, where p=1.0 and $\Delta p = -0.05$. The stopping rule for both algorithms is when a specified number of class 2 customers are served at server 3 at both the nominal and the perturbed systems. To compare the efficiency of the algorithms, we define the following measure of efficiency

$\alpha = $ measure of efficiency $= \dfrac{n_b}{n_b + n_c + n_i}$, where

$n_i = $ total number of customers served when only the perturbed sample path is active (insert operation is used).

$n_c = $ total number of customers served when only the nominal sample path is active (cut operation is used).

$n_b = $ total number of customers served when both sample paths are active.

This measure of efficiency is the proportion of the time in which the nominal and the perturbed sample paths are simultaneously active. Clearly, the number of states is always greater than or at best equal to the number of event lists (or equivalently, the number of equivalence classes defined by the event matching algorithm). More importantly for our discussion, the number of event lists is independent of N_1 and N_2 if $N_1, N_2 > 0$, whereas the number of states grows combinatorially with increasing N_1 and N_2. Therefore, intuitively, for large N_1 and N_2, when a mismatch of event lists (in the case of event matching) or states (in the case of state matching) occurs, the event matching algorithm searches through a much smaller set for an event list match than the state matching algorithm does for a state match. Hence, for large N_1 and N_2 we expect smaller n_c and n_i for the event matching algorithm than for the state matching algorithm and therefore we expect more efficiency.

Our experiment consists of a number of smaller experiments where we increase the number of customers of class 1 and 2 in each experiment and compute the efficiency of the algorithms. Table 1 below shows the nominal and perturbed average performances obtained from state matching and event matching algorithms, and the measures of efficiency for these algorithms for each experiment (in the last two experiments, only event matching is used because the measure of efficiency of the state matching algorithm is almost zero). Fig. 10 shows the efficiency of the event matching and of the state matching algorithms as the number of customers in the system increases. The simulation is stopped when the nominal and the perturbed systems serve 10000 customers of class 2 altogether. Note that the two paths are constructed simultaneously.

Table 1 Event Matching (EMPA) vs. State Matching (EPA)

(10 run/experiment, 10,000 customers of class 2 at server 3/run, 1,000 customers warm up/run)

	Event	Matching		State	Matching	
Number of customers $N_1=N_2$	TP_{32}	$TP_{32}+$ ΔTP_{32}	α	TP_{32}	$TP_{32}+$ ΔTP_{32}	α
2	0.45755	0.46138	0.88181	0.45728	0.46182	0.84882
3	0.48305	0.48850	0.76820	0.48564	0.48869	0.63897
4	0.49413	0.49870	0.61629	0.49472	0.49677	0.36268
5	0.49741	0.50286	0.47486	0.49858	0.50149	0.16986
6	0.49819	0.50348	0.37992	0.49933	0.50102	0.06311
8	0.49996	0.50261	0.30526	0.49870	0.50180	0.00711
10	0.50000	0.50384	0.27985	0.50153	0.50224	0.00215
16	0.50047	0.50220	0.26638	—	—	—
32	0.49995	0.50176	0.25944	—	—	—

Note that for a relatively large number of customers the efficiency of the state matching algorithm decreases rapidly and soon the proportion of time that this algorithm constructs both sample paths simultaneously becomes almost zero. In comparison, with the increase of the number of customers in the system, the efficiency of the event matching algorithm decreases more slowly until it levels off at a non-zero value. With a large number of the customers in the system, a mismatch of event lists occurs essentially because the customers being served at server 3 at the nominal and the perturbed systems have different classi-fications (servers 1 and 2 are almost always busy at both systems). Therefore, the proportion of the time that both sample paths are simultaneously active at the event matching algorithm is the proportion of the time that the class of customers being served at server 3 at the two systems are the same. With an argument similar to the one used in Section 6.2.1, it can be easily shown that this quantity is nonzero.

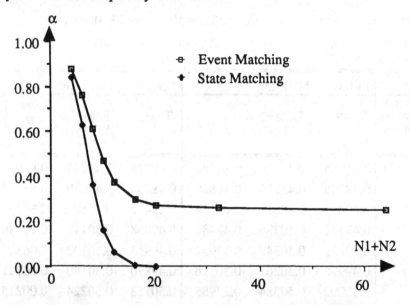

Fig. 10 Experimental Comparison of State vs. Event Matching EPA

Remark Actually, for the above system we can consider an even more efficient event matching algorithm using a slightly different GSMP model, whereas the state matching algorithm will have the same behavior. Note that for the above system the set of event lists is {the end of service at server 1, the end of service at server 2, the end of service of class 1 at server 3, the end of service of class 2 at server 3}. Since the distributions of service times of class 1 and class 2 customers are equal, it is possible to identify the events of the end of service of the two classes of customers at server 3 as a single event: i.e., the end of service at server 3. Given this identification, for a large number of customers, there is rarely a mismatch of event lists between the nominal and the perturbed systems and the efficiency of the event matching algorithm in this case will approach 1 as the number of customers in the systems approaches infinity. It is clear from this remark that when different classes of customers have the same distribution of service times at a server, this alternative GSMP model is better suited for the event matching algorithm. This observation further underscores the importance of an appropriate "model" in simulation and PA application (see Chapter 5).

Lastly, the cut-and-paste idea depends basically on the Markov property. For the GSMP model of this chapter, this implies that the mathematical state of the DEDS is discrete and can be specified by a finite number. However, once we deviate from the Markov property, say to the case of an M/G/1/K queue, the specification of the state of the system now involves a real number representing the age of the service lifetime. In principle, of course, we can convert to the Markov case by approximately discretizing this real number and apply "cut-and-paste" to the resultant enlarged state space. However, this is highly inefficient since matching the discrete age will be difficult and structural information about the problem is not utilized. We present an extension which circumvents this difficulty.

Let us use the example of generating the sample path of an M/G/1/K queue from that of an M/G/1 queue once again. Suppose at an event time t, the queue length is K and an arrival occurs. The remaining service time of the server is x. For the sample path of the M/G/1/K queue we must now cut (or freeze) and wait for a state or event match along the path of the M/G/1 queue. For event match we must find a state transition time with a remaining service time equal to x for which the next event is a service completion. The probability of finding such an exact match is of course zero, and even that of an approximate match is very small. However, this difficulty is not necessary. Note that with the exception of the general service time, the rest of the system is Markov. All we need is to find a state transition time with a remaining service time y ≥ x for which the next event is a service completion. Instead of pasting the sample path beginning at such a state transition time, we can merely wait until the remaining service time has decreased to x and commence the "pasting" at that time. The fact the rest of the system is Markov permits such a shift in time. This idea can obviously be extended to other non-Markov DEDS so long as only one non-Markov event is involved.

Exercise Formalize the extension of this idea to a general GSMP model with one source for non-Markov event lifetimes (see [Strickland and Cassandras 1989]).

6.3 Cut-and-Paste as a Generalization of the Rejection Method

In this section, we formalize the idea that the cut-and-paste method can be viewed as a generalization of the rejection method as was suggested in the introduction of this chapter. To illustrate our point we consider the simplest form of the rejection method [Brately, Fox, and Scharge 1987]:

Consider a random variable X with probability density function f(x). Assume that f(x) is bounded and is nonzero only on some finite interval [a, b) and define

$c = \max\{f(x) : a \le x \le b\}$

To generate samples of this random variable the following procedure, called rejection method, is used

1) generate u, a sample of a uniform random number U, on (a, b);

2) generate v, a sample of a uniform random number V, on (0, c);

3) if v ≤ f(u) accept u as a sample of X; otherwise go to step 1.

The above method can be formalized in the following form, which makes the connection between this method and the cut-and-paste procedure apparent: Let $A = \{(U_1, V_1), (U_2, V_2), \ldots \}$ be an infinite sequence of couples of random variables on a probability space Ω, where $\{U_1, U_2, \ldots \}$ is an i.i.d. sequence of random variables distributed as the random variable U and $\{V_1, V_2, \ldots\}$ is an i.i.d. sequence of random variables distributed as the random variable V.

The rejection method defines random times $\tau_1, \tau_2, \ldots$ as follows

$$\tau_1, \tau_2, \ldots : \Omega \to \{1, 2, \ldots\}$$

$$\tau_1(\xi) = \min\{i ; V_i(\xi) \le f(U_i(\xi))\}.$$

In other words, $\tau_1(\xi) = m$ means that at the mth try of the rejection method the first sample for X is accepted and all the samples prior to m were rejected and discarded. Similarly

$$\tau_j(\xi) = \min\{i ; V_i(\xi) \le f(U_i(\xi)) \text{ and } i > \tau_{j-1}(\xi)\}$$

or, in other words, $\tau_j(\xi)$ is the index of the jth accepted sample for X.

The assertion of the rejection method is that the sequence $\{U_{\tau_1}, U_{\tau_2}, \ldots\}$ is an i.i.d. sequence of random variables distributed as the random variable X (Fig. 11).

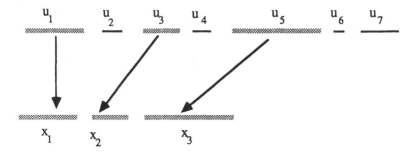

Fig. 11 The Rejection Method

Now we formalize the cut-and-paste method in a similar fashion. It is illustrative to consider this method in its simplest form. We consider the problem introduced in Section 2 and assume that, to construct the perturbed path, the state matching method and the cut-and-paste operations are used (no insertion). Using a different notation than that used in Section 2, we define

$X_p(t)$ = number of customers at server 1 at time t for the nominal system.

$X_{p'}(t)$ = number of customers at server 1 at time t for the perturbed
system (operating with routing parameter p').

$X_p(t)$ and $X_{p'}(t)$ define two stationary Markov processes on $t \geq T_0$. If we assume that $X_p(T_0) = X_{p'}(T_0)$, in other words the two processes start from the same initial value, then the cut-and-paste operations define random times P_0, $C_1, P_1, C_2, P_2, \ldots$ as follows

$$P_0 = T_0$$

$$C_1(\xi) = \inf \{t : X_p(t, \xi) \neq X_{p'}(t, \xi) , t > P_0\},$$

or in other words, C_1 is the random variable of the first instant when the two processes become different.

$$P_1(\xi) = \inf \{t + C_1(\xi): X_p(t + C_1(\xi), \xi) = X_{p'}(C_1(\xi), \xi) , t > 0\}.$$

P_1 is the first passage time of the process $X_p(t)$ to the state $X_{p'}(C_1)$ after C_1. The part of the sample path of $X_p(t)$ between C_1 and P_1 is not used to construct the sample path of $X_{p'}(t)$ and is "rejected." C_i and P_i are similarly defined as follows

$$C_i = \inf \{ t + P_{i-1} : X_p(t + P_{i-1}) \neq X_{p'}(\sum_0^{i-1}(C_j - P_{j-1}) + t) , t > 0 \} \text{ and}$$

$$P_i = \inf \{ t + C_i : X_p(t + C_i) = X_{p'}(\sum_0^{i}(C_j - P_{j-1})) , t > 0 \}$$

The assertion of the cut-and-paste method is that the sequence

$$\{ X_p(t) : P_0 \leq t < C_1, X_p(t) : P_1 \leq t < C_2, X_p(t) : P_2 \leq t < C_3, \ldots \}$$

is a sequence of random processes defined on random intervals of which each term is statistically indistinguishable (as a random process) to the terms of the following sequence (see Fig.12)

$$\{ X_{p'}(t) : P_0 \leq t < C_1, X_{p'}(t) : C_1 \leq t < C_1 + (C_2 - P_1), X_p(t) : C_1 + $$

$$(C_2 - P_1) \leq t < C_1 + (C_2 - P_1) + (C_3 - P_2). \ldots \}$$

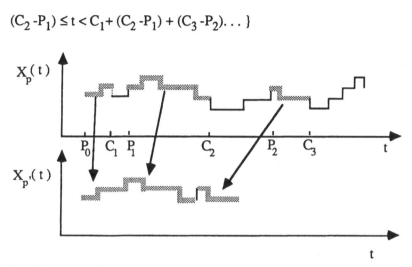

Fig. 12 Cut and Paste

To summarize, the above discussion shows that in the rejection method samples of a random variable are generated; samples satisfying a specified condition are accepted, otherwise they are rejected. The accepted samples constitute samples of the desired random variable. On the other hand, in the cut-and-paste method sample path segments of a random process are generated (in this case we are viewing a long sample path as a number of shorter sample path segments between cut and paste points); sample path segments satisfying a specified condition are accepted, otherwise they are rejected. The resultant path assembled from the accepted sample segments represents a sample path of the desired random process. In this sense, we can view the cut-and-paste method as a generalization of the rejection method for random processes.

6.4 First Order Propagation Rules and Approximate EPA

To implement EPA, we need two practical ingredients:

(i) *Efficiency of the Cut-And-Paste Operation:* As discussed in Section 2, the event matching approach described is a form of aggregated EPA in which the cut-and-paste operations are performed only when the aggregated state of the NP and the PP starts to differ and to re-join. More generally, we can view event matching as one of the many possible ways of aggregation. All these aggregation methods are attempts to increase the efficiency of the EPA method.

(ii) *Detection of State Sequence or Event Order Change:* In order to know when to "cut," we must have ways to detect the violation of deterministic similarity or state sequence difference between the NP and the PP. In general, this can always be accomplished by simulating both the NP and the PP in parallel. But this is precisely what PA proposes to avoid. Thus, the question is how to use the accumulated perturbations directly to determine such deviations.

The purpose of this section is to detail the procedures for accomplishing (ii), which is called finite PA or FPA for short. In FPA, we assume that a parameter (e.g., a mean service rate) is changed by a small but finite amount. The finite change of a parameter induces a finite change of event times, i.e., the perturbations considered in the analysis are finite. The finite perturbations will accumulate and eventually will lead to event order-changing, causing possible discontinuities in the sample performance function. In what is denoted as the first order FPA, we consider only the possibilities of adjacent event order change and estimates the performance measure as a result of such changes. The perturbation generation rule in first order analysis is just the same as that for infini-tesimal analysis (see Chapter 3); however, the perturbation propagation rules incorporates the effect of the adjacent event order-changing. If the perturbation is small so that event order changing does not occur, applying simply the infini-tesimal propagation rules is enough to predict the perturbed path. When the perturbation accumulates so that two events change their order in the perturbed path, the first order propagation rules can be used to predict (at least in the short run) the events and event times after the order changing. Note that when an event order change first occurs, an event usually only changes the order of occurrence with its adjacent event; but as the perturbation accumulates, it is entirely possible that an event will change its order with another event which is two or more events ahead of or behind it in the nominal path. First order analysis deals with only the order change between two adjacent events (for this reason, the technique is called "first order" analysis). If order-changing happens between events which are not adjacent to each other, errors may occur in estimating the perturbed path using first order propagation rules. Thus, first order analysis is an approximate technique. It provides the estimate of the performance measure for a system with a finitely perturbed parameter rather than providing the performance sensitivities.

Furthermore, similar to Chapter 3, first order propagation rules should be defined with respect to a class of problems we wish to consider. Let us consider the same queueing networks as in Chapter 3, i.e., closed single-class networks with finite or infinite buffers. These rules can be grouped into two sets: those pertaining to "No Input (NI)" and those pertaining to "Full Output (FO)." To illustrate the principle, we shall discuss only the "NI" case. The propagation rules for FO can be considered as the dual of that for the "NI" case; readers are referred to [Ho, Cao, and Cassandras 1983] for details of the FO propagation rules.

The first order perturbation propagation in an "NI" case can be described in the following three rules. The first rule deals with the situation where an idle period ("NI") in the nominal path is eliminated in the perturbed path because of a finite perturbation. Fig. 13a shows the nominal path and Fig. 13b shows the perturbed path. The start of the jth service period is denoted by $t_{j,s}$. In the nominal path, server j meets an idle period at time t_j, and the idle period is terminated by server i at t_i, $t_i > t_j$. Suppose that the two servers obtain perturbations Δ_i and Δ_j, respectively. Then in the nominal path the service completion times of the two customers in servers i and j will be $t'_i = t_i + \Delta_i$, and $t'_j = t_j + \Delta_j$. If the two perturbations are such that $t_i + \Delta_i > t_j + \Delta_j$, then the idle period still exists in the perturbed path. The perturbation of server i is propagated to server j, i.e., $\Delta_j = (t_i + \Delta_i) - t_i = \Delta_i$. However, if $t_j + \Delta_j > t_i + \Delta_i$, then the two service completion times of servers i and j change order because of the perturbations. In the perturbed path, the customer coming from server i arrives at server j before the idle period starts. Therefore, as illustrated in Fig. 13b, no idle period exists in the perturbed path. In this case, only partial perturbation is propagated from server i to server j: After the propagation, server j obtains a perturbation $\Delta_j = (t_j+\Delta_j) - t_i = \Delta_j+(t_j - t_i)$. In the case shown in Fig. 13b, server j is said to have a "Potential No Input (PNI)." To summarize, we have

First order propagation rule 1:
If server i terminates an NI of server j ($t_i > t_j$), and

  (i).   if $t_i +\Delta_i > t_j + \Delta_j$, then after the propagation $\Delta_j = \Delta_i$, i.e., IPA applies;

  (ii). if $t_j + \Delta_j > t_i +\Delta_i$, then an NI becomes a PNI, and $\Delta_j = \Delta_j+(t_j - t_i)$ after the propagation.

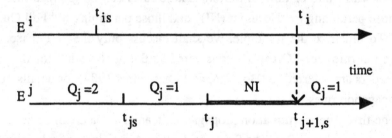

Fig. 13a An No Input Case

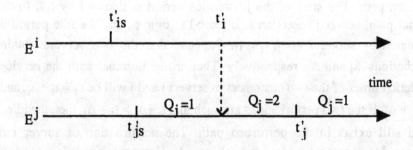

Fig. 13b A Potential No Input Case

The second propagation rule studies the situation where a PNI in the nominal path becomes an NI in the perturbed path, i.e., an NI is created. This propagation rule can be explained by using Fig. 13b as the nominal path and Fig. 13a as the perturbed path. Keep in mind that now the t'_i and t'_j shown in Fig. 13b should be considered as t_i and t_j, while the t_i and t_j in Fig. 13a are the perturbed values t'_i and t'_j. From the same discussion as for propagation rule 1, we have

First order propagation rule 2:
If server j meets a PNI shown in Fig. 13b ($t_j > t_i$), and

(i). if $t_j + \Delta_j > t_i + \Delta_i$, then no propagation takes place (PNI remains a PNI);

(ii). if $t_i + \Delta_i > t_j + \Delta_j$, then a new NI is created because of the perturbation, and $\Delta_j = \Delta_i - (t_j - t_i)$ after the propagation.

The third propagation rule applies when the server terminating an NI of another server changes because of the perturbation. This is explained in Fig. 14. The figure shows that in the nominal path an idle period of server j is terminated by server i at time t_i, and at time $t_k > t_i$ a customer from server k arrives at server j. Suppose that in the perturbed path $t'_k = t_k + \Delta_k < t'_i = t_i + \Delta_i$, then in the per-turbed path the idle period of server j is terminated by server k instead of server i. As a result, server j will obtain the perturbation of server k in the perturbed system. This is stated as the following propagation rule:

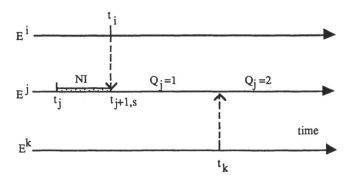

Fig. 14 The Changing of Critical Servers Terminating an NI

First order propagation rule 3:
If server i terminates an NI of server j at t_j and afterwards server j receives a
customer from server k at $t_k > t_j$, then

(i).   if $t_k + \Delta_k > t_j + \Delta_j$, then no propagation takes place;

(ii).   if $t_j + \Delta_j > t_k + \Delta_k$, then in the perturbed path server k becomes the
        server that terminates the NI of server j, and $\Delta_j = \Delta_k - (t_k - t_j)$ after
        the propagation.

Entirely similar propagation rules have been developed for the Full
Output (FO) case in simple queueing networks with finite capacity buffers.

Exercise Work out the corresponding three rules governing the "Full Output"
case. (Hint: see [Ho, Cao, and Cassandra 1983]).

Various examples indicate that the first order perturbation analysis rules
yield the exact prediction for performance measures of short sample paths of
finitely perturbed networks. This should not be surprising since FPA is
basically a thought experiment to carry out brute force simulation in the short
term. The fact that the first order perturbation analysis provides the exact
results for short sample paths is useful. In fact, this feature makes the first order
perturbation analysis an accurate tool for estimating the performance
sensitivities of systems whose sample paths consist of short regenerative
periods. For an example see [Cao 1987c] where the first order analysis is
applied to a multiclass finite source queue, resulting an accurate estimate of the
sensitivity of the system throughput. It is known that infinitesimal perturbation
analysis does not work for the system. For long sample paths, FPA is of course
only an approximation. Its accuracy depends on the validity of the heuristic
"stochastic similarity" assumption mentioned in Chapter 2. It can also be
viewed as the limit of approximate EPA where the entire state space has been
aggregated into one state for matching purposes. Consequently, we never "cut"
at all (in other words, we "paste" immediately after we cut). Between the two
extremes (EPA or EMPA which are exact algorithms at one end and the FPA at

the other), a spectrum of approximate cut-and-paste possibilities exists depending on the amount of "matching" required. The central idea is that the error of approximate matching when piecing together trajectory segments tends to average out statistically[4]. Some experimental evidence of this belief can be found in [Li 1988]. Furthermore, an extensive statistical experiment comparing FPA and IPA on a large number of randomly generated queueing networks [Dille 1987] shows that the errors for long sample paths are consistently better when using FPA.

Exercise Consider the same example of Fig. 10 of Section 5.5. Solve it using FPA rules.

Finally, For more complex queueing networks and DEDS, more complex FPA rules must be worked out. We can visualize general FPA rules as simply a short term "forward-looking" thought experiment of the perturbed path used solely to detect deviation in the event or state sequence of the system. Once a deviation is detected, "cut-and-paste" is then used to avoid the inefficiency that comes from generating two parallel trajectories.

---

[4] Note this issue of state matching occurs also in regenerative simulation and various approximations to regeneration.

CHAPTER SEVEN

Generalized Sensitivity Analysis

Chapters 3, 4, and 5 primarily discussed a straightforward approach to the problem of constructing sensitivity estimates of a DEDS performance from a single sample path. The principal issue dealt with is the IPA algorithm, the consistency of the IPA estimate, and some extensions of the applicability of the IPA algorithms to situations where it does not appear to be applicable at the first glance. Chapter 6 by way of the "cut-and-paste" idea begins to address the broad issue of computing the performance of a general DEDS, $J(\theta)$, at different values of the system parameter(s), θ, or the response surface exploration in the experimental design terminology. In this chapter, we would like to continue this viewpoint and treat PA not so much as a particular algorithm or a group of algorithms but as an approach or mind-set towards the efficient processing and analysis of DEDS performance and to view PA as the first step in the eventual construction of a theory of DEDS very much in parallel with the control theory for CVDS (see Appendix C). We submit that PA-like techniques complement extant simulation and queueing network methodologies. The first three sections of this chapter are basically independent from each other and can be read in any order. However, they mutually reinforce one another by providing many connections and share perspectives with each other and with other parts of this book.

7.1 Efficient Sample Path Generation

In this section, we introduce one more variation on algorithms for simultaneously generating sample paths of two or more GSMPs. In this case no cut-

and-paste of the sample paths is used and the sample paths are active at all times. Our objective in using this algorithm is to avoid duplication as much as possible during parallel simulation experiments of structurally similar DEDS. We assume that all event lifetimes are exponentially distributed.

7.1.1 The Standard Clock and the M/M/1 Example

To help fix ideas of the approach, we consider the simple M/M/1 example with arrival and service rates λ and μ, respectively. We start with the following steps:

(i) mark a series of event instants on the time axis by generating a series of exponentially distributed random variables with unit mean. This series of event instants is called the a **standard clock** and is illustrated in Fig. 1.

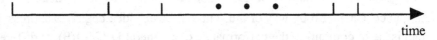

time

Fig. 1 A Standard Clock

For the M/M/1 case, we need a sample sequence of event times at the rate of $\lambda+\mu$. This sequence can be obtained by re-scaling the time unit by $1/(\lambda+\mu)$. Next,

(ii) determine the type of the successive event times on this standard clock sequence according to

an arrival event	if $u_i < \lambda/(\lambda+\mu)$,
a departure event	if $u_i \geq \mu/(\lambda+\mu)$ and $n>0$,
ignore the event	otherwise,

where u_i, $i=1, 2, \ldots$, is a sequence of u.i.i.d random variables on $[0,1)$, and n is the content of the queue (Fig.2a)

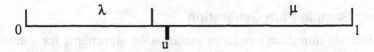

Fig. 2a Determination of Event Type on the Nominal Path

(iii)   update the state (n in this case), according to the results of (ii).

Steps (i - iii) constitute a simulation of the M/M/1 queue.

Now suppose we wish to simulate the same example with only a change in the arrival rate from λ to $\lambda+\Delta\lambda$. Conventional approach requires us to repeat steps (i - iii) with λ replaced by $\lambda+\Delta\lambda$. However, this is not necessary. We only need to rescale the standard clock sequence by $1/(\lambda+\Delta\lambda+\mu)$ and modify step (ii) by

(ii.p) determine   the   successive   event   times   on   this   standard   clock sequence   according to

an arrival event	if $u_i < \lambda+\Delta\lambda/(\lambda+\Delta\lambda+\mu)$,
a departure event	if $u_i \geq \mu/(\lambda+\Delta\lambda+\mu)$ and n>0,
ignore the event	otherwise,

where u_i, i=1, 2, . . . is the **same** sequence of u.i.i.d random variables on [0,1) as in step (ii) and n' is the content of the perturbed queue (Fig.2b)

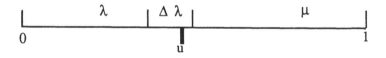

Fig. 2b Determination of Event Type on the Perturbed Path

Remark 1 Note that except for the requirement of a separate memory register to keep track of the perturbed state n', the modification of $\lambda+\mu$ to $\lambda+\Delta\lambda+\mu$ in any time scaling, and steps (ii) and (ii.p), the simulation of the nominal and the perturbed trajectories share everything else. There is a minimal duplication and the method is ideally suited to be carried out in parallel! In fact, this saving becomes even more dramatic as the number of different $\Delta\lambda$'s that we wish to explore increases. Fig. 3 shows a comparison of the relative CPU times of a series of parallel simulation experiments (number of different $\Delta\lambda$'s) using conventional brute force repetitions of simulations vs. the above standard clock approach. For the conventional approach we also used both an existing simulation language and a customized program which took maximal advantage of this simple example. The performance measure used is the mean first passage time of the queueing length exceeding K starting from an empty queue.

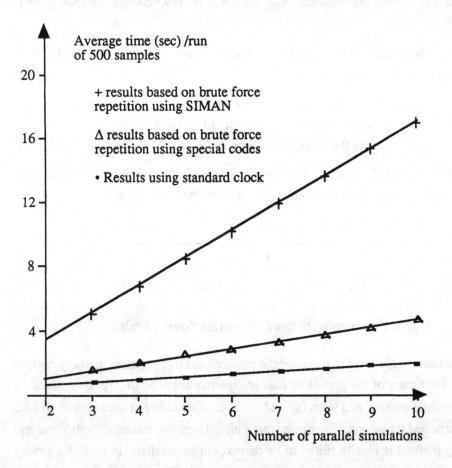

Fig. 3 Comparison of Brute Force vs. Standard Clock Approach to Simulation

Remark 2 Note also that instead of sharing segments of trajectories as in "cut-and-paste" we share computational efforts at the more detailed level of the random variable generation. Like the event list matching case in Chapter 6, we incur some waste when the state sequence becomes incompatible with the event list, i.e., a departure event is inconsistent with n or n' =0. Such event must be discarded.

Remark 3 Step (i), and to a lesser extent step (ii) or (ii.p), is completely independent of step (iii). Step (i) can be performed in nonreal time if required.

Remark 4 Note also in this approach the timing of events (i.e., the clock mechanism) are never perturbed even in thought as in IPA. Instead, we rely on the probabilistic event type determination and event list feasibility to provide proper identification and thinning of specific events.

The above illustration shows how to use standard clock to simulate structurally similar but parametrically different systems. It is also possible, within reason, to extend the approach to cover parametrically identical but structurally perturbed systems and other complex perturbations. For example,

(a)   adding the finite buffer constraints to the M/M/1 queue: This will only require a change in step (iii) on the updating of the state making n=buffer limit and arrival event incompatible. Both nominal and perturbed paths use the same step (iii). Similar statements applies to other variations involving

(b)   complex queueing disciplines

(c)   history dependent state transitions

Furthermore, for performance measures that are state dependent only, we even do not have to generate the standard clock. For example, consider the case in Fig. 3. We can eliminate step (i) completely and work with only the imbedded Markov chain and determine the mean number of transitions to queueing lengths exceeding K. To do this we only require steps (ii), (ii.p), and (iii). At the conclusion of the experiment we can multiply the result (the mean number of transitions) by $\lambda+\mu$ to get the mean first passage time back. This is because step (i) is independent of steps (ii - iii). This idea of eliminating the event times generation is known as "discrete time conversion" and in fact was first used by [Hordijk, Iglehart, and Schassaberger 1976].

It is instructive to compare the features of the standard clock vs. the brute force simulation approach to the study of parametrically different but structurally similar simulations.

BRUTE FORCE	STANDARD CLOCK
co-generation of event time and event type via a given distribution.	complete separation of event time and event type determination.
no sharing of anything in event time determination.	timing of events is shared among all parallel simulations.
no sharing of anything in event type determination.	one random number is used in all event-type determinations.
maintains a future event list.	standard clock replaces future event list and is a one time setup effort.
requires separate simulation experiment.	requires only separate memory for the state of each parallel simulation.
no wasted events.	some wasted event-type determinations due to state incompatibility, e.g., this method may not be advantageous during light traffic.
not easy to switch to discrete time conversion.	easy to switch to discrete time conversion.

The extension of and the variations on the above simple example to other Markov networks is more or less obvious. Again we use an example to illustrate this. Consider the multiclass network of Section 6.2 (Fig. 6.9 repeated below) and assume that we want to construct the sample paths of N different versions of this queueing network that only differ in their routing parameters.

For example, assume that all the routing parameters of these systems are equal except for $p_{13}^{(1)}$, and $p_{13}^{(1)} = p_i$ for the ith version of the system. Using a GSMP model for each system, we denote these models by $\Sigma_i = (X_i, \Gamma_i, P_i)$. Note that an obvious brute force way to construct the sample paths of these systems simultaneously is to consi-

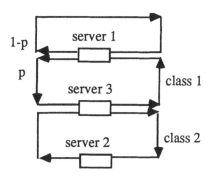

der the collection of the N GSMPs as a new GSMP, $\Sigma = (X, \Gamma, P)$, and simulate this new and much larger GSMP (the state space of this larger GSMP is the cross product of the state spaces of individual GSMPs and the set of possible events is the union of all such sets of individual GSMPs). In fact, this is the most inefficient way of simulating these systems simultaneously. Following the idea in the M/M/1 example, considerable simplification is possible because of the structural similarities of these GSMPs. At the ith system the set of events Γ_i is:

$\Gamma_i = \{ e_{1i} =$ the end of service at server 1 at system i, $e_{2i} =$ the end of service at server 2 at system i, $e_{31i} =$ the end of service of a class 1 customer at server 3 at system i, $e_{32i} =$ the end of service of a class 2 customer at server 3 at system i $\}$, $i = 1, \ldots, N$.

Since the event lifetimes of all the corresponding events happened at the same server at all the systems (e.g. all e_{1i} 's) have the same distribution we can in fact identify these events among different systems and consider one event (e.g.,

e_1) corresponding to all of them. This process defines a new GSMP, $\overline{\Sigma} = (\overline{X}, \overline{\Lambda}, \overline{P})$; the state space of this GSMP, Ξ, is the cross product of state spaces of Σ_i 's, X_i, and the set of events, Γ, is : $\Gamma = \{ e_1, e_2, e_{31}, e_{32} \}$. Note that for an event, e.g., e_1, to be active it suffices to have at least one customer at server 1 at one of the systems 1 to N. At the occurrence instant of this event we update the states of all the individual GSMPs that have a customer at server 1 according to their transition rules. We do not update the GSMPs that have an idle server 1. It is clear that by simulating Σ we avoid considerable duplication of simulation efforts compared to when these systems are simulated individually. The GSMP Σ can be formally described as the disjunction of the GSMPs Σ_i 's (see Section 7.2).

A further simplification of the above scheme is possible if we keep all the events in Γ active at all times or equivalently always simulate the maximal event list. This approach can be viewed as an application of the uniformization method [Keilson, 1979] in this context. Note that the occurrences of e_1, e_2, e_{31} and e_{32} can be viewed as arrivals of Poisson processes. Let $\lambda_1, \lambda_2, \lambda_3$ and λ_4 be the rates of these Poisson processes respectively. We can then simulate the superposition of these Poisson processes which is itself a Poisson process with rate $\lambda = \lambda_1 + \lambda_2 + \lambda_3 + \lambda_4$ (step (i) of the M/M/1 example in Section 7.1.1) and use an independent probabilistic device to decide separately which of the above four events occurred; e.g., the occurred event is e_{31} with probability λ_3 / λ. Similar to the above, upon occurrence of an event from the maximal event list, states of all the systems for which the occurrence of the occurred event is possible are updated and the others are ignored (steps (ii) and (iip)). This approach simplifies the updating and generation of the future event list. There is, of course, the possibility of extra effort in this case since it is possible that an event from the maximal event list would occur where no state would be updated.

Exercise Devise a standard clock simulation for an M/M/1 two-class network involving λ_1, λ_2, μ_1, and μ_2 using only **three** event types, e_1=C1 arrival, e_2=C2 arrival, and e_3=service completion. In what sense is this more efficient than the obvious use of **four** event types? (HINT: you may use additional u.i.i.d. random numbers to sub-divide e_3.)

Whereas the cut-and-paste method attempts to eliminate duplications at the trajectory level as was pointed out in Chapter 6 (i.e., use segments of the trajectory of either the nominal or the perturbed systems when the segments are deterministically similar or incrementally similar), this new approach attempts to eliminate duplications at the level of simulation effort (i.e., generation of random numbers, finding the next event type and event time, etc.). Furthermore, this approach is also similar to the Augmented Chain analysis to be discussed in Section 7.2. The GSMP framework and the identification of events chosen in our approach will simplify the explanations in Section 7.2.2.

Viewed in this light of efficient sample path generation for performance analysis purposes, conventional discrete event simulation of structurally similar systems appears to involve considerable duplications that can be eliminated. The ideas of infinitesimal perturbation analysis, cut-and-paste, standard clock, and the likelihood ratio approach to be discussed below are merely the beginning of what appear to be enormous opportunities for the development of parallel simulation techniques when combined with hardware advances in computer technology.

7.1.2 IPA via the Standard Clock

While the standard clock method is primarily a technique for finite perturbation analysis, one can nevertheless ask the question: "can it be used for the calculation of performance derivative estimates?" or "what is the relationship to IPA if $\Delta\lambda$ in Section 7.1.1 approaches zero?" Useful insight can be gained by considering a slight variation of the M/M/1 example above. Let us generate the standard clock in step (i) at a rate $\nu > \lambda + \mu$ and carry out step (ii) by

$$\text{event type} = \left\{ \begin{array}{ll} \text{arrival} & \text{w.p. } \lambda/\nu \\ \text{departure} & \text{w.p. } \mu/\nu \\ \text{ignore} & \text{w.p. } 1-(\lambda+\mu)/\nu \end{array} \right\} \qquad (1)$$

It is clear that except for some wasted event time generation for the "ignore" cases, a legitimate M/M/1 trajectory can be generated using the above steps (i - ii) in Section 7.1.1 and Eq.(1). A typical sample path may look like that of Fig. 4a. Now **imagine** we perturb λ by $\Delta\lambda\rightarrow 0$ and use Eq.(1) to generate the perturbed path. The only difference in this case is that some "arrival" event in Fig. 4a will now be re-classified as "ignore" events (assuming $\Delta\lambda\rightarrow 0$).

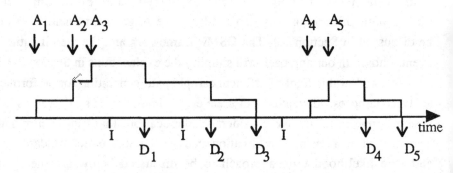

Fig. 4a Standard Clock Generated Sample Path for an M/M/1 Queue

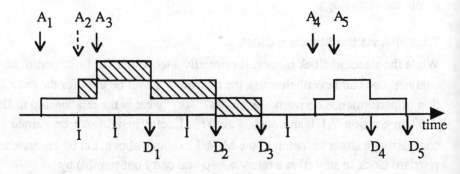

Fig. 4b Effect of changing an Arrival event to Idle

Typically, for example, in Fig.4b, the second arrival A_2 may become an "ignored" event, I. The effect of changing A_2 to I on the system performance, the response time of customers, is represented by the shaded area in Fig. 4b, or the amount $(D_3 - A_2)$. Such a change occurs with probability $\Delta\lambda/\lambda$. Now summing over all possible changes of arrival events to "ignored" events over the busy period, we get

$$\Delta WT\big|_{\text{ith b.p.}} = \frac{\Delta\lambda}{\lambda}[(D_3\text{-}A_1)+(D_3\text{-}A_2)+(D_3\text{-}A_3)]. \tag{2}$$

Eq.(2) is interesting if we reverse time. Then $(D_3\text{-}A_3)$ is recognized as the service time of the first arriving customer in reverse time, i.e., s_1. Similarly, $(D_3\text{-}A_2) = s_1 + s_2$, and so forth. The result is that Eq.(2) can be rewritten as

$$\Delta WT\big|_{\text{ith b.p.}} = \Delta\lambda\sum_{i=1}^{3}\sum_{j=1}^{i}\frac{s_j}{\lambda}. \tag{3}$$

Summing over M busy periods and dividing by $\Delta\lambda$, we finally get

$$\frac{\Delta WT}{\Delta\lambda} = \frac{1}{N}\sum_{n_m=1}^{M}\sum_{i=1}^{n_m}\sum_{j=1}^{i}\frac{s_j}{\lambda},$$

which is recognized as the valid formula for the derivative estimate of the mean response time with respect to arrival rate for an M/M/1 queue (Sections 3.4 and 4.4). Since M/M/1 queues are known to be reversible [Kelly 1979][1], we have thus established the correctness of this approach (Eq. (2)) to derivative estimation. The interesting point to note here is that this standard clock approach to derivative estimation is in fact the reverse of the techniques of IPA extension discussed in Chapter 5. Instead of smoothing out infrequent finite perturbations into frequent infinitesimal ones, we convert frequent infinitesimal

---

[1] An M/M/1 queue is reversible in the following sense: A movie is shot tracking the arrivals and departures of this queue. We then run the movie backward where arrival becomes departure and vice versa; service distribution becomes that of inter-arrival. Statistically, we cannot tell the difference whether we are watching the movie in normal or reverse time.

perturbations in the timing of events (the usual IPA procedure) into infrequent finite perturbations in event types with the attendant finite perturbations in system performance. However, these finite perturbations are smoothed out by the probability of these perturbations, a kind of smoothed PA discussed in Section 5.4. For more details see [Vakili 1990a]. Also worth repeating is the earlier Remark 4 where we note that in standard clock there are no event timing perturbations but only event type perturbations.

7.1.3 The Alias Method of Choosing Event Types

One efficient way of obtaining a random variable distributed over the integers 1,2,...,n with probabilities $p(i)$, i=1,2,...,n, is the alias method [Bratley et al. 1987]. This method can be used to further reduce the computation effort of the standard clock simulation approach in determining the event type at every transition instant (i.e., step (ii) in Section 7.1.1). The method requires only one uniformly distributed variable, one comparison, and at most two memory references per sample. It is thus **independent** of the size of the possible event list, an important advantage in the simulation of large systems via the standard clock approach. However, this method requires pre-computing two tables of length n, which is a one-time effort.

The alias method uses two tables, $R(i)$ and $A(i)$, i=1,2,...,n , $0 \le R(i) \le 1$, and $A(i)$ is a mapping from the set $\{1,2,...,n\}$ to itself. The description of the algorithm generating the random variable with distribution $p(i)$ below essentially follows [Bratley et al. 1987].

(1)   Generate a uniformly distributed random variable $u \in [0, 1)$.

(2)   Let v=nu (v is uniform on [0, n)).

(3)   Set $I = [v]$; I is the smallest integer which is bigger than v. (I is uniform on integers 1,2,..,n.)

(4)   Set w=I-v (note that w is uniform on [0, 1) and independent of I).

(5)   If $w \le R(I)$, then output e = I; otherwise, output e= A(I).

In the algorithm, we first generate a uniformly distributed integer I on 1,2,...,n; then we adjust the probabilities by replacing the number $I \in \{1,2,...,n\}$ by its "alias" A(I), with a certain "aliasing" probability 1-R(I). If we choose the aliases and the aliasing probabilities properly, then the random variable generated by this algorithm, e, has the desired distribution. From the algorithm, we have

$$P[w \leq R(I), I = i] = \frac{R(i)}{n},$$

which says that the probability of getting "i" without aliasing is smaller than 1/n. On the other hand,

$$P[w > R(I), I=j] = \frac{1-R(j)}{n}.$$

Summing the probabilities of the mutually exclusive ways to get e=i, we obtain

$$P[e = i] = \frac{R(i)}{n} + \sum_{\{j : A(j)=i\}} \frac{1-R(j)}{n},$$

which provides a means of increasing the probability of getting "i" to above 1/n. Thus, if we choose A(i) and R(i), i=1,2,...,n, such that the above quantity equals p(i), then the random variable e has distribution p(i), i=1,2,...,n.   it is worthwhile to note that the above relation does not uniquely specify the values of A(i) and R(i).   There may be many tables which can be used as the aliases and aliasing probabilities. The following is an algorithm generating a proper set of A(i) and R(i), i=1,2,...,n.

(1) Set $H = \emptyset$, and $L = \emptyset$. ($\emptyset$ is the null set.)

(2) For i=1 to n:
     (a) set R(i) = np(i);
     (b) if R(i) > 1, then add i to the set H;
     (c) if R(i) < 1, then add i to the set L.

(3) (a) if $H = \emptyset$, stop;
     (b) otherwise select an index j from L and an index k from H.

(4) (a) Set A(j) = k; (A(j) and R(j) are now finalized.)

  (b) Set R(k) = R(k) + R(j) -1;

  (c) if R(k) ≤ 1, remove k from H;

  (d) if R(k) < 1, add k to L;

  (e) remove j from L.

(5) Go to step 3.

This algorithm runs in $O(n)$ time because at each iteration the number of indices in the union of H and L goes down by at least one. Also, the total "excess" probability in H always equals the total "shortage" in L; i.e.,

$$\sum_{i \in H} [R(i) - 1] = \sum_{i \in L} [1 - R(i)]$$

This shows that in step 2(a) if H = ∅, then L = ∅. At the last iteration, there is just one element in each of H and L. Hence R(j) -1 = 1- R(k) at step 3(b) and at step 4(b) R(k) = 1; this leaves both H and L empty. The proof of the algorithm is left as an exercise.

Exercise Work through the Alias algorithm for the case of p(1)=.1, p(2)=.4, and p(3)=.5

Exercise Prove that in the algorithm generating A(i) and R(i), we have

  (a) If np(i) < 1, then $\{j : A(j) = i \} = \emptyset$.

  (b) If np(i) > 1, then $R(i) + \sum_{\{j : A(j) = i\}} [1 - R(j)] = np(i).$

7.2 Trajectory Projection and State Augmentation

The purpose of this section is to present two other approaches to the generation of DEDS trajectories (sample paths) which are closely related to the cut-and-paste method discussed in Chapter 6. They are the trajectory projection [Cao 1989e] and the state augmentation [Cassandras and Strickland 1989b,c] methods. The trajectory projection approach uses an automaton model of a discrete event dynamic system and will provide a connection to the Wonham-Ramadge model discussed in Appendix D. We use this approach to formulate the problem of constructing a sample path of the perturbed system as a problem of trajectory projection. The interesting feature of this approach is that it sets up some relationship between the two recently developed areas in the discrete event system theory, the automaton model and perturbation analysis. The state augmentation method studies the stochastic feature of both the nominal and perturbed paths of a Markov process and provides a mapping of state transitions which leads to an augmented Markov process, from which the perturbed path can be predicted. It is a different view of "cut and paste."

7.2.1 Trajectory Projection of the Automata Model

In Chapter 6, we presented a way to formalize the description of the cut-and-paste method as a generalization of the rejection method for random number generation. The material in this section can be viewed as a conceptual description of "cut-and-paste" via the automaton model of Ramadge and Wonham. At the same time, this section serves as an additional introduction to the notations and the descriptive power of the logical models. Recall from Appendix D that a discrete event dynamic system (DEDS) in this approach is described by a generator which generates a language. A sample path is a sequence of events represented by a string of symbols in the language. This approach does not consider the time elapsed between two events (i.e., it provides an untimed model), nor does the approach emphasize the stochastic nature of the system. However, if we wish to visualize the trajectories as that of a Markov process, then we can take the viewpoint that the average times between any two events are given or can be determined for such processes and thus we can derive most average performance measures from a knowledge of the event sequences plus the average inter-event times. (See also the discussion with the state augmentation method later in this section.) For readers who are not interested in these various connections, this subsection can be skipped.

Let us first review some notations in the automaton and formal language theory. An event of a DEDS is represented by a symbol. An *alphabet* Σ is a finite non-empty set of symbols, which represents a set of possible events. A *string* over Σ is a finite sequence of the elements of Σ. Thus, a string represents a sequence of events of a DEDS. Since the state transitions are determined by the events which trigger the transitions, a string may be, under some feasible conditions which will be made clear later, translated into a sequence of states for a given initial state. The string consisting of zero symbols is denoted by 1. The set of all finite strings over Σ is denoted by Σ^*. A juxtaposition of any two strings s and t, denoted as st, is also a string. In a DEDS, the trajectory repre-sented by the string st can be obtained by pasting the trajectory represented by t to the end of that represented by s. A string s is a *substring* of a string t if there are strings u and v such that t=usv. If u=1, then s is called a *prefix* of t; if v=1, then s is called a *suffix* of t. For convenience, we also say that t is a prefix or a suffix of itself.

A *language* over an alphabet Σ is any subset of Σ^*. That is, a set of event sequences forms a language. The *closure* of a language $L \subseteq \Sigma^*$ is defined as

$$\overline{L} := \{\ s: s \in \Sigma^* \wedge (\exists\ t \in \Sigma^*)\ st \in L\}$$

This means that if $t \in L$, then all the prefixes of t belong to $\overline{L}$. (In words, this reads that $\overline{L}$ consists of all strings, s, of Σ^* and for which there exists string $t \in \Sigma^*$ such that $st \in L$.)[2]. A language is called a *closed* one if it is the same as its closure. A closed language contains all the prefixes of the strings in the language. If a trajectory is to be represented by a string of events, then clearly the string and its prefixes must belong to a closed language which contains all the trajectories.

The concept of a language describes only the sequences of events. Another important element of a DEDS is the state. Let Σ be the set of all events and Q be the set of all states. The state transitions are triggered by events and

---

[2] We use the symbol "$\wedge$" to mean "AND" and "$\vee$" for "OR."

can be determined by transition functions which are mappings $\delta: \Sigma \times Q \rightarrow Q^3$. The meaning of the transition functions is this: If a DEDS is at state $q \in Q$, then an event $\sigma \in \Sigma$ triggers a transition from state q to state $\delta(\sigma,q)$. Therefore, a DEDS can be modeled as an *automaton* or a *generator* defined as a 4-tuple:

$$G = (\Sigma, Q, \delta, q_0)$$

where Σ is the set of all events (the alphabet of the automaton), Q is the set of *states*, $\delta: \Sigma \times Q \rightarrow Q$ is the *transition function*, and q_0 is the *initial state*. In general, $\delta(\sigma,q)$ is a partial function. (i.e., for any q, the domain of σ is a subset $\Sigma_\delta(q) \subseteq \Sigma$.) The transition function is also extended to $\delta : \Sigma^* \times Q \rightarrow Q$ according to

$$\delta(1, q) = q$$

and

$$\delta(s\sigma, q) = \delta(\sigma, \delta(s, q))$$

whenever $q'=\delta(s,q)$ is defined and $\sigma \in \Sigma_\delta(q')$.

To study the predictability of a DEDS, we need to extend the transition function to $\Sigma - \Sigma_\delta(q)$. An event belongs to $\Sigma - \Sigma_\delta(q)$ if and only if it belongs to Σ but does not belong to $\Sigma_\delta(q)$. We define

$$\delta(\sigma,q) = q, \quad \text{if } \sigma \in \Sigma - \Sigma_\delta(q).$$

Thus, the extended transition function is defined on set Σ. To distinguish the original domain $\Sigma_\delta(q)$ and the extended domain Σ, we say that $\delta(\sigma,q)$ is *properly defined* if $\sigma \in \Sigma_\delta(q)$. Note that $\delta(\sigma,q)=q$ may also hold for some $\sigma \in \Sigma_\delta(q)$. Furthermore, for any $s \in \Sigma^*$, we say that $\delta(s, q_0)$ is properly defined if for any prefix u, suffix v, and event σ satisfying $s=u\sigma v$, $\delta(u,q_0)$ is properly defined and $\sigma \in \Sigma_\delta(q')$, $q'= \delta(u,q_0)$. We use $\delta(s, q_0)!$ to denote the fact that

---

[3] This deterministic state transition function can be made to be equivalent to the stochastic state transition function of a GSMP, ψ, in Section 3.3 or $p(x'; x, \iota^*)$ in Appendix B.4. by

$\delta(s, q_0)$ is properly defined. From the definition, if $\delta(s, q_0)$ is properly defined, then the string s corresponds to a trajectory of the DEDS with the initial state q_0; on the other hand, if $\delta(s, q_0)$ is not properly defined, then there is a prefix u of s such that $\delta(u, q_0)$ is properly defined bu t $\sigma \notin \Sigma_\delta(q')$, $q' = \delta(u, q_0)$, meaning event σ cannot occur at state $q' = \delta(u, q_0)$. Such a string s cannot be translated into a trajectory of the DEDS.

With these definitions, the language *generated* by G is defined as

$$L(G) := \{ s : s \in \Sigma^* \wedge \delta(\sigma, q_0) \, ! \}$$

$L(G)$ contains all the strings $s \in \Sigma^*$ such that $\delta(s, q_0)$ is properly defined or, equivalently, $L(G)$ contains all the strings that correspond to trajectories of the DEDS. For this reason, we also call $s \in L(G)$ a trajectory of the DEDS, G. By the definition, $L(G)$ is a closed language. Note that because of the extension of the transition function $\delta(s, q)$ to set $\Sigma - \Sigma_\delta(q)$, for any $s \in \Sigma^*$, $\delta(s, q_0)$ always has a value. However, $\delta(s, q_0)$ may not be properly defined since there may exist some u, σ, and v, satisfying $s = u\sigma v$, for which $\sigma \notin \Sigma_\delta(q')$, $q' = \delta(u, q_0)$.

Let $\Pi \subseteq \Sigma^*$ be a closed language. A projection from Σ^* on Π is defined by $P_\Pi(1) = 1$, and for any string $s \in \Sigma^*$ and any event $\sigma \in \Sigma$,

$$P_\Pi(s\sigma) = \begin{cases} P_\Pi(s)\sigma & \text{if } P_\Pi(s)\sigma \in \Pi, \\ P_\Pi(s) & \text{if } P_\Pi(s)\sigma \notin \Pi. \end{cases}$$

Roughly speaking, the projection of s on Π is a string in Π that can be constructed from s by sequentially removing the symbols which make the prefixes of s outside of Π. For instance, let $\Pi = \{1, \alpha, \alpha\alpha, \alpha\alpha\beta, \alpha\gamma, \alpha\gamma\alpha, \alpha\gamma\alpha\beta\}$ and $s = \alpha\beta\alpha\gamma\beta$. Then $P_\Pi(s) = \alpha\alpha\beta$. In the definition, Π must be closed. Otherwise, we may have $P_\Pi(s) \neq s$, even for $s \in \Pi$. For instance, if $\Pi = \{1, \alpha,$

---

redefining events and enlarging the event set, Γ. Similarly, the definition of $\Sigma_\delta(q)$ below is the same as $\Gamma(x)$, the feasible event set.

$\alpha\alpha\beta$} and s=$\alpha\alpha\beta$, then according to the definition of projection, $P_\Pi(\alpha)=\alpha$, $P_\Pi(\alpha\alpha)=\alpha$, $P_\Pi(s) =P_\Pi(\alpha\alpha\beta) =\alpha \neq s$.

Consider a DEDS G=(Σ, Q, δ, q_0). The operation of projection onto L(G) is denoted by $P_{L(G)}$. For s $\in \Sigma^*$, let s'= $P_{L(G)}(s)$ be the projection of s on L(G). By the definitions of projection and language L(G), we have

$$\delta(s,q_0) = \delta(s', q_0)$$

Note that $\delta(s', q_0)$ is properly defined, while $\delta(s, q_0)$ may not be. Thus, projecting a string s onto L(G) is simply removing some events from s so that the remaining string s' corresponds to a trajectory of the DEDS G under the transition functions δ. In other words, s' is the biggest sub-sequence of s which belongs to L(G).

Now, let us consider two DEDSs: G = (Σ, Q, δ, q_0) and H = (Σ, R, χ, r_0). Note that the sets of events are the same for both systems. In case they have different event sets Σ_1 and Σ_2, we can choose $\Sigma=\Sigma_1 \cup \Sigma_2$. We now define the predictability of a string and a language.

Definition 7.2.1.

a. A string t $\in$ L(H) is said to be *predictable* from a string s $\in$ L(G) if and only if t=$P_{L(H)}(s)$.

b. A language K $\subseteq$L(H) is said to be *predictable* from L(G), denoted by L(G)$\rightarrow$K, if and only if $(\forall$ t$\in$ K) $(\exists$ s$\in$ L(G))t = $P_{L(H)}(s)$. In words, this expression reads "for all t$\in$ K, there exists an s$\in$ L(G) such that t = $P_{L(H)}(s)$."

For any language L and any closed language Π, we define

$$P_\Pi(L) = \{t : (t\in \Pi)\wedge (\exists s\in L)t = P_\Pi(s)\}.$$

$P_\Pi(L)$ is called the *projection* of L on Π. With this notation, it is easy to see that for any K $\subseteq$L(H), L(G)$\rightarrow$K if and only if K$\subseteq P_{L(H)}(L(G))$.

The intuition of the definitions is the following: In a DEDS, if $t \in L(H)$, $s \in L(G)$, and $t = P_{L(H)}(s)$, then for any prefix u of $s \in L(G)$, we can check whether the transition function $\chi(u, r_0)$ of H is properly defined, and then we can determine the string $t \in L(H)$. In other words, t can be constructed from s, or equivalently, a trajectory of the discrete event system H can be obtained by removing all the nonfeasible events in a trajectory of the discrete event system G. Projection in this sense is just the same as the cut-and-paste idea presented in the last chapter.

Next, we shall introduce the disjunction of two DEDSs. This concept will also be used in the state augmentation method discussed in the next section.

Definition 7.2.2. The *disjunction* of the two DEDS G and H is defined by $H \vee G = (\Sigma, Q \times R, \delta \vee \chi, (q_0, r_0))$, where $Q \times R$ is the set of states of the disjunction DEDS, (q_0, r_0) is the initial state of the DEDS, and $\delta \vee \chi$ is the transition function of $H \vee G$, which is the *disjunction* of the two transition functions δ and χ defined as follows:

$$(\delta \vee \chi)(\sigma, (q, r)) = \begin{cases} (\delta(\sigma, q), \chi(\sigma, r)) & \text{if } \sigma \in \Sigma_\delta(q) \wedge \sigma \in \Sigma_\chi(r), \\ (\delta(\sigma, q), r) & \text{if } \sigma \in \Sigma_\delta(q) \wedge \sigma \notin \Sigma_\chi(r), \\ (q, \chi(\sigma, r)) & \text{if } \sigma \notin \Sigma_\delta(q) \wedge \sigma \in \Sigma_\chi(r), \\ \text{undefined} & \text{if } \sigma \notin \Sigma_\delta(q) \wedge \sigma \notin \Sigma_\chi(r). \end{cases}$$

By the extension of the transition functions δ and χ, this equation can be rewritten as

$$(\delta \vee \chi)(\sigma, (q, r)) = \begin{cases} (\delta(\sigma, q), \chi(\sigma, r)) & \text{if } \sigma \in \Sigma_\delta(q) \vee \sigma \in \Sigma_\chi(r), \\ \text{undefined} & \text{if } \sigma \notin \Sigma_\delta(q) \wedge \sigma \notin \Sigma_\chi(r). \end{cases}$$

The domain of $(\delta \vee \chi)$ is $s \in (\Sigma_\delta(q) \cup \Sigma_\chi(r))$. As mentioned above, we extend the definition of $(\delta \vee \chi)$ by defining $(\delta \vee \chi)(s, (q, r)) = (q, r)$ for $s \notin (\Sigma_\delta(q) \cup \Sigma_\chi(r))$. The definition implies that $(\delta \vee \chi)$ is properly defined if and only if either δ or χ

is properly defined. $(\delta \vee \chi)$ can also be defined for any string $s \in \Sigma^*$. The language generated by $H \vee G$, $L(H \vee G)$, is called the *disjunction* of the two languages generated by H and G, $L(H)$ and $L(G)$. As we can see, the state space of the disjunction is the product of the two state spaces of the two DEDS; each trajectory of the disjunction is a shuffle of two trajectories of the two DEDS. (For any two strings $s_1 = \alpha_1\alpha_2,...,\alpha_n$ and $s_2 = \beta_1\beta_2,...,\beta_n$, a string consisting of all α_i, i=1,2,...,n and β_j, j=1,2,...,m, in any possible order is called a shuffle of the two strings s_1 and s_2.). In short, we can think of the disjunction of two DEDSs as the natural combination of two DEDS running autonomously. For a simple example see Example 7.2.1 of the next section.

If $s \in L(G)$, then $\delta(s,q)$ is properly defined. By the definition of the disjunction of two transition functions, this implies that $(\delta \vee \chi)(s,(q,r))$ is properly defined. Therefore, $s \in L(H \vee G)$. Thus, we have

$$L(G) \subseteq L(H \vee G).$$

Similarly,

$$L(H) \subseteq L(H \vee G).$$

Now, we can state a sufficient condition for the predictability as follows:

If $P_{L(H)}(u) = P_{L(H)}P_{L(G)}(u)$ holds for any $u \in L(H \vee G)$, then $L(G) \rightarrow L(H)$.

The proof of this condition can be found in [Cao 1989e]. The condition can be intuitively explained as follows. For any string $s \in \Sigma^*$, $P_{L(H)}(s)$ can be viewed as all the information about s contained in language $L(H)$. The sufficient condition claims that if the information contained in $L(H)$ can be obtained from that contained in $L(G)$, then $L(H)$ is predictable from $L(G)$, an intuitively reasonable statement. However, the converse of this statement is not true. This can be shown by a counter example. Let $\Sigma = \{\alpha,\beta,\gamma\}$, $L(G) = \{1,\alpha,\alpha\beta,\alpha\gamma\}$, and $L(H) = \{1,\alpha,\alpha\gamma\}$. We have $P_{L(H)}(\alpha) = \alpha$, $P_{L(H)}(\alpha\gamma) = \alpha\gamma$. Thus, $L(G) \rightarrow L(H)$.

However, for $s=\alpha\beta\gamma$, we have $P_{L(H)}(s)=\alpha\gamma$, and $P_{L(H)}P_{L(G)}(s)$ $=P_{L(H)}\{P_{L(G)}(s)\}$ $=P_{L(H)}(\alpha\beta)=\alpha$. This implies that $P_{L(H)}\neq P_{L(H)}P_{L(G)}$.

As pointed out in Chapter 1, the primary purpose of the formal language / finite state machine approach above is not quantitative performance analysis but the construction of a formal mathematical machinery to state or to prove precisely various qualitative properties of DEDS. Its starting point is the event strings. Together with some description of the "rules of operations" of the system, a language (sets of event strings) is created to meet system specifications. Conceptually, the formal language model is most closely related to the generalized semi-Markov scheme (GSMS), i.e., the GSMP without the specification of the event lifetimes. However, in a GSMS (or a GSMP) the feasible event list, $\Gamma(s)$ for all s, together with the state transition function, is used to summarize the operational logic of a DEDS. The event string is a derived object. This does not present much of a problem in simple examples for either approach. In general, however, the step from a high level verbal description to finding a $\Gamma(s)$ that generates the required behavior becomes a problem to be solved. In fact, the so-called supervisor synthesis problem of the formal language approach can be viewed as the construction of $\Gamma(s)$. Qualitative concepts, long familiar in control and systems theory, such as controllability, observability, and stability, then arise naturally in similar fashion for DEDS. For some recent work along these lines beyond what was illustrated in Appendix D, see [Wonham 1989, Ozveren and Willsky 1991]. An interesting connection along these lines between the qualitative performance analysis of formal language models and Petri nets, and the commuting conditions (CU) of Chapter 5 were studied by [Glasserman and Yao 1991]. Since the commuting condition basically says that the state reached is independent of the **order** of occurrence of the event, its satisfaction implies certain qualitative performance property of the DEDS.

7.2.2 The State Augmentation Method for Markov Chains

The state augmentation method was proposed in [Cassandras and Strickland 1989b,c]. The basic idea is that, based on a nominal sample path of a Markov process, a sample path of an augmented Markov process (similar to the disjunction of the two DEDS discussed in the last subsection) can be constructed. The state space of the augmented Markov process can be partitioned into subsets in such a way that each state of the perturbed Markov process corresponds to one of the subsets, and the steady-state probability of the state of the perturbed process is proportional to that of the corresponding subset.

We use a simple example to explain the basic idea of the method. Let G_1 be an M/M/1/1 queue and G_2 be an M/M/1/2 queue. Our purpose is to predict the performance of G_2 by using a sample path of G_1. The states of the M/M/1/1 queue can be denoted as $x_{1,0}=0$ and $x_{1,1}=1$, those of the M/M/1/2 queue $x_{2,0}=0$, $x_{2,1}=1$, and $x_{2,2}=2$. The event sets are $\Sigma_1 = \Sigma_2 = \{\alpha, \beta\}$ for both processes with α denoting an arrival and β a departure. As the first step of the state augmentation method, we construct an augmented Markov process G of the two processes G_1 and G_2. The states of the Markov process G are (0, 0), (0, 1), (0, 2), (1, 0), (1, 1), and (1, 2). The state transition diagrams for G_1, G_2, and G are shown in Fig. 5. The first (second) number in a state of G corresponds to a state of G_1 (G_2). The event set of G is the same as those of G_1 and G_2. An event is feasible for state (x_1, x_2), if it is feasible for either x_1 or x_2. The augmented Markov chain G thus defined is, in fact, the disjunction of the two processes G_1 and G_2. The feasible event sets for every state are represented by the arrows going out the particular state in the diagrams. The figure indicates (by dashed circles) that the states (1, 0) and (0, 2) are transient states, since they cannot be reached from other states.

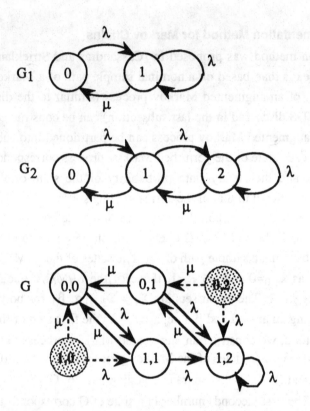

Fig. 5 State Transition Diagrams of G1, G2, and G

An interesting and easily verified feature of the augmented Markov process G is: if we partition its state space into $c_0 = \{(0, 0), (1, 0)\}$, $c_1 = \{(0, 1), (1,1)\}$, and $c_2 = \{(0, 2), (1, 2)\}$ (i.e., partition the the state space of G vertically), then the steady-state probabilities of these three subsets are just the same as those of $x_{2,0}$, $x_{2,1}$, and $x_{2,2}$ in the Markov process G_2, i.e., $P(c_i) = P(x_{2,i})$, $i = 0,1,2$. Therefore, if we have an **ergodic** sample path of G^4, we can

---

[4] In an ergodic process the time average of a sample path equals with probability one the ensemble average in steady state; such a sample path is called an ergodic sample path. Roughly speaking, an ergodic sample path exhibits all possible behavior of the process with the appro-

obtain the steady-state probability of G_2 and predict its performance. Given a sample path of G_1, i.e., given an event sequence of process G_1, we can construct a sequence of states for process G by following the transition rules specified in Fig. 5. Unfortunately, from a sample path of the M/M/1/1 system G_1, it is not possible to directly construct an ergodic sample path of G because some events that are feasible for a state (x_1, x_2) in process G may not be feasible for the corresponding state x_1 in process G_1. For example, event β (a departure) is feasible for (0, 1) in process G but is not feasible for state 0 in G_1. This means that, based on the event sequence of a sample path of G_1, we can never construct a sample path of G which contains a transition from state (0, 1) to (0, 0). Therefore, the sample paths obtained from the directly construction of the sample paths of G_1 are not ergodic and cannot be used to predict the performance of G_2. Note that (0, 1) is the only state that has a feasible event which is not feasible for the corresponding state (0 in this case) in G_1 in this example.

Now consider a sample path of G. Let us start from the initial state and follow the path to its first entry to state (0, 1). We know that before the sample path enters (0,1), it can be constructed from a corresponding sample path of G_1 without worrying about the ergodicity. However, once the path enters (0, 1), the next event may not be feasible in the sample path of G_1. This may cause a problem in constructing a representative sample path for G using a sample path of G_1. To overcome this difficulty, we start to cut the sample path of G when G enters (0, 1) until G reaches one of the other three states (0, 0), (1, 1), and (1, 2). Once G reaches one of these states, the sample path can again be constructed from that of G_1. Thus, by cutting the sample path, we can obtain segments of the sample path of G. Each of them can be constructed from a sample path of G_1 without the trouble caused by the undesirable state (0, 1).

Recall that our purpose is to predict the performance of G_2. Since the process G is cut when its state is at (0, 1), at the end of each segment after

---

priate frequency. However, there may exist some sample paths with probability zero on which the time averages do not equal the ensemble average.

cutting the corresponding state of G_2 is always 1. Therefore, we have to wait until the sample path to reach state $(1, 1)$ to guarantee that the next segment starts with state 1 for G_2. Now, since the cut part of the sample path of G is of no interest to the prediction of the performance of G_2, when constructing a sample of G from that of G_1, we may force the sample path of G to move to $(1,1)$ immediately after it reaches $(0,1)$. This is done by changing the transition rule of G at state $(0, 1)$ from the one specified in Fig. 6a to the one in Fig. 6b. This is called "transformation of the state transition."

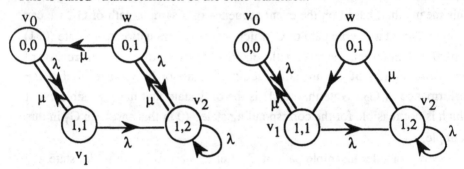

Fig. 6a State Transitions of G   **Fig. 6b** State Transitions After Transformation

Finally, of course, it is necessary to prove that the sample path obtained this way does provide the steady-state probability of G_2. This can be formally discussed in a general setting [Cassandras and Strickland 1989c]. Now we can summarize the procedure of obtaining the steady-state probability of G_2 from a sample path of G_1 as follows:

(i) Define the augmented Markov process with state space $\{v_0, v_1, v_2, w\}$, in which $v_0 = (0, 0)$, $v_1 = (1, 1)$, $v_2 = (1, 2)$, and $w = (0, 1)$.

(ii) Define the state transition function for G: $\delta(\alpha, v_0) = v_1$, $\delta(\alpha, v_1) = v_2$, $\delta(\alpha, v_2) = v_2$, $\delta(\alpha, w) = v_1$, $\delta(\beta, v_1) = v_0$, and $\delta(\beta, v_2) = w$. (Recall α is an arrival, β a departure, and $\delta(\alpha, w) = v_1$ is the transition function after the transformation.)

(iii)   Choose an initial state of G_2. Based on a sample path of G_1, apply the transition function defined in step ii to construct a sample path of G.

(iv)   The steady-state probability of G_2 is then $P(x_2 = i) = P(v_i)/\{\Sigma_i \, P(v_i \,)\}$, $i=0,1,2$.

In this sense, the performance of G_2 can be obtained from a trajectory of G_1. The augmented chain approach is also in the spirit of the standard clock since "cut-and-paste" is done using pieces of the underlying clock mechanism and no event timing perturbation is contemplated.

7.3 The Likelihood Ratio Method

In the introduction to Chapter 6 we have already alluded to another approach to the single-run gradient estimation technique known as the likelihood ratio method. The earliest discussion of the method apparently is due to [Aleksandrov et al. 1968] but independently re-discovered by [Glynn 1987, Rubinstein 1986b, and Reiman and Weiss 1989b]. Let us motivate this by way of a simple example.

Example Consider the exponentially distributed random variable x with density function $f(x;\theta)=(1/\theta)\exp(-x/\theta)$. Pretend for the moment that we cannot compute analytically the quantity $dE[x]/d\theta$ and must use a simulation experiment. Thus,

$$\frac{dE[x]}{d\theta} \equiv \frac{d}{d\theta} \int_0^\infty xf(x;\theta)dx \approx \frac{d}{d\theta}\{\frac{1}{N}\sum_{n=1}^N x_n\},$$

where x_n, $n=1,2,...,N$ are N independent realizations of x. Assuming the interchangeability of integration with differentiation (see Section 7.3.4 and Eq. (21) below for a discussion), we get

$$\frac{dE[x]}{d\theta} = \int_0^\infty xf'(x;\theta)\frac{f(x;\theta)}{f(x;\theta)}dx = \int_0^\infty x\frac{d\ln f(x;\theta)}{d\theta}f(x;\theta)dx$$

$$= E\{x\frac{d}{d\theta}[(-\ln\theta)-\frac{x}{\theta}]\} = -1 + \frac{2\theta^2}{\theta^2} = 1,$$

which agrees with the known answer. This result says that we can obtain an estimate of the derivative of $E[x]$ by taking a sample of x at $f(x;\theta)$ and re-scaling it by the factor $d\ln f(x;\theta)/d\theta$ instead of by the usual method of using another sample at $f(x;\theta+\Delta\theta)$ and differencing That is, we have

$$\frac{dE[x]}{d\theta} \approx \frac{1}{N}\sum_{n=1}^N \{x_n\frac{d\ln[f(x_n,\theta)]}{d\theta}\}.$$

The scale factor can be considered as the limit of

$$\frac{1}{\Delta\theta}\{E_{\theta+\Delta\theta}[x]-E_\theta[x]\} = \int_0^\infty x\{\frac{1}{\Delta\theta}\frac{f(x;\theta+\Delta\theta)-f(x;\theta)}{f(x;\theta)}\}f(x;\theta)dx$$

as $\Delta\theta\to0$ (see Section 7.3.2 below) and the ratio $f(x;\theta+\Delta\theta)/f(x;\theta)$ can be explained as the answer to the question *"what is the likelihood that a sample x from f(x;θ) is in fact generated by f(x;θ+Δθ)?"* Now if instead of the derivative of $E[x]$ we wanted to compute that of $E[L(x)]$, the same argument applies. We merely have to change x to $L(x)$ in the above development. The scale factor times $L(x)$ is then denoted as the likelihood ratio (LR) estimate of $dE[L(x)]/d\theta$. In the below subsections we shall apply this basic idea to the sample path of general DEDS and analyze some of the basic properties of LR estimates.

7.3.1 A Markov Chain Example and Variance Reduction

A good insight into the LR method can be obtained by considering its application to a Markov chain model. Consider a sample path represented by the state sequence $s_0, s_1, s_2, \ldots$ and the sample performance measure

$$L(\xi) = \frac{1}{N+1} \sum_{i=0}^{N} F(s_i) ==> J(\theta) \equiv E[L(\xi)] = \int \{ \frac{1}{N+1} \sum_{i=0}^{N} F(s_i) \} P(\xi) d\xi, \quad (4)$$

where
$$P(\xi) \equiv p(s_0) \prod_{i=0}^{N-1} p(s_i, s_{i+1}),$$

and we have assumed that $P(\xi)$ depends on a parameter "θ." A similar manipulation, as in the previous example via differentiation, interchanging the operators "$\int$" and "$d/d\theta$," yields

$$\frac{dJ(\theta)}{d\theta} = \int L(\xi) \frac{\partial \ln P(\xi)}{\partial \theta} P(\xi) d\xi,$$

where the term $\partial \ln P(\xi)/\partial \theta$ can be written as

$$\frac{p'(s_0)}{p(s_0)} + \sum_{i=0}^{N-1} \frac{p'(s_i, s_{i+1})}{p(s_i, s_{i+1})} \equiv \frac{p'(s_0)}{p(s_0)} + \sum_{i=0}^{N-1} d(s_i, s_{i+1}). \quad (5)$$

Ignoring the initial condition terms $p(s_0)$ for the moment and combining (4) and (5), we find an estimate of $dJ(\theta)/d\theta$ given by

$$\frac{1}{N} \left[\sum_{i=0}^{N} F(s_i) \right] \left[\sum_{j=0}^{N-1} d(s_j, s_{j+1}) \right].$$

This product of two sums can be represented by the sum of the contents of the n^2 cells in a two-dimensional array (See Fig. 7). However, the variance of the LR gradient estimator grows with "N" which makes steady-state performance sensitivity estimation difficult. The reason for this is as follows. Ordinarily when computing a sample average, we use the form of

$$\frac{1}{N}\sum_{i=1}^{N}[\text{random variable}]_{\text{ith sample}} \qquad (6)$$

It is well-known that the variance of this quantity goes to zero as N goes to infinity because of the presence of the 1/N term (see Appendix B.4.1). Thus, by increasing the length of the simulation, N, we can achieve convergence. However, for our current estimate we are dealing with the sample average which has the form

$$\frac{1}{N}\left[\sum_{i=1}^{N}[\text{random variable}]_{\text{ith sample}}\right]\left[\sum_{j=1}^{N}[\text{random variable}]_{\text{jth sample}}\right]. \qquad (7)$$

Because of the N^2 terms of the product of two sums, the variance of the sample average actually increases with N. This is the main inherent difficulty with the LR method. The intuitive reason is this. As discussed at the beginning of this section, we essentially ask the question what is the likelihood that a trajectory of the DEDS, $x(t;\theta,\xi)$, can actually be generated by a process characterized by $\theta+\Delta\theta$. The answer requires us to "re-scale" L(x) accordingly by the ratio $f(\theta+\Delta\theta,x)/f(\theta,x)$. It is also clear, for $x(t;\theta,\xi)$ of a long duration it will become increasingly unlikely that $x(t;\theta,\xi)$ is in fact generated by $f(\theta+\Delta\theta,x)$. The re-scaling ratio can become very large or very small leading to numerical difficulties in computation. This is another explanation of the variance problem associated with the LR method (see also the discussion in Appendix B.4.4 on importance sampling).

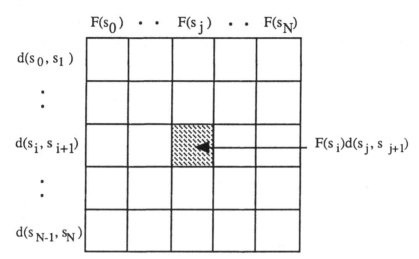

Fig.7 Representation of Likelihood Ratio Estimate

One way to ameliorate the problem is through the use of the regenerative technique. Suppose we can identify a state, say s_0, for the Markov chain which has the regenerative property, i.e., the sequence of states, $s_0, \ldots s_i, \ldots . s_0, \ldots s_k, \ldots . s_0, \ldots .$ can be decomposed into segments, each starting with s_0 and ending with the first return of the sequence to s_0, that are independent from each other. We denote these segments as r-segments. Since the r-segments are independent, they can be regarded as separate experiments or simulation runs. Using these ensembles of r-segments, we can construct an asymptotically unbiased LR estimate which has variance decreasing to zero as the number of r-segments in the ensemble increases. In terms of Fig. 7, what regenerative simu-lation accomplishes is illustrated in Fig. 8.

In other words, instead of one experiment, now we have an ensemble of experiments each of shorter length. Sample average is now computed using only the sum of the small black rectangles, each corresponding to one r-segment or experiment. Provided the variances of each of the small rectangles are manageable, then we are back in the situation of Eq.(6)[5]. Unfortunately, except for simple example problems, the mean length of regenerative segments for real problems is often uncomfortably long. For example, even for a small multiclass queueing network with 3 servers and 24 customers of 2 classes, the mean length of regenerative cycles can reach 17 million transitions [Zhang 1990]. In other words, the smaller rectangles (upper triangles) in Fig.8 still have variances that are too large (due to the effect of Eq.(7) on these smaller rectangles). Further attempts to reduce the variance problem for the LR method can be found in [Zhang and Ho 1991a] and Section 8.4. In general, the straight-forward applica-tion of the method produces likelihood ratio estimates that have a big variance, and hence is suitable only for systems with short regenerative sample paths.

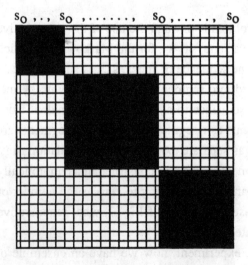

$$S_O, \ldots, S_O, \ldots\ldots, S_O, \ldots\ldots, S_O$$

Fig. 8 Regenerative LR Method

---

[5] Actually, [Glynn 1987] showed that only the upper triangle of the rectangular regions are needed. This will further reduce the variance somewhat.

7.3.2 Basic Analysis

Consider a stochastic system with a parameter $\theta \in R$. Suppose the performance measure of the system is

$$J(\theta) = E\{L(x(\theta))\}$$

where x is a random variable whose c.d.f. (cumulative distribution function) is $F(\theta,x)$ depending on θ. To simplify the discussion, we assume that x is a scalar. Thus, we have

$$J(\theta) = \int_{-\infty}^{\infty} L(x)dF(x,\theta). \tag{8}$$

Let $\xi = F(\theta, x)$. Then ξ is a random variable uniformly distributed on [0,1), and x is a function of ξ:

$$x = x(\theta,\xi) = F^{-1}(\theta,\xi) = \inf\{x: F(\theta,x) \geq \xi\}. \tag{9}$$

Eq.(8) is equivalent to

$$J(\theta) = \int_0^1 L[x(\theta,\xi)]d\xi. \tag{10}$$

On the other hand, let $f(\theta,x)$ be the p.d.f. (probability density function), then (8) can be written as

$$J(\theta) = \int_{-\infty}^{\infty} L(x)f(\theta,x)dx. \tag{11}$$

Taking the derivatives of the both sides of (10) and (11), we obtain

$$\frac{\partial J(\theta)}{\partial \theta} = \frac{\partial}{\partial \theta}\int_0^1 L(x(\theta,\xi)d\xi = \int_0^1 \frac{\partial L[x(\theta,\xi)]}{\partial \theta}d\xi$$

$$= \int_0^1 \frac{\partial L(x)}{\partial x} \frac{\partial x(\theta,\xi)}{\partial \theta} d\xi = E[\frac{\partial L(x)}{\partial x} \frac{\partial x(\theta,\xi)}{\partial \theta}] \tag{12}$$

and

$$\frac{\partial J(\theta)}{\partial \theta} = \frac{\partial}{\partial \theta} \int_{-\infty}^{\infty} L(x) f(\theta,x)dx$$

$$= \frac{\partial}{\partial \theta} \int_{-\infty}^{\infty} L(x) \frac{\partial f(\theta,x)}{\partial \theta} dx = \frac{\partial}{\partial \theta} \int_{-\infty}^{\infty} L(x) \frac{\partial \ln f(\theta,x)}{\partial \theta} f(\theta,x)dx$$

$$= E[L(x) \frac{\partial \ln f(\theta,x)}{\partial \theta}], \tag{13}$$

respectively, provided that the two operations of differentiation and integration are interchangeable in each case. Eq.(13) also requires $f(\theta, x) \neq 0$. If the interchangeability in (12) or (13) holds, then $(\partial L(x)/\partial x)(\partial x/\partial \theta)$ or $L(x)[\partial \ln f(\theta, x)/\partial \theta]$ is an unbiased estimate of $\partial J/\partial \theta$. $(\partial L(x)/\partial x)(\partial x/\partial \theta)$ corresponds to the perturbation analysis estimate, and $L(x)[\partial \ln f(\theta, x) /\partial \theta]$ is called the *likelihood ratio estimate* (LR) of $\partial J/\partial \theta$.

Both the perturbation analysis and the likelihood ratio estimates can be considered as the limiting value of the difference of two Monte Carlo estimates of the performance measure divided by the parameter increment. Using two uniformly distributed random variables ξ and η to determine the random variables x in the system with $\theta+\Delta\theta$ and the system with θ, we obtain the following expression for the finite difference estimate of $\partial J /\partial \theta$:

$$e(\Delta\theta,\eta,\xi) = \frac{L[x(\theta+\Delta\theta,\eta)] - L[x(\theta,\xi)]}{\Delta\theta}. \tag{14}$$

Now, letting $\eta = \xi$ and $\Delta\theta \to 0$ in Eq.(14), we get $(\partial L(x)/\partial\theta) = (\partial L(x)/\partial x)$ $(\partial x/\partial\theta)$. This clearly shows that the perturbation analysis estimate $(\partial L(x)/\partial x)$ $(\partial x/\partial\theta)$ corresponds to (14) with a common random variable used for both the systems with $\theta+\Delta\theta$ and θ.

To study the relation between the LR estimate and the estimate (14), we rewrite (13) in the following form:

$$L(x)\frac{\partial \ln f(\theta, x)}{\partial\theta} = L(x)\frac{1}{f(\theta, x)}\frac{\partial f(\theta, x)}{\partial\theta}$$

$$= \lim_{\Delta\theta\to 0}\frac{1}{\Delta\theta}\{ L[x(\theta,\xi)]\frac{f(\theta+\Delta\theta,\xi)}{f(\theta,\xi)} - L[x(\theta,\xi)]\}. \tag{15}$$

This shows that the estimate first takes $L[x(\theta, \xi)]\{f(\theta+\Delta\theta, \xi)/f(\theta, \xi)\}$ as an estimate of $L[x(\theta+\Delta\theta, \xi)]$, then calculates the limit value as $\Delta\theta \to 0$. Fig. 9 gives a geometrical explanation of both estimates. The underlying mechanism is actually as follows. Suppose that $L[x(\theta,\xi)]$ is the output of a simulation for the system with θ, then this same value would be the output of a simulation for the system with $\theta+\Delta\theta$, if the realization of the random variable is $x(\theta+\Delta\theta,\eta) = x(\theta,\xi)$. From this and Eq.(9), η can be determined by

$$F^{-1}(\theta+\Delta\theta, \eta) = F^{-1}(\theta, \xi),$$

or

$$\eta = F(\theta+\Delta\theta, F^{-1}(\theta,\xi)). \tag{16}$$

From Eq.(16), the measure of $\Delta\eta$ corresponding to $\Delta\xi$ is

$$\Delta\eta = \frac{\partial F[\theta+\Delta\theta, F^{-1}(\theta, \xi)]}{\partial\xi}\Delta\xi$$

$$= \frac{\partial F(\theta+\Delta\theta, x)}{\partial x}\frac{\partial x}{\partial\xi}\Delta\xi.$$

where $x=F^{-1}(\theta, \xi)$. Since

$$\frac{\partial x}{\partial \xi} = \frac{1}{\frac{\partial F}{\partial x}},$$

we have

$$\Delta\eta = \frac{\frac{\partial}{\partial x}F(\theta+\Delta\theta, x)}{\frac{\partial}{\partial x}F(\theta, x)}\Delta\xi = \frac{f(\theta+\Delta\theta, x)}{f(\theta, x)}\Delta\xi.$$

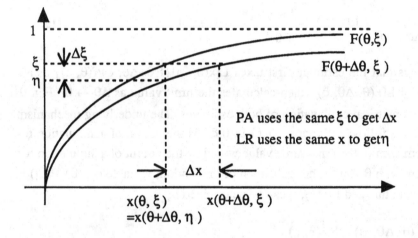

Fig. 9 Determining η in Eq.(13)

Therefore, when taking the expectation, the value of $L(x)$ should be modified by a factor $f(\theta+\Delta\theta,x)/f(\theta,x)$. More precisely, the expected performance for the system with $\theta+\Delta\theta$ can be obtained by taking $x(\theta, \xi)$ as the value of $x(\theta+\Delta\theta, \eta)$ and changing the probability measure accordingly:

$$E\{L[x(\theta+\Delta\theta, \eta)]\} = E\{L[x(\theta+\Delta\theta, F(\theta+\Delta\theta, F^{-1}(\theta,\xi))]\times \frac{\frac{\partial}{\partial x}F(\theta+\Delta\theta, x)}{\frac{\partial}{\partial x}F(\theta, x)}\}.$$

This explains the first term on the right-hand side of Eq.(15). The essential idea is to find the p.d.f. of $L[x(\theta+\Delta\theta,\eta)]$ at the value of $L[x(\theta, \xi)]$, which is obtained by a simulation for the system with θ. Since the realizations for both $x(\theta+\Delta\theta,\eta)$ and $x(\theta, \xi)$ are assumed to take the same value, the likelihood ratio method is called the *common realization (CR) estimate* in [Cao 1987b].

We have explained that in the perturbation analysis estimate (12) $\eta =\xi$, and in the likelihood ratio estimate Eq.(13) η is determined by Eq.(16). Now, consider the more general case. Let $\eta = g(\theta,\phi,\xi)$ be a differentiable function which maps $[0,1)$ to itself. $\eta = g(\theta, \theta+\Delta\theta, \xi)$ is chosen for η in (16). Assume $g(\theta, \theta, \xi) = \xi$. Then when taking expectation, the probability measure of η should be modified. That is,

$$e(\Delta\theta,\eta, \xi) \equiv \frac{1}{\Delta\theta} \{L[x(\theta+\Delta\theta,g(\theta,\theta+\Delta\theta,\xi))]\frac{\partial g(\theta,\theta+\Delta\theta,\xi)}{\partial\xi} - L[x(\theta,\xi)]\}$$

is an unbiased estimate of $\Delta J(\theta)/\Delta\theta$. Note that $\partial[g(\theta, \theta,\xi)]/\partial\xi =1$, and the second term can be written as $L[x(\theta,\xi)]\{\partial[g(\theta, \theta,\xi)]/\partial\xi\}$. Taking the limit of $\Delta\theta\to 0$, we obtain an unbiased estimate of $\partial J(\theta)/\partial\theta$:

$$\frac{\partial}{\partial\phi} \{L[x(\phi,g(\theta,\phi,\xi))]\frac{\partial g(\theta,\phi,\xi)}{\partial\xi}\}_{\phi = \theta}. \tag{17}$$

The perturbation analysis and likelihood ratio estimates are two special cases of the estimate (17) with $g(\theta, \phi ,\xi) = \xi$ and $g(\theta, \phi ,\xi) = F[\phi, F^{-1}(\theta, \xi)]$, respectively.

Exercise Devise another unbiased sensitivity estimate by using form (17).

7.3.3 An Example of the Variances of PA and LR Estimates
Since the PA estimate corresponds to the common random number (CRN) estimate of $\Delta J(\theta)/\Delta\theta$, we can use the properties of the CRN estimate to study that of the PA estimate. Since both $x=F^{-1} (\theta+\Delta\theta, \eta)$ and $x = F^{-1}(\theta, \xi)$ are non-decreasing functions of η and ξ, $L[x(\theta+\Delta\theta, \eta)]$ and $L[x(\theta, \xi)]$ are monotonic

and in the same direction. Under this condition, it was proved in [Rubinstein and Samorodnitsky 1985] that the CRN estimate of $\Delta J(\theta)/\Delta\theta$ has the smallest variance, i.e.,

$$\text{var}\,[e(\Delta\theta,\,\xi,\,\xi)] \leq \text{var}[e(\Delta\theta,\,\eta,\,\xi)]\,,$$

for all ξ and η having an arbitrary joint c.d.f. whose marginals are uniform distributions. Letting $\Delta\theta\rightarrow0$ and assuming that the two operations "var" and "$\lim_{\Delta\theta\rightarrow0}$" are interchangeable, we have

$$\text{Var}[\lim_{\Delta\theta\rightarrow0}\,e(\Delta\theta,\xi,\xi)] \leq \text{Var}[\lim_{\Delta\theta\rightarrow0}\,e(\Delta\theta,\eta,\xi)]\,.$$

Note that $\lim_{\Delta\theta\rightarrow0}e(\Delta\theta,\,\xi,\,\xi) = [\partial L(\theta,\,\xi)/\partial\theta]$ is the PA estimate. Thus, we conclude that under the above assumptions the PA estimate has the smallest variance. There is a wide class of systems whose performance measures are monotonic functions. For instance, in queueing networks the mean system throughput is a non-decreasing function of the mean service time of a customer at a server.

As an example of the comparison of the variances of the PA and the likelihood ratio estimates, we consider a simple cyclic queueing network consisting of two servers and only one customer. The mean service times of the two

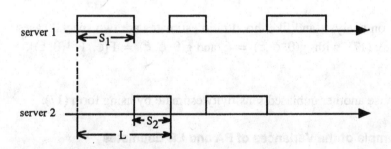

Fig. 10 Regenerative Periods in the Example

servers are s_1 and s_2, respectively. The state process of this system can be decomposed into regenerative periods, each consisting of one busy period and one idle period of each server (see Fig. 10). The length of the kth regenerative period is $L(s_1, s_2) = s_{1,k} + s_{2,k}$, where $s_{1,k}$ and $s_{2,k}$ are the service times of the kth customer at servers 1 and 2, respectively. The mean length of a regenerative period is

$$E[L] = E(s_{1,k} + s_{2,k}) = s_1 + s_2.$$

Its derivative with respect to s_1 is $\partial E[L(s_1)]/\partial s_1 = 1$. For exponential service distribution, (9) takes the form

$$s_{1,k} = - s_1 \ln (1 - \xi).$$

The PA estimate of $\{\partial E[L(s_{1,k}+s_{2,k})]\}/\partial s_1$ is

$$e_1 = \frac{\partial L(s_{1,k}, s_{2,k})}{\partial s_{1,k}} \frac{\partial s_{1,k}}{\partial s_1} = - \ln (1 - \xi) = \frac{s_{1,k}}{s_1}.$$

On the other hand, since $f(s_{1,k}, s_1) = (1/s_1) \exp \{-s_{1,k}/s_1\}$, we have

$$\frac{\partial \ln f(s_1, s_{1,k})}{\partial s_1} = \frac{1}{s_1} (\frac{s_{1,k}}{s_1} - 1).$$

Thus, the likelihood ratio estimate of $\{\partial E[L(s_{1,k}+s_{2,k})]\}/\partial s_1$ is

$$e_2 = L(s_{1,k}+s_{2,k}) \frac{\partial \ln f(s_1, s_{1,k})}{\partial s_1} = \frac{s_{1,k}+s_{1,k}}{s_1} (\frac{s_{1,k}}{s_1} - 1).$$

Both estimates are unbiased, since

$$E[e_1] = E[e_2] = 1.$$

Exercise Verify this, particularly $E[e_2]=1$.

However, the variances of thes two estimates differ significantly. In fact,

$$\text{var} [e_1] = E[e_1^2] - [E(e_1)]^2 = 1,$$

and

$$\text{var}[e_2] = E[e_2^2] - \{E[e_2]\}^2 = E[\frac{s_{1,k} + s_{2,k}}{s_1}(\frac{s_{1,k}}{s_1} - 1)]^2 - 1$$

$$= 2\rho^2 + 6\rho + 13, \qquad \rho = \frac{s_2}{s_1}.$$

This shows that the variance of the likelihood ratio estimate is at least 13 times as big as that of the PA estimate. Note that $L(s_{1,k}, s_{2,k})$ is an increasing function of $s_{1,k}$, and the required interchangeability holds. Thus, the PA estimate has the smallest variance in the example.

Exercise[6] Let $x_1, x_2, ..., x_n$ be n random variables with the joint probability density function $f_\theta(x_1, x_2, ..., x_n)$ depending on a parameter θ and $J(\theta) = L_n(x_1, x_2, ..., x_n)$ be a performance measure. $L_n(x_1, x_2, ..., x_n)$ is an unbiased estimate of $J(\theta)$ and $L_n(x_1, x_2, ..., x_n) d[\ln f_\theta(x_1, x_2, ..., x_n)]/d\theta$ is the likeli-hood ratio estimate of $d[J(\theta)]/d\theta$.

(a)   Using the Schwartz inequality prove

$$\text{Var}[L_n \frac{d\ln f_\theta(x_1,...,x_n)}{d\theta}] \geq \frac{[\frac{d}{d\theta}E(L_n^2) - E(L_n)\frac{d}{d\theta}E(L_n)]^2}{\text{Var}(L_n)}.$$

(b)   Explain the meaning of the inequality in (a).

(c)   Let $x_1, x_2, ..., x_n$ be exponentially distributed random variables with mean θ, calculate the lower bound of the variance for the LR estimate of $d[J(\theta)]/d\theta$ by using the inequality.

7.3.4 Comparison of the PA and Likelihood Ratio Estimates

We have shown via the example of the above subsection that the likelihood ratio estimate usually has a bigger variance than the PA estimate. In this section, we shall discuss some other aspects of the two estimates.

---

[6] The authors are indebted to W. B. Gong for this exercise.

Recall that the most important condition for these two estimates is the inter-changeability. For the PA estimate, it is

$$\frac{\partial}{\partial\theta}\int_0^1 L[x(\theta,\xi)]d\xi = \int_0^1 \frac{\partial}{\partial\theta}L[x(\theta,\xi)]d\xi. \tag{18}$$

The corresponding equation for the likelihood ratio estimate is

$$\frac{\partial}{\partial\theta}\int_{-\infty}^{\infty} L[x]f(\theta,x)dx = \int_{-\infty}^{\infty} L[x]\frac{\partial f(\theta,x)}{\partial\theta}dx. \tag{19}$$

Eq.(18) is equivalent to (see Chapter 4)

$$\lim_{\Delta\theta\to0} \frac{E[r(\theta,\Delta\theta,\xi)]}{\Delta\theta} = 0 \tag{20}$$

where $r(\theta, \Delta\theta, \xi)$ is defined in Eq. (4.3). As discussed in Chapter 4, Eq. (20) usually requires that the probability that $L[x(\theta, \xi)]$ is discontinuous in $[\theta,\theta+\Delta\theta]$ is of order $o(\Delta\theta)$.

Assume that the first and second derivatives of $f(\theta, x)$ with respect to θ exist and are continuous. We have the Taylor expansion:

$$L(x)\{\frac{f(\theta+\Delta\theta, x) - f(\theta, x)}{\Delta\theta}\} - L(x)\frac{\partial f(\theta, x)}{\partial\theta} = L(x)\{\frac{1}{2}\frac{\partial^2 f(\theta', x)}{\partial\theta^2}\}\Delta\theta,$$

where $\theta'\in [\theta, \theta+\Delta\theta]$. From this, we find a sufficient and necessary condition for Eq.(19) as follows:

$$\lim_{\Delta\theta\to0}\int_{-\infty}^{+\infty} \{L(x)\frac{\partial^2 f(\theta', x)}{\partial\theta^2}\Delta\theta\}\,dx=0,$$

or

$$\int_{-\infty}^{\infty} L(x) \frac{\partial^2 f(\theta, x)}{\partial \theta^2} dx < \infty \tag{21}$$

in a neighborhood of θ[7]. Note that Eq. (20) is much more restrictive than Eq. (21). Eq. (20) requires basically the continuity of $L(x)$, while (21) requires almost only the boundedness of $L(x)$. Thus the LR methodology can be expected to be more applicable when the distributions are explicitly known.

Another important issue is, for a DEDS, does the estimate based on one sample path converge with probability one as the length of the sample path goes to infinity? The answer, of course, depends on the particular DEDS. However, we can conclude that, except for some DEDS whose state processes are regenerative processes, the likelihood ratio approach does not provide a convergent estimate. This statement can be verified as follows: Suppose that random variables $x_1(\theta, \xi_1), x_2(\theta, \xi_2), ..., x_n(\theta, \xi_n)$ appear in the sample path with a finite length. As the length of the sample path goes to infinity, n also goes to infinity. In this multivariable case, the likelihood ratio estimate based on a finite sample path takes the form

$$\lim_{\Delta\theta \to 0} \frac{1}{\Delta\theta} \{ L[x_1(\theta, \xi_1), ..., x_n(\theta, \xi_n)] \times [\prod_{i=1}^{n} \frac{f(\theta+\Delta\theta, \xi_i)}{f(\theta, \xi_i)} - 1] \}$$

Letting $n \to \infty$ and assuming the interchangeability holds for "$\lim_{n \to \infty}$" and "$\lim_{\Delta\theta \to 0}$," we obtain the value of the likelihood ratio estimate when the sample path goes to infinity as follows:

$$\lim_{\Delta\theta \to 0} \frac{1}{\Delta\theta} \{ \lim_{n \to \infty} L[x_1(\theta, \xi_1), ..., x_n(\theta, \xi_n)] \times \lim_{n \to \infty} [\prod_{i=1}^{n} \frac{f(\theta+\Delta\theta, \xi_i)}{f(\theta, \xi_i)} - 1] \}$$

---

[7] The authors are indebted to Prof. S. Marcus for pointing out an inaccuracy in the proof of this condition in an early draft of this book.

If the system is ergodic, then the first $\lim_{n\to\infty}$ in the above expression converges to $J(\theta)$. However, there is no compelling reason why the second $\lim_{n\to\infty}$ in the above expression should exist. This means that the likelihood ratio method usually can not apply to a DEDS for the estimate of the steady-state performance measure unless regenerative property is used.

In short, the IPA estimate has a smaller variance, enjoys the nice property of convergence with probability one and a certain amount of robustness with respect to uncertainties in the underlying distribution (Section 3.4), but requires more restrictive conditions than the LR estimate. On the other hand, the LR estimate usually has a bigger variance, but applies to a wider range of systems since the requirement for interchangeability is much milder.

There is another way of viewing the LR approach to gradient estimation. Going back to the example at the beginning of this section where we wish to compute the derivative with respect to θ of the quantity $E[x]$ when x is exponentially distributed. Once again we have

$$\frac{dE[x]}{d\theta} = \int_0^\infty xf'(x;\theta)\frac{f(x;\theta)}{f(x;\theta)}dx = \int_0^\infty x\frac{d\ln f(x;\theta)}{d\theta}f(x;\theta)dx \ .$$

However, instead of defining $x[d\ln f(x;\theta)/d\theta]$ as a new (or re-scaled) integrand, we can view $[d\ln f(x;\theta)/d\theta]f(x;\theta)$ as a new (or derivative) density function with which we shall use to evaluate the expectation of x. We call this density function the **weak derivative** of $f(x;\theta)$. Of course, since density functions are always positive, we must be careful to define and normalize this term. For the expo-nential density function $f(x;\theta)=[1/\theta][\exp(-x/\theta)]$, it can be shown that the weak derivative is given by [Pflug 1989].

$$\frac{1}{\theta e}[\frac{e}{\theta}\left(\frac{x}{\theta}-1\right)\exp(-\frac{x}{\theta})1(x\geq\theta) - \frac{e}{\theta}\left(1-\frac{x}{\theta}\right)\exp(-\frac{x}{\theta})1(x\leq\theta)]$$

where $1(\cdot)$ is the unit step function. Direct calculation again yields the result that $dE[x]/d\theta = E_f[x]=1$. Of course this approach requires us to evaluate either analytically or numerically an expectation with respect to a new distribution, i.e., a new simulation. For each new parameter, θ, there will be one extra simu-lation. Hence it can be argued that this is not in the spirit of single-run gradient estimations.

Lastly, it is appropriate to recapitulate the conceptual basis of the PA and LR methods of single run performance gradient estimations. Both methods rely fundamentally on the inherent randomness in the system. Suppose the system parameter θ of interest had been a deterministic service time, D, of a server. If we observe the sample path of the system at $\theta=D$, then it makes no sense to ask the question what is the probability that such a path could have been generated by an identical system with $\theta=D+\Delta D\neq D$. The probability will always be zero. The PA method via "cut-and-paste" as described in Chapter 6 may run into similar difficulties. All that is required is for the system under $\theta=D+\Delta D$ to have behaviors that are patently impossible if $\theta=D$. Then no amount of cut and wait will yield a match for pasting. In this sense, the superficial first impression of the impossibility of single-run performance gradient estimations need not be so counter-intuitive after all. Although we do not perturb the system in the sense of introducing a deliberate intrusion, we instead let natural randomness do the job. In the realm of continuous variable dynamic systems there is a well-known precedent. The impulse response, hence the complete characterization, of an arbitrary linear dynamic system can be determined by auto-correlating the output when the input is subjected to Gaussian white noise [Davenport and Root 1958]. It is not necessary to inject impulse or step inputs to the system.

7.4 DEDS Models Revisited

It is appropriate at this point to review the models of DEDS and the sample path based techniques that we have developed so far in order to provide some perspectives. As discussed in Chapter 1 and Appendix D, the emerging view of a general DEDS appears as follows. A DEDS is an automaton or a finite state machine consisting of

(i) A finite state space X composed of a set of discrete states $x_0, x_1, x_2, \ldots,$

(ii) A finite set of events, $\alpha \in A$,

(iii) An enabled event list for each state $\Gamma(x)$, and

(iv) A next state transition function $x_{next} = f(x_{now}, \alpha)$.

Given an initial state, (i-iv) specify the dynamical evolution of the DEDS. The trajectory is then described by the sequence of (state,event) pair called **trace** while the event sequence alone is denoted as a **string**. The set of all feasible event strings from a DEDS constitutes a **language**, L. This view is representative of the finite state machine, finitely recursive process, the Petri net model of Appendix D and that of the GSMS of Appendix B. Absent from this description is the timing information on the duration of various states along the trajectory and the timing mechanism for determining the occurrence and order of events. From the viewpoint of GSMP, we can visualize an |A|-dimensional vector sequence of samples of random variables $\omega(n) = [\ . \ . \ \omega_\alpha(n) \ . \ .]$ which gives the lifetime of the nth occurrence of the α-event. Given a state x, the enabled event list $\Gamma(x)$ schedules the lifetimes of the enabled events using the clock mechanism $\omega_\alpha(n)$. Once scheduled, the event clock of each event runs down until it reaches zero. At that time the event is said to have occurred. The event occurrence times are denoted similarly as $\tau_\alpha(n)$'s. The state now transits and the process repeats. Any event scheduled but have not occurred and is not enabled in the new state is simply interrupted. Thus, t he occurrence of an event is determined by the various competing lifetimes of enabled events in each state. This mechanism in turn determines the duration of each state. The important point to be stressed here is the *independence of the timing mechanism from the automaton structure* of the DEDS. The $\omega(n)$ may be specified or determined separately from the dynamic evolution of the automaton. A glimpse of this viewpoint is already present in the standard clock approach discussed in Section 1 of this chapter. We may now pursue this further. Consider the graphical depiction of a GSMP in Fig.11 .

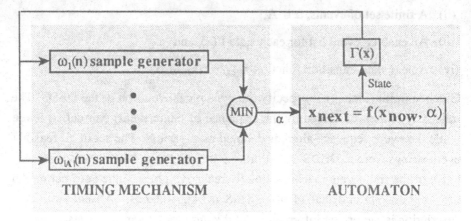

List of enabled events

$\omega_1(n)$ sample generator

$\omega_{1A}(n)$ sample generator

MIN

α

$\Gamma(x)$

State

$x_{next} = f(x_{now}, \alpha)$

TIMING MECHANISM   AUTOMATON

Fig. 11 The Structural and the Timing Parts of a DEDS

Fig. 11 clearly illustrates the independence of the structural and the timing parts of the system. This view also clarifies the cut-and-paste (Chapter 6), the standard clock (Section 7.1), and the augmented chain approach (Section 7.2) to generating perturbed trajectory from the nominal. Suppose the perturbation is in a parameter of the structural part of the DEDS. Typically this may change (let us say, reduce) the state space. However, the event sample generating processes remain unchanged. Thus, we need only compare the event lists $\Gamma(x)$ and $\Gamma(x')$ to determine if the mechanisms for generating the next piece of trajectory are identical. Here a "piece" means the duration of a state or one sample from the $\omega_\alpha(n)$ generators. We cut and paste "pieces" entirely on this basis and only at $\tau_\alpha(n)$ points. There is no perturbation to the timing part of the DEDS in Fig. 11. This is the basic idea behind the standard clock and the augmented chain approach. On the other hand, suppose we perturbed the timing (stochastic) part of the DEDS, say the service rate represented by one the $\omega_\alpha(n)$ generators. Now, if the perturbation is very small and we assume the event list between the perturbed and the nominal will always be the same (deterministic similarity), then IPA essentially says that it is not necessary to devise separately a perturbed $\omega_\alpha(n)$ generator. A little bit of calculus and bookkeeping using the automaton part of the DEDS can in fact enable us to generate the perturbed trajectory. This is fine so long as the state sequence remains the same (or essentially the same under condition (CM) of Chapter 5). We must cut when the state sequence starts to differ. Now the trajectory "pieces" here are generally much longer lasting over many state transitions. However, in order to simplify the subsequent pasting we make the assumption that the $\omega_\alpha(n)$ generating processes are Markov. However, from this viewpoint the Markov assumption is purely a matter of convenience. Cut-and-paste is not fundamentally restricted to Markov systems. Using event matching for cut-and-paste, just how much extra computation must be done on the timing portion in order to generate perturbed trajectory for non-Markov system is still an open question. For example, [Strickland and Cassandras 1989] showed that if there exist only one non-Markov generator, the extra work is minimal. This discussion also classifies PA techniques into two major categories (Fig. 12):

(i) Cut-and-paste only at $\tau_\alpha(n)$ points — In this category we have the augmented chain approach and the standard clock method of Chapter 7.

(ii) Cut-and-paste at non-$\tau_a(n)$ points — Here we include the large class of IPA-applicable problems where no cutting is ever necessary, the various extension of IPA in Chapter 5, and the extended PA of Chapter 6.

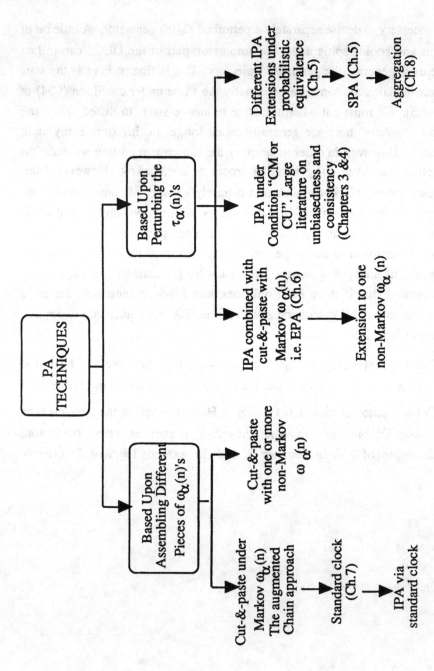

Fig.12 Taxonomy of PA Techniques

CHAPTER EIGHT

Aggregation, Decomposition, and Equivalence

The idea of aggregation is a fundamental concept in engineering. Whenever we make a model of a real system for analysis or optimization purposes, we are consciously or unconsciously making aggregation decisions as to what features to include and what details to gloss over in our model. Of course, depending on the problem, the actual aggregation methodology may be intuitive, informal, approximate, formal, precise, or exact. However, there are two universal desirable characteristics for any aggregation technique. We denote the first characteristic as **External Independence**. By this we mean that whatever aggregated equivalent we synthesize to represent a complex part of a system, it can only depend on the part to be aggregated and should not depend on anything external to the part. From the practical and computational viewpoint, this is a necessary requirement. The idea is that the aggregation effort is similar to a one-time set-up charge. Once carried out, the simpler equivalent model can be used over and over in analysis when other parts of the system change. Only through external independence, can one realize any practical computational saving. If each time the external part of the system changes, the aggregated equivalent, whether exact or approximate, must be re-computed, then no saving results. The second less important but still desirable aggregation property is what we shall call **Closure**. The idea here is that each time we aggregate a system into a simpler equivalent, the resultant system should contain no new constructs that do not present in the original system. For example, A complex queueing network is aggregated to form another simpler queueing network. Both networks contain only customers and servers. We do not require the creation of any new element. Similarly, a Markov system when aggregated should retain the Markov property. Otherwise, much of the advantage is lost.

This closure property is particularly important when a hierarchical view of aggregating a complex system into many levels is adopted. We shall in this chapter judge various aggregation and PA techniques using the external independence and closure yardsticks and measure the extent to which they exactly or approximately achieve these requirements. After a review of the known results on aggregation of queueing networks in Section 1, we treat in Section 2 the relationship between aggregation and perturbation analysis. The principal idea here is that aggregation properly used actually extends the applicability of IPA. In Sections 3 and 4 we take a rather broad view of aggregation and demonstrate a new approach to the computation of the performance measures and their derivatives of very large arbitrary Markov chains.

8.1 Equivalent Server and Aggregation

8.1.1 Product-form Networks

The concepts of equivalent server and network aggregation have been established for product-form networks ([Chandy et al. 1975], [Brandwajn 1985], Appendix A, and Chapter 3). The concepts have the obvious appeal of reducing a complex system to a simple one. The basic idea can be explained as follows. Consider an irreducible closed network of single-server exponential nodes with a single-class of customers. Let N be the number of customers and M be the number of servers. The system state is a vector $\mathbf{n} = (n_1, n_2, ..., n_M)$, where n_i denotes the number of customers waiting and in service at server i. Suppose that the system can be partitioned into two sub-networks S_1 and S_2 with the condition that upon entering into S_2, the destination of a customer is independent of its origin. This condition is necessary because once S_1 is aggregated, there is only one input to S_2; thus, in the aggregated system, the origins of all customers arriving at S_2 are indistinguishable[1]. Without loss of

---

[1] For example, if the routing probability of the network is such that after partitioning the two nodes in S_2, say, nodes 3 and 4, can receive customers only from nodes 1 and 2 in S_1 respectively, then the condition is violated.

generality, we can write $\mathbf{n} = (\mathbf{n}_1, \mathbf{n}_2)$, where $\mathbf{n}_1 = (n_1, ...,n_k)$ and $\mathbf{n}_2 = (n_{k+1}, ...,n_M)$. It was shown in [Chandy et al. 1975] that S_1 can be replaced by a load-dependent exponential server with specified service rates without changing the marginal steady-state distribution of $\mathbf{n}_2$. That is, S_1 can be aggregated into an equivalent server. The load-dependent service rates of the equivalent server, $\mu(1), ..., \mu(N)$, can be determined by "shorting" the sub-network: If all service times in S_2 are set to zero, $\mu(n)$ is the throughput through S_2 when there are a total of n customers in the shorted network. This is called the Norton theorem[2] and can be extended to other product-form networks such as the so-called BCMP networks [Kritzinger et al. 1982]. One important property of the equivalent server of a subnetwork S_1 of a product-form network is that its service rates are independent of both the structure of the complimentary subnetwork S_2 and the service rates of the servers in S_2; i.e., the external independence property is satisfied.

Another feature of the equivalent server of a product-form network is that $\mu(1), ..., \mu(N)$ can also be interpreted as the instantaneous departure rates from S_1, conditional on the number of customers in S_1. Let

$$\mu_{n_1} = \sum_{i \in S_1} \mu_i \, \varepsilon(n_i) \, p_i, \tag{1}$$

where $\varepsilon(n_i)=1$ if $n_i>0$, $\varepsilon(n_i)=0$ if $n_i=0$, and p_i is the probability that upon leaving server i a customer leaves the sub-network S_1. Then, for n=1,2,...,N we have

$$\mu(n) = E[\mu_{n_1} \mid \sum_{i \in S_1} n_i = n]. \tag{2}$$

Using the product-form formulas, we can check that Eq.(1) and (2) yield the same service rates as the shorted throughput method [Hsiao and Lazar 1990].

Exercise Verify this assertion in terms of a simple 3 server network.

---

[2] In analogy to the Norton theorem of electric circuit theory.

Of course, for product-form networks it has been pointed out that aggrega-tion does not offer any computational savings ([Vantilborgh 1978] and [Balsamo and Iazeolla 1984]) and that the effort required to determine the equivalent load-dependent server is no different from that required to evaluate the product-form formula directly. However, the concept of the so-called Norton's equivalent server is very useful. The ideas of the shorted throughput through S_2 or the conditional departure rate from S_1 given by (1) and (2) provide directions for generalization to non-product form networks.

Exercise Consider an M/M/1 queue with two classes of inputs with rates λ_1 and λ_2. The service rates for both classes of customers are the same and equal to $\mu > \lambda_1 + \lambda_2$. Show that the IPA estimate of the average system time for both customer classes with respect to, say λ_1, is equivalent to that of a single class (λ_1 arrival only) M/M/1 server with reduced service rate $\mu - \lambda_2$.

8.1.2 General Networks

The aggregation of general queueing networks reviewed here is based on the discussion in [Glasserman and Ho 1989]. As stated in the previous subsection, there are two characterizations of the equivalent server, one is based on the shorted subnetwork and the other on the conditional departure rates. These two characterizations suggest two ways of constructing aggregated nodes for sub-networks of a general network.

Consider a general network S consisting of two complimentary sub-networks S_1 and S_2 which satisfies the required routing condition for aggre-gation. The service times of the servers in the network have general distributions. Suppose we replace subnetwork S_1 by a load-dependent expo-nential server with $\mu(1), \ldots \mu(N)$. We get an equivalent network of S. We choose $\mu(n)$ to be the throughput through S_2 in a network in which there are n customers and all the service times in S_2 are set to zero. This "aggregation" is only approximate, in the sense that the marginal steady-state distribution of S_2 in the equivalent network is not the same as that in the original network. However, since both distributions are the same when the aggregation is applied to product-form networks, we hope that the equivalent network may provide an approximate value of the marginal distribution for S_2 in S. An heuristically more general approximation would be to replace the sub-network S_1 by a load-dependent server whose service distribution at each value of the load is given by the interarrival distribution along the "shorted" path [Ho and Yang 1986]. Experimentally, there is evidence that such approximations behave adequately for the purpose of performance analysis using PA methodology [Glasserman and Ho 1989, Ho and Yang 1986].

In the other approach, we choose the conditional departure rates from S_1 as the service rates of the equivalent server. In the product-form network case, these rates are the same as those obtained by shorting the subnetwork S_2. But for non-product-form networks, the two sets of rates obtained from these two methods are different. Note that the rates obtained by shorting S_2 are independent of the structure of S_2 and the service distributions of all servers in S_2. The service rates obtained by the instantaneous departure rates, however, usually depend on the parameters of S_2, since in Eq.(1) $\varepsilon(n_i)$'s are random variables the distributions of which depend on the parameters of S_2. The fact that these two sets of rates are the same for product-form networks is a consequence of the product-form formulas. Therefore, the equivalent network using the conditional departure rates as the service rates of the load-dependent server is different from that based on a shorted network. In fact, the following discussion shows that the aggregation based on the conditional departure rates provides an "exact" result.

To simplify the exposition, we consider a class of Markovian queueing networks. We suppose that the service time at every station can be represented by the method of stages, i.e., the service time can be represented as the passage time through a finite Markov chain. This type of service distribution is called the Coxian distribution (see Appendix A). Any distribution with a rational Laplace transform is a Coxian distribution. Thus, Coxian distribution is general enough to approximate any general distribution (just by approximating the

Laplace transform of the distribution by a rational function). We denote the state of the queueing network S with such service time distributions as

$$X = ((n_1, c_1),..., (n_M, c_M)),$$

where n_i is the number of customers in server i, and c_i records the stage of service of the customer being served at server i. We take $c_i = 0$ if and only if $n_i = 0$. We also assume that the Markov process X is irreducible and ergodic. Thus, there exists a unique steady-state distribution $P(X)$ of the process.

Let X_1 and X_2 be the states of the two sub-networks S_1 and S_2, i.e., $X_1 = ((n_1, c_1),...,(n_k, c_k))$ and $X_2 = ((n_{k+1}, c_{k+1}),...,(n_M, c_M))$. Our goal is to replace these two subnetworks with two load-dependent servers without changing certain marginal distributions. Let $\mu_i(c)$ be the rate of the exponential holding time in stage c at server i, $p_i(c)$ be the probability that upon exiting from stage c a customer actually leaves the sub-network containing server i for the other subnetwork. (Note that $p_i(c)$ equals the product of the probability that a customer completes its service at server i after exiting from stage c and the probability that a customer leaves the subnetwork containing server i after the completion of its service at server i.) The instantaneous departure rates from the two subnetworks are

$$\mu_{S_1}(X_1) = \sum_{i \in S_1} \sum_{c_i} \mu_i(c_i) p_i(c_i)$$

and

$$\mu_{S_2}(X_2) = \sum_{i \in S_2} \sum_{c_i} \mu_i(c_i) p_i(c_i).$$

Let $n_{S_1} = n_1 + ... + n_k$ and $n_{S_2} = n_{k+1} + ... + n_M$ be the numbers of customers in subnetworks S_1 and S_2, respectively. Then the conditional departure rate from S_1, given that the number of customers in S_1 is $n_{S_1} = n$, is

$$\mu_{S_1}(n) = E[\mu_{S_1}(X_1) | n_{S_1} = n] = \sum_{X_1 : n_{S_1} = n} \frac{P(X_1)}{P(n_{S_1} = n)} \mu_{S_1}(X_1). \tag{3}$$

Similarly, the conditional departure rate from S_2, given that the number of customers in S_2 is $n_{S_2}=n$, is

$$\mu_{S_2}(n) = E[\mu_{S_2}(X_2) \mid n_{S_2}=n] = \sum_{X_i\,n_{S_i}=n} \frac{P(X_2)}{P(n_{S_2}=n)} \mu_{S_2}(X_2). \qquad (4)$$

Now, we construct a simple cyclic network consisting of two load-dependent exponential servers. The service rate of server 1 when there are n customers in its queue is $\mu_{S_1}(n)$; the service rate of server 2 for n customers is $\mu_{S_2}(n)$. Denote the numbers of customers in servers 1 and 2 by n_1^* and n_2^*, respectively. We assert that in steady-state the marginal distribution of n_{S_2}, the number of customers in subnetwork S_2 of S, is the same as the steady-state distribution of n_2^* in the equivalent two-server cyclic queue. Namely,

$$P(n_{S_2}=n) = G^{-1} \prod_{i=1}^{n} \frac{\mu_{S_1}(N-i)}{\mu_{S_2}(N)},$$

where G is a normalizing constant. To see this, partition the state space into sets $\{x: n_{S_2}=n\}$, $n=0,1,\ldots,N$, and write $x \in n$ if x is in the nth such set. In steady-state, the flow of probability in and out of each set is balanced. Thus,

$$\sum_{x\in n}\sum_{x'\notin n} \pi(x)\, q(x, x') = \sum_{x'\notin n}\sum_{x\in n} \pi(x')\, q(x', x), \qquad (5)$$

where $\pi(x)$ is the equilibrium probability of x and $q(x,x')$ the transition probability from x to x'. With the one step transition assumption (only one customer can move from S_1 to S_2 or vice versa at one time), the summation over $x'\notin n$ need only be carried out over the cases $x'\in$ n-1 and n+1. Equation (5) then becomes

$$\sum_{x\in n} \pi(x)\, [\mu_{S_2}(x) + \mu_{S_1}(x)] = \sum_{x'\in n-1} \pi(x')\mu_{S_1}(x') + \sum_{x'\in n+1} \pi(x')\mu_{S_2}(x'). \,(6)$$

Multiplying and dividing the three terms in (6) by $P(n_{S_2}=n)$, $P(n_{S_2}=n-1)$, and $P(n_{S_2}=n+1)$, respectively, we get from (3) and (4)

$P(n_{S_2}=n)[\mu_{S_2}(n) + \mu_{S_1}(N-n)]$

$= P(n_{S_2}=n-1)\mu_{S_1}(N-n+1) + P(n_{S_2}=n+1)\mu_{S_2}(n+1),$

which is recognized as the balance equations for a birth-death process with birth rates $\mu_{S_1}(N), \ldots, \mu_{S_1}(1)$ and death rates $\mu_{S_2}(1), \ldots, \mu_{S_2}(N)$ and the solu-tion given above in terms of a cyclic two-server system.

The result indicates that a subnetwork with general service time distributions can be replaced by a load-dependent exponential server without changing the steady-state marginal queue length distributions for the rest of the network. Note, in particular, that when S_2 consists of a single exponential server with rate μ, our result says that the steady-state performance of this server could actually be evaluated by replacing its complement by a load-dependent exponential server - even if the original network is not product-form. The catch here is that the parameters $\mu_{S_1}(n)$ are at least as difficult to obtain *analytically* as the perfor-mance itself. Put it another way, the "exact" equivalent of S_1 is *not* externally independent, each time S_2 changes, it must be re-determined.

The above result has been treated in many different forms [Kemeny and Snell 1960], [Courtois 1978, 1981], [Balbo and Denning 1979], and [Brandwajn 1985]. It is in fact a special case of a more general result in [Bremaud 1981] which can be stated as follows: For any queue Q having time varying arrival rate λ_t and departure rate μ_t, there exists a *first-order equivalent* queue with a load-dependent arrival $\lambda_n(t) = E[\lambda_t \mid Q_t =n]$ and a load-dependent service rate $\mu_n(t) = E[\mu_t \mid Q_t =n]$, where Q_t is the queue length of Q at time t. If Q_t and its first-order equivalent queue have the same initial distribution, then they have the same distribution at any time $t \geq 0$. In our case, $\mu_{S_2}(i)$ is the service rate of the first-order equivalent server of S_2, and $\mu_{S_1}(N-i)$ is the arrival rate to the server.

8.2   Perturbation Analysis on Aggregated Systems and Aggregated Perturbation Analysis

Consider the situation in Fig.1.Suppose that we can replace part A by part A' for the purpose of analyzing the performance behavior of part B (represented here by a single "queue-server" combination for simplicity).   Then three possible advantages of aggregated PA present themselves:

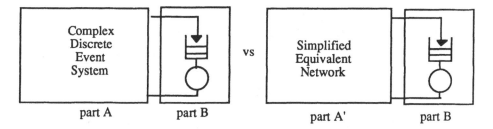

Fig. 1 Equivalent Networks

(i) *saves simulation/experimentation time*

It is clear that in any simulation /experiment involving the system, it will be easier to program and faster to run the simulation by using parts A' and B vs. A and B.

(ii) *extending the applicability of PA techniques*

A particular PA algorithm, such as the Infinitesimal Perturbation Analysis (IPA) algorithm, may not be applicable to parts A-B and yet may be provably applicable to parts A'-B. Recall that if in some part of the system a small change in a parameter may cause event order changes which result   in   discontinuity in performance, then infinitesimal perturbation analysis may fail to provide an unbiased estimate. Now, if that part of a network where the event order changes occur could be aggregated into a single node, then the event order changes may not occur in the aggregated system, and the infinitesimal perturbation analysis could be applied to the aggregated system and could provide an unbiased sensitivity estimate for the original system.

(iii)   *applies to real systems*

This advantage is potentially the most significant. By designing a PA algorithm based on parts A'-B, which are simple, we can apply this simple PA algorithm to the real complex system (i.e., parts A-B). The implication of this in real-time real-world applications is obvious. PA is not an all-or-nothing tool. It is not necessary to develop or to use PA algorithms which are applicable to each system at the most detailed level. Depending on the problems and questions posed, one can use PA algorithms of the appro-priate complexity regardless of the actual complexity of the underlying system. As pointed out in (ii) above, this is important for real-world real-time applications

In this section, we shall discuss the above mentioned approaches in most cases by showing experimental results. Limited theoretical analysis will be given for special cases.

8.2.1 Perturbation Analysis of Aggregated Systems

Suppose we want to estimate the sensitivity of the performance of a queueing network S with respect to a parameter of the service distribution of one of the servers in S. Denote the server as server 1 and the parameter as μ_1. According to the discussion in Section 8.1, we can aggregate servers 2 to M into an equivalent load-dependent server with service rates $\mu_2(1)$, ..., $\mu_2(N)$, such that the queue length distribution of the network consisting of server 1 and the equivalent server is the same as the marginal queue length distribution of S. We can apply the infinitesimal perturbation analysis for load-dependent servers (Section 3.6) to the aggregated network and obtain the estimate of the performance sensitivity for the aggregated network. Note that although the two networks, the original one and the aggregated one, have the same steady-state performance, the sensitivity of the steady-state performance may be different for both networks. The reason is, in general, $\mu_2(1)$, ..., $\mu_2(N)$ depend on the steady-state probability of S (see Eq.(4)) and hence implicitly depend on μ_1. Applying perturbation analysis to the aggregated network gives us the sensitivity of performance with respect to μ_1 while $\mu_2(1)$, ..., $\mu_2(N)$ are fixed. Thus, the dependency of $\mu_2(1)$, ..., $\mu_2(N)$ on μ_1 is lost in the perturbation analysis estimate obtained from the aggregated network. This approach gives the correct estimate only for networks for which the service rates of the equivalent server of a subnetwork do not depend on the parameters of the complimentary sub-network, e.g., the product-form networks. In general, our estimates obtained by applying PA to the aggregated system can only be approximations of the true values of the original system.

Several examples of load-dependent Jackson networks were studied in [Ho and Yang 1986]. Simulation results showed that the perturbation analysis estimate of the aggregated system are very close to the sensitivity given by the product-form formula, and that for the same length of the simulation the pertur- bation analysis applied to the aggregated system yields more accurate estimates than the perturbation analysis applied to the original network. This is not sur- prising because of rationale (i) mentioned in the introduction of this section and since the external independence property is satisfied here.

A more interesting set of aggregation experiments was performed in [Ho and Yang 1986]. Consider a general non-product form network S consisting of two sub-networks S_1 and S_2. In order to obtain an equivalent server of S_1, we first short the subnetwork S_2, i.e., set the service times of all servers in S_2 to zero. Then, we simulate the shorted network with a population of 1,2,...,N and for each population record the interarrival time distributions of the customers passing through the shorted path. The service time distribution of the equiva- lent server with n customers in its queue is chosen to be the distribution of the interarrival time on the shorted path in the shorted network with a population of n. Simulation experiments were carried out. In all these examples the two sensitivity estimates, one from the original network and the other from the aggregated one, are quite close. Of course, obtaining the interarrival distributions requires considerable of computation. Unless much parametric study is contemplated for the subnetwork S_2, this approach to aggregating S_1 is not practical; it just provides a concept of aggregation and experimental evidence for the advantages (i) mentioned above.

8.2.2 Aggregated Perturbation Analysis

The more useful advantages (ii) and (iii) mentioned above call for an approach we denoted as aggregated PA. Consider a queueing network $S = (S_1, S_2)$, where S_1 consists of servers 1 to k, and S_2 consists of servers k+1 to M. Assume that we are estimating the performance sensitivity with respect to one of server M's parameters. We first determine the service rates of the equivalent server of S_1, $\mu(1)$, ..., $\mu(N)$ (by simulation or analytical formula). A PA algorithm for the derivatives of performance defined in S_2 with respect to the parameter of server M is derived based on the aggregated network. This aggregated perturbation analysis algorithm is applied to the original network instead of the aggregated network; but when perturbations are propagated between S_1 and S_2, it works as if subnetwork S_1 had been replaced by an exponential server with rates $\mu(1)$, ..., $\mu(N)$. The idea is that we not only use a simpler PA algorithm but also possibly extend the applicability of the algorithm. The algorithm is as follows.

Aggregated Perturbation Analysis Algorithm
Initialize : $\Delta_1 = \Delta_{k+1} = ... = \Delta_M = 0$;
If a customer moves from server i to server j, then

$n_i := n_i - 1, n_j := n_j + 1$;

If i=M, do perturbation generation;

If $i \in S_2$ and $j \in S_2$ and if $n_j = 0$, then $\Delta_j = \Delta_i$;
(perturbation propagation in S_2)

If $i \in S_2$ and $j \in S_1$, then

$n_{S_1} = n_{S_1} + 1$, and

$$\Delta_1 = (1 - \frac{\mu(n_{S_i})}{\mu(n_{S_i}+1)})\Delta_i + \frac{\mu(n_{S_i})}{\mu(n_{S_i}+1)}\Delta_1 \; ;$$

(perturbation propagation from server i in S_2 to the aggregated server)

If $i \in S_1$ and $j \in S_2$, then

$$n_{S_1} = n_{S_1} -1, \text{ and if } n_j = 0, \text{ then } \Delta_j := \Delta_1.$$

(perturbation propagation from the aggregated server to server j in S_2)

The aggregated perturbation analysis algorithm records the perturbations of the servers in the non-aggregated subnetwork S_2, $\Delta_{k+1},..., \Delta_M$, and the perturbation of the equivalent server, Δ_1. When a customer moves from one server in S_2 to another server in S_2, the algorithm works in the same way as the ordinary perturbation analysis. When a customer moves from a server in S_2 to a server in S_1, or from a server in S_1 to a server in S_2, the algorithm propagates perturbation as if S_1 is aggregated into an equivalent server. When a customer moves from one server in S_1 to another server in S_1, no perturbation is propa-gated even if an idle period is terminated.

The aggregated perturbation analysis algorithm permits us to work on a sample path of the original network rather than on a sample path of the aggre-gated network. Although both sample paths are stochastically equivalent in the sense that the steady-state distributions of both networks are the same, the sample path of the original network should contain more information about the original network than the aggregated network does.

It should be emphasized at the beginning that strictly speaking there is no assurance that applying the PA algorithm developed on the basis of the aggre-gated system equivalent will necessarily yield the correct estimate when applied to the original system. Two issues are involved here. First of all, as pointed out above, in general our aggregation result do not possess the property of "external independence." If the parameters change even infinitesimally in the S_2 subnetwork, then the equivalent server of S_1 will also change. This effect is not captured when we develop a PA estimate using the aggregated equivalent. Secondly, even if the aggregated and the original networks had the same performance derivatives, it is by no means clear that the aggregated perturbation analysis applied to the original network would yield the same performance derivative as that obtained by the ordinary perturbation analysis applied to the original network. That is, the correctness of the aggregated perturbation analysis is not theoretically proved. However, we can consider two limiting cases of a product-form network and show that at least in those cases the aggregated algorithm yields exact results (Experimentally, we have observed that this is often sufficient to ensure that the intermediate values are close as well.) In the network, S_1 consists of any number of servers and customer classes, and S_2 consists of only one exponential server with service rate μ. Only one class of customers, say C_1, can visit S_2. Now as $\mu \to \infty$, class C_1 customers spend less and less time outside S_1 to the point where customers arriving at S_2 virtually find it idle. In other words, the algorithm in this case only cycles between the cases of zero or one customer at the outside server. Conversely, as $\mu \to 0$, we only need to consider the cases of N and N-1 customers at the outside server. Direct calculation of the limiting value of Δ_1 in the above algorithm yields the value of "1" and "1- $(\lambda_{N-1} / \lambda_N)$," respectively, for d[TP]/dμ where TP is the mean departure rate from S_1. On the other hand, under aggregation we view the network as a single class two server closed network with N (class 1) customers. Then the above results agree with that of the equivalent single-class two-server cyclic network.

Exercise Carry out this calculation .

[HINT: In the two limiting cases, the system behaves essentially as a 2-server cyclic queue with only one customer (for $\mu \to \infty$) and with a never idle load-dependent server (for $\mu \to 0$). See [Glasserman and Ho 1989].

8.2.3 Aggregation of Multiclass Queueing Networks—An Example

One important application area of the aggregated algorithm is the sensitivity estimation of multiclass queueing networks. As discussed in Chapter 5, infinitesimal perturbation analysis does not provide an unbiased estimate of the sensitivity of performance for many multiclass queueing networks. This is because a small change in a parameter may induce changes in the order of events, which results in the discontinuity of the sample performance function with respect to the parameter. However, the order change may happen only within the events (e.g., customer transitions) in a subnetwork. In this case, if we aggregate the subnetwork as an equivalent server, the order change may not show up in the aggregated network. Thus, if we apply the aggregated perturbation analysis to a sample path of a multiclass network, the effect of order change may be buried in the subnetwork and is "aggregated" away. In this sense, the aggregated perturbation analysis is similar to the idea of "smoothed perturbation analysis" (Section 5.5 and [Gong and Ho 1987]) which applies perturbation analysis to a conditional mean of a performance measure. In the conditional mean, the discontinuity of the sample performance function is "averaged" out.

As an example, we consider a simple multiclass network shown in Fig. 2 (This example is from [Glasserman and Ho 1989]). There are two classes of customers, two customers of each class. Both classes visit server 2 but only one class visits each of the other two servers. All servers are exponential, the means at servers 1 and 2 are 1.0, and that at server 3 is μ. We choose the average time required by server 3 to serve one customer in steady state (including idle times), T, as the performance. It is known that the infinitesimal perturbation analysis does not yield a consistent estimate for $dT/d\mu$ (see [Cao 1987c] and [Heildelberger et al. 1988]).

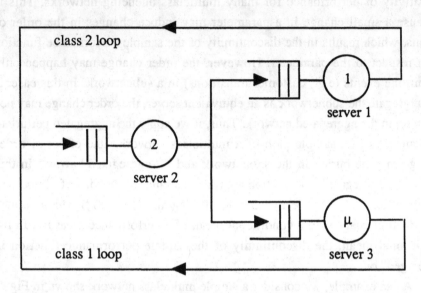

Fig. 2 A Two-Class Product-Form Network

To apply the aggregated perturbation analysis to the network, we decompose the network into two subnetworks, S_1 and S_2. S_1 consists of servers 1 and 2 and S_2 consists of server 3. Since the class 2 customers never leave S_1, the service rate of the equivalent server for class 2 customers is zero. The service rates of the equivalent server for class 1 customers depend on the number of the class 1 customers in it. By Norton's theorem, we can determine the rates as $\mu(1) = 0.5$ and $\mu(2) = 0.6$. Both the aggregated perturbation analysis estimates and ordinary perturbation analysis (applied to the original network) estimates are listed in Table 1. The results are based on the average of twenty replications of 10,000 services each. The numbers in parentheses are the estimated half-widths of 95% confidence intervals.

μ	Exact	Aggregated PA	Simple IPA
0.01	0.175	0.174 (0.004)	0.003
0.1	0.244	0.242 (0.006)	0.031
0.25	0.342	0.339 (0.008)	0.062
0.5	0.467	0.471 (0.10)	0.106
1.0	0.630	0.642 (0.014)	0.157
2.0	0.792	0.795 (0.016)	0.224
4.0	0.907	0.905 (0.019)	0.315
5.0	0.972	0.976 (0.019)	0.465

Table 1 Aggregated PA and IPA Estimates of $dT/d\mu$ in the Example.

Several additional examples were also shown in [Glasserman and Ho 1989]. They involve hyperexponential service times, servers with a last-come, first-served (LCFS) pre-emptive resume discipline, and more classes and servers. In short, more complex nonproduct-form features. Since analytical results are not available for $dT/d\mu$ for such networks, the aggregated perturbation analysis estimates were compared with the finite difference estimates obtained by repeated simulations. Both the "shorted S_2" and the "conditional departure" methods of aggregation were used. Experimental evidence suggests that these methods are reasonably accurate and potentially very useful. Indeed our consideration of the product-form networks suggests that the main difficulty lies in finding a good equivalent server and not in the aggregated perturbation analysis algorithm itself. The particular aggregation techniques for equivalent servers reported here are interesting and plausible ones. But other possibilities exist. The work in this area has just begun.

8.3 Aggregation and Decomposition of Markov Chains

The Markov chain model of DEDS is extremely general in the sense that no structural assumption about the system is required. So long as we are willing to increase the size of the state space, any kind of dynamic dependence on the finite past can be captured by redefining what is meant by the word "state." Mathe-matically, the model is also very clean and susceptible to analysis. A large amount of literature exists on the subject. The principal drawbacks of this model are computational and conceptual. Because of the combinatorial explosion of the size of the discrete state space, any realistic problem other than the simplest academic examples soon outstrips even the capability of a supercomputer. We can easily be required to invert numerically matrices of size billions by billions. Furthermore, by essentially one-dimensionalizingthe state space in the sense of listing one state after another, all structural information of the problem are suppressed[3]. For these reasons, serious performance evaluations on and insights about arbitrary real problems cannot be easily acquired by using the Markov chain model. In this section we shall address attempts at alleviating this computational burden and the extent to which they are successful.

The simplest form of aggregation for Markov chains is called "lumpability." We partition the state space, S, into non-intersecting sets $\{A_1, A_2, \ldots, A_N\}$ and consider each set as a distinct state. In a sense, this can be thought of as **state** aggregation in contrast with **time** aggregation to be described separately below. As the Markov chain evolves, it will travel through these sets. We shall denote the set in which the state lies at time t as A(t). Of course, it is important that the Markov property is retained for the sets visited by the Markov chain; i.e.,

$$P[A(t+1) / A(t), A(t-1), A(t-2), \ldots] = P[A(t+1) / A(t)] \qquad (7)$$

---

[3] Just compare the effort and memory size required for simulating a simple network vs. that of the same network modeled as an arbitrary Markov chain.

In general, we cannot expect this to be true for arbitrary chains and arbitrary partitions of the state space. It is not difficult to convince oneself that the necessary and sufficient condition for (7) is [Kemeny and Snell 1960, Theorem 6.3.2]

$$P(A_j/s_k \in A_i) = P(A_j/s_i \in A_i) \text{ for any } j \text{ and all } s_k, s_i \in A_i \qquad (8)$$

In other words, knowing the current state as opposed to merely knowing the set in which the current state lies does not enable us to predict the next set any better. Any partition of states into sets A_i, i=1,2, . . ., N, satisfying (8) will be considered an "exact" aggregation in the sense of exactly retaining the Markov property. However, it is also clear that this is a rather restrictive condition. Thus, in general, exact state aggregation is not often possible and cannot be considered as a general method of reducing the computational burden. In terms of the criteria set forth at the beginning of this chapter, such aggregations do not satisfy the "closure" property.

We shall now propose another method of exact aggregation which shall extend the applicability of the Markov chain approach by enabling an effective computational solution of some very large Markov chains, e.g., with a state space of one billion states. The basic idea can be described by an analogy to continuous variable dynamic systems (CVDS). For a general dynamic system governed by finite order linear differential equations, it is well-known that we can, to a very large extent, study its behavior in terms of a finite set of basis solutions of these differential equations. Different behaviors or performance are simply represented by different linear combinations of these basis solutions. The problem of determining the basis solutions can be separated from that of performance evaluation. The latter can be a static (non-dynamic) problem of a finite dimension. Our proposed method employs a similar decomposition: (i) performance evaluation of system sample paths starting from a given set of initial states and (ii) the probabilistic distribution of this set of states and the performance based on this distribution. We shall show effective computational techniques for handling (i) and (ii) involving 10,000 and 1 00,000 states respectively[4]. This will give a total state space size of one billion.

The method begins by choosing a subset set, A, of the state space. The sample paths of the Markov chain are segmented according to the times when the state transits through the set A, as illustrated in Fig. 3.

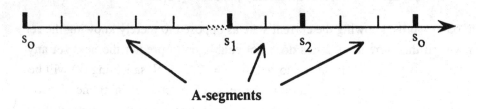

Fig.3 Example of A-Segments: $A = \{s_0, s_1, s_2\}$

Each of the A-segments is independent from the others on account of the Markov property. All A-segments starting from the same initial state, say s, can be grouped and analyzed together using whatever performance analysis techniques such as likelihood ratio or perturbation analysis methods. The result will be the **characteristic** of the A-segments. These represent the lower level behaviors. On the other hand, the states within A form another imbedded Markov chain with equilibrium distribution $v(s)$ for all $s \in A$. This imbedded chain has a smaller state space of size $|A|$, which roughly has the order of $|S|$ / L, where L is the average length of the A-segments. Each transition of this imbedded chain constitutes an A-segment. Assuming performance results of the A-segments can be determined, then the remaining problem is to merge these results according to the distribution $v(s)$ which must in general be experimentally determined. Note here the A-segments play the role of the basis solutions and $v(s)$ the linear combinations in analogy to the CVDS framework. More importantly, this approach to aggregation via the A-segments satisfies the "closure" property since the high level set A forms another Markov chain.

---

[4] By these numbers we mean the A-segments are approximately 10,000 transitions long and $|A|=100,000$.

Furthermore, as is typical with all approaches to decomposition, the effort involved in solving |A| lower level problems of dimension L can be expected to be much less than that of solving a problem of dimension L*|A|=|S|. The paragraphs below outline such an approach.

Consider a sample path represented by the state sequence $s_0, s_1, s_2, ...$ Following the same notation we adopted at the beginning of Section 7.3, we denote the sample and mean performance as

$$L(\xi) = \frac{1}{N+1} \sum_{i=0}^{N} F(s_i) \implies J(\theta) \equiv E[L(\xi)] = \frac{1}{N+1} \int \sum_{i=0}^{N} F(s_i) P(\xi) d\xi \tag{9}$$

$$\text{where} \quad P(\xi) = p(s_0) \prod_{i=0}^{N} p(s_{i+1}, s_i)$$

$L(\xi)$ is the sample performance measure, and $P(\xi;\theta)$ the probability of the sample path $\xi = (s_0, s_1, s_2, ..., s_N)$. Now we propose to compute the mean value of L in two steps. First, find the mean value of L over each A-segment starting with state s, and then average those mean values over all s∈ A, i.e.,

$$J(\theta) = \sum_{s \in A} [\int_{\Omega_s} L dP_s] \, v(s), \tag{10}$$

where we use LdP_s as a shorthand notation for the integration with respect to the probability space Ω_s of the A(s)-segment (starting with state s). Taking deriva-tives on both sides of (10), we get

$$\frac{\partial J(\theta)}{\partial \theta} = \sum_{s \in A} \{\frac{\partial}{\partial \theta} \int_{\Omega_s} L dP_s\} v(s) + \sum_{s \in A} \{\int_{\Omega_s} L dP_s\} v'(s). \tag{11}$$

Eq.(11) shows that the performance sensitivity problem can be decomposed into two smaller problems: (i) summing the derivatives of the mean value of the A(s)-segments over A, weighted by v(s), - the first summation on the right-hand side of (11), and (ii) summing the mean values of A(s)-segments over A, weighted by v'(s) the second summation of (11). The first sum can be carried out in principle by any methods described so far in this book. In particular, the likelihood ratio method seems to be particularly appropriate since by proper choice of the set A, the mean length of A-segments ($\approx |S|/|A|$) can be made fairly short (say, less then ten thousand transitions) to minimize the variance problem with the LR method. Furthermore, as will be made clear shortly, using the LR method will enable a certain amount of computations be shared by the calculation of the second summation of (11). As for the derivative $v'(s)$ of the equilibrium distribution involved in the second summation, several possibilities exist. First, $v'(s)$ can be zero in which case there is no contribution from the second term. Second, $v(s)$ and hence $v'(s)$ may be known analytically in closed form in which case the second summation may be directly evaluated. Lastly, the most frequent case in practice, we must somehow numerically solve for $v'(s)$. These three cases will be discussed separately below and in Section 8.4.

8.3.1 The Case of $v'(s)=0$

The easiest example is when the set A is a singleton. In this case, the method obviously reduces to the case of regenerative LR method discussed in Section 7.2. There is nothing new to add. However, it is worthwhile to point out that this is not the only example satisfying $v'(s)=0$ nor does general regenerative simulation necessarily imply $v'(s)=0$.

Example 1 Regeneration does not imply $v'(s)=0$.

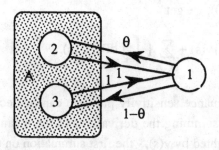

Fig.4 Example of a Regenerative System That Does Not Have n'(s)=0

Consider the system shown in Fig. 4. The state space = {1,2,3}. Let A={2, 3} and the transition probability among the states be as indicated with 0<θ<1. Each return to A clearly causes the system to regenerate and yet v(2)=θ and v(3)=1-θ. Therefore dv / dθ ≠ 0.

Example 2 v'(s)=0 does not imply regeneration.

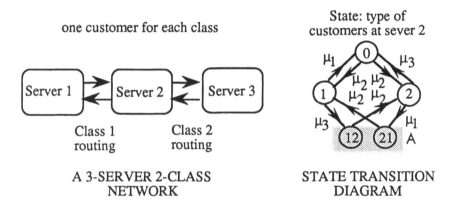

one customer for each class

State: type of
customers at sever 2

A 3-SERVER 2-CLASS
NETWORK

STATE TRANSITION
DIAGRAM

Fig. 5 Example of a Markov Chain With A={12, 21} and Non-Regenerating.

Consider the queueing network in Fig. 5, in which the class 1 customer circulates between servers 1 and 2, and the class 2 customer between servers 2 and 3. Let μ_1 and μ_3 be the service rates of servers 1 and 3, respectively, and μ be the service rate of server 2 for both customers. There are five possible states characterized by the customer position and class at server 2. Let A = {12, 21}. By symmetry, if $\mu_1 = \mu_3$ then the conditional steady-state probabilities of the states v{12} and v{21} over A are equal and independent of parameters such as μ_2. Hence $dv(s)/d\mu_2 = 0$ for s over A. Yet the segment starting in {12} is probabilistically different from the segment starting in {21}; hence the system is not regenerative from A.

A sufficient condition for v'(s)=0 for s∈ A is given in [Zhang and Ho 1989, 1991b]. However, the condition is quite restrictive. The more important point made here is the more general view of regenerative segments. It need not always be defined as starting from a fixed state but from a fixed set of states. However, in the latter case, we must be concerned with $v'(s)$ or the second term on the right-hand side of Eq.(11).

8.3.2 The Case Where $v'(s)$ Can Be Calculated In Closed Form

In this section, we give an example to show that we might be able to compute $v'(s)$ if there exists an imbedded Markov chain the distribution of which is somehow available. This example shows another possibility that analytical results can be utilized in simulation to control the length of A-segments and hence the variance of the estimates.

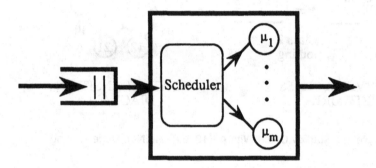

Fig.6 The Modified G/M/m Server With Scheduler

Consider a service station (see Fig. 6) with servers $S_1, S_2, \ldots, S_m$, which may not be identical. Customers arrive singly with i.i.d. interarrival times according to some given distribution. Let $1/\lambda$ be the mean interarrival time. Service times at S_i are exponentially distributed i.i.d.'s with mean $1/\mu_i$. $\mu=\mu_1+ \ldots +\mu_m$ is the full service rate of the station. The system is stable if $\lambda < \mu$. The queueing discipline is FCFS. There is a scheduler dispatching customers from the head of the queue to the idle servers. The only constraint on the scheduler is that no server will be kept idle if there are customers waiting in the queue. This is the so-called the*work conserving* requirement. We call the system in Fig.6, which can be quite complicated, the *modified* GI/M/m queue because of the existence of the scheduler and the nonidentical servers. There is always an imbedded GI/M/1 queue with service rate μ in the system whenever all the servers are busy. Let us see how we can take advantage of this property. Let T_i be the instant immediately prior to the ith arrival that queues (i.e., the customers immediately served upon arrival are not counted). We shall use these T_i instants to define our A-segments. In other words, the set A is defined by A={(m+k, 0-)| k≥0} where 0- means immediately prior to an arrival and k is the length of the queue. It turns out that the imbedded Markov chain at the instants {T_i | i=0,1,2, . . } has a geometric steady-state distribution   [Kleinrock 1975 p. 247] $v((m+k, 0-)) = (1-\sigma)\sigma^k$, where σ is the root of $\sigma=A*(\mu-\mu\sigma)$ and A* the Laplace transform of A(t). Knowing this, we can develop a scheme to calculate v'(s). Details are omitted here (see [Zhang 1990]). The case where v'(s) must be calculated numerically is separately treated in Section 8.4.

8.4 Decomposition via the A-Segment Algorithm for Very Large Markov Chains

In this section we shall describe a scheme for the numerical calculation of v'(s). This is of course the most frequently encountered case for Eq.(11). Indeed, if the above aggregation / decomposition method is going to be anything other than a limited curiosity, we must develop an effective computational procedure for v'(s) of the imbedded Markov chain, which can still have a reasonably large size (say 100,000 if the total |S|= one billion and the average length of A-segments L=10,000). Now the steady-state distribution of the imbedded chain obeys v=Qv, where Q is its transition probability matrix. Differentiating, we get

$$v'[Q\text{-}I] + vQ' = 0. \tag{12}$$

In principle, we can attempt to experimentally determine Q, Q', v, and v' by any combinations of simulations and numerical solutions of the linear equations (12) However, this is hardly a practical approach when the size of the chain is at 100,000. Since Q will be a 100,000 ×100,000 matrix thus requires a memory size of 10 billion storage locations; there is also the associated numerical problem of inverting a matrix of such magnitude. Instead, some iterative but convergent approach to directly estimate v'(s) will be devised. Furthermore, we shall take advantage of the fact that it is the expected value of some performance measure with respect to v'(s) (the second term on the r.h.s. of (11)) that we are ultimately interested in. v'(s) need not be estimated equally accurately for all states; it is only those states that occur most frequently that are important[5]. As we shall see, simulation and iterative calculations provide an automatic means for allocating more computational effort to estimating v' for those states. It is these combi-nations (A-segment decomposition and estimating v'(s) of the imbedded chain) that enable us to solve the performance evaluation problem of very large Markov chains.

To solve (12) for v' using noisy versions of Q, Q', and v, we apply the idea of stochastic approximation - finding roots of h(x)=0 when only noisy estimates of h(x) are available (see Appendix C). Briefly, one theoretical and two practical problems must be dealt with: theoretically, we must show the convergence of any algorithm for estimating v'(s); practically, any algorithm we propose must first avoid storing matrices of estimates of Q and Q' since they have the dimensions of |A| x |A|[5], and second, we should avoid schemes that require updating v'(s) for all s's each time we terminate an A-segment. Ideally, to save computational effort, only v'(s) for a particular s need to be revised upon the termination of an A(s)-segment. The algorithm to be described below satis-fies all these requirements.

---

[5] What we mean by the remarks that no storage of Q is required and that no structural assump-tions are made about Q is this: During the course of the iterative computation for v' no use is made of any structural knowledge of Q nor is there any requirement for the storage of Q. *The only data required are what a real system would generate as it evolves* . However, if a simulation is attempted, then there will be the separate question of how to generate the state transitions of an arbitrary Q without requiring the storage of Q itself. Usually, some structural assumption is made, e.g., the system is a queueing network with given topology, in order to generate sample paths without the need of the storage of Q. We submit that the algorithm we are about to describe cannot be held responsible to such assumptions.

Recall that the general stochastic approximation algorithm for finding the roots of $h(z)=0$ is (Appendix C Section 4)

$$z_{k+1} = z_k + a_k \hat{h}(z_k) \tag{13}$$

where $\hat{h}(z_k)$ is a noisy estimate of $h(z_k)$ with $E[\hat{h}(z_k)]=h(z_k)$, $Var[\hat{h}(z_k)]<\infty$, $a_k \to 0$ and $k a_k \to \infty$. Let us now apply the algorithm in (13) to find the solution $v'(s)$ to Eq.(12). Assume that an A-segment starts with $s=Z_k \in A$ and terminates with $s'=Z_{k+1} \in A$. We shall see that the particular version of (13) for the solution $v'(s)$ to Eq.(12) is

$$z_{k+1}(s') = z_k(s') - a_k z_k(s') \qquad \text{for } s' \neq Z_{k+1} \tag{14}$$

$$z_{k+1}(Z_{k+1}) = z_k(Z_{k+1}) + a_k \{ z_k(Z_k)\frac{k}{i_k(Z_k)} - z_k(Z_{k+1})+D_{Z_k Z_{k+1}} \}$$
$$\text{for } s' = Z_{k+1} \tag{15}$$

where $i_k(s)$ is the number of visits to $s \in A$ in the first k transitions of the imbedded chain, and $D_{ss'}$ is a random variable whose conditional mean, given that the A-segment starts with s and terminates with s', equals the ratio of the derivative of the transition probability from s to s' in the imbedded Markov chain to the transition probability itself, i.e., $q'(s,s')/q(s,s')$ and $E[|D_{ss'}|^2]<\infty$.

To avoid updating all state $s \in A$ each time we encounter a state transition of the imbedded chain, we take advantage of the simplicity of (14) and (15) and define a simple change of variables $Y_k \equiv p_k z_k$ with $(1-a_k)p_{k+1} \equiv p_k$. Such a p_k and a_k sequence is always possible, e.g., $a_k = 1/(k+1)$, $p_k = k$. In terms of Y_k, (14) and (15) become

$$Y_{k+1}(s') = Y_k(s') \qquad \text{for } s' \neq Z_{k+1} \tag{16}$$

$$Y_{k+1}(Z_{k+1}) = Y_k(Z_{k+1})+a_k p_{k+1}\left[\frac{k}{p_k} \frac{Y_k(Z_k)}{i_k(Z_k)} + D_{Z_k Z_{k+1}}\right]$$
$$\text{for } s' = Z_{k+1} . \tag{17}$$

Eqs.(16) and (17) now have the required property of selectively updating only one value of v'(s) for each transition of the imbedded chain. If we choose $a_k = 1/(k+1)$ and $p_k = k$, then the algorithm simply becomes

$$Y_{k+1}(s') = Y_k(s') \qquad\qquad \text{for } s' \neq Z_{k+1}$$

$$Y_{k+1}(Z_{k+1}) = Y_k(Z_{k+1}) + \frac{Y_k(Z_k)}{i_k(Z_k)} + D_{Z_k Z_{k+1}} \qquad \text{for } s' = Z_{k+1}.$$

Next, we must demonstrate a particular $D_{ss'}$ with $E[D_{ss'}] = q'(s,s')/q(s,s')$. Let X_i be the ith state of the original Markov chain, and denote the first state of the kth A-segment as the n_kth state of the original Markov chain. Thus, $Z_k = X_{n_k}$ and $Z_{k+1} = X_{n_{k+1}}$. Define

$$D_{X_{n_k} X_{n_{k+1}}}(\omega_k) = \frac{\partial}{\partial \theta} \log[P_{X_{n_k}}(\omega_k)]$$

$$= \sum_{i=n_k}^{n_{k+1}-1} \frac{p'(X_i, X_{i+1})}{p(X_i, X_{i+1})} = \sum_{i=n_k}^{n_{k+1}-1} d(X_i, X_{i+1}) \qquad (18)$$

where ω_k denotes the sample path of the A-segment starting in state $s = X_{n_k}$ and terminating in state $s' = X_{n_{k+1}}$. Eq.(18) shows that $D_{ss'}$ is in fact a term used in the likelihood ratio estimate of the A-segment (see Section 7.2 and 8.3). In (18),

$$P_{X_{n_k}}(\omega_k) = P_s(\omega_k) = \prod_{i=n_k}^{n_{k+1}-1} p(X_i, X_{i+1}), \qquad X_{n_k} = s, \; X_{n_{k+1}} = s'$$

is the probability of the A-segment ω_k. Let $\Omega_{ss'}$ be the space of the A-segments starting from s and terminating in s'. Then we have

$$q(s, s') = \sum_{\omega_k \in \Omega_{ss'}} P_s(\omega_k).$$

Given that the A-segment begins with s and terminates with s', the conditional probability of ω_k is simply $P_s(\omega_k)/q(s, s')$. Thus, the mean value of $D_{ss'}$ defined in (18), given that $X_{n_k}=s$ and $X_{n_{k+1}}=s'$, is

$$E[D_{s\,s'}] = \sum_{\Omega_k \in \Omega_{ss'}} \frac{\partial \log[P_s(\omega_k)]}{\partial \theta} \frac{P_s(\omega_k)}{q(s,s')} = \frac{q'(s,s')}{q(s,s')}. \tag{19}$$

The second equation is by virtue of $\Sigma \{\partial \log[P_s(\omega_k)]/\partial \theta\} P_s(\omega_k) = q'(s, s')$. Eq.(19) justifies that the $D_{ss'}$ defined in (18) can be chosen as that in algorithm (15).

We also need to show that $\hat{h}(z_k)$ as defined by (13) - (15) has the property that $E[\hat{h}(z_k)] = h(z_k)$. Note that both $\hat{h}(z_k)$ and $h(z_k)$ are now $|A|$-dimensional vectors. Their components are denoted as $[\hat{h}(z_k)]_s$ and $[h(z_k)]_s$, respectively. Examining (13) - (15) component-wise, for states s' = 1, 2, . . . , $|A|$, we note

$$[\hat{h}(z_k)]_{s'} = \begin{cases} -z_k(s') & s' \neq Z_{k+1}, \\ z_k(Z_k)\dfrac{k}{i_k(Z_k)} - z_k(s') + D_{Z_k s'} & s' = Z_{k+1}. \end{cases} \tag{20}$$

Taking the conditional expectation of (20) for a given Z_k (Z_{k+1} is random), we have

$$E\{[\hat{h}(z_k)]_{s'} | Z_k\} = -z_k(s') + z_k(Z_k)\frac{k}{i_k(Z_k)} q(Z_k, s') + E[D_{Z_k s'} | Z_k, s'] q(Z_k, s').$$

By virtue of the property of $D_{ss'}$ just shown in (19), this equation reduces to

$$E\{[\hat{h}(z_k)]_{s'} | Z_k\} = -z_k(s') + z_k(Z_k)\frac{k}{i_k(Z_k)} q(Z_k, s') + q'(Z_k, s'). \tag{21}$$

Now let us take the expectation of (21) with respect to Z_k. Note that $P(Z_k=s) \equiv v(s)$ and $\lim_{k \to \infty} E[k/i_k(Z_k)] = 1/v(s)$. Denoting $\lim_{k \to \infty} z_k(s) = z(s)$, we get

$$\lim_{k \to \infty} E[\hat{h}(z)]_{s'} = \sum_{s \in A} v(s)\{z(s)\frac{1}{v(s)} q(s, s') - z(s')\} + \sum_{s \in A} v(s) q'(s,s')$$

$$= \sum_{s \in A} z(s) q(s, s') - z(s') + \sum_{s \in A} v(s) q'(s,s'),$$

where z is a vector with components $z(s')$, and z_k given by the algorithms (14) and (15) approaches z as $k \to \infty$. Writing the above equation in a vector form, we have

$$\lim_{k \to \infty} E[\hat{h}(z)] = z [Q - I] + vQ' = h(z). \tag{22}$$

Thus, $E[\hat{h}(z_k)] = h(z_k)$ holds asymptotically.

Finally we need to show that (14) and (15), or (16) and (17), satisfy the convergence conditions of stochastic approximation of (13). The task is compli-cated by the fact that the "noise" in the estimate $\hat{h}(z_k)$ is correlated for different k. The usual simple conditions for convergence of stochastic approximations are not applicable. The convergence proof utilizing some results of [Kushner and Clark 1976] is fairly complex and not particularly enlightening. We omit the details [Zhang 1990].

To demonstrate the effectiveness of this approach we perform two sets of experiments:

(i) Performance comparison with respect to the LR method without A-segmentation and

(ii) Performance on Markov chains with very large state spaces but with known analytical solutions.

For the first set of experiments, we choose the network of Fig. 7.

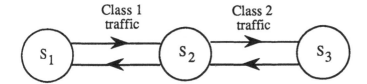

Fig. 7 A Two-Class Three-Server Network

The data for the network is: $\mu_{1,1} = 2$, $\mu_{2,1} = 1.2$, $\mu_{2,2} = 1.8$, $\mu_{3,2} = 2.5$, where $\mu_{i,j}$ denotes the mean service time of class j customers at server i, and $N_1 = N_2 = N$ is the number of customers of each class. For A, we choose the set of states that has all customers at server S_2 and the first k customers in the queue at S_2 alternate in classification, e.g. the state $[C_1, C_2, C_1, C_2, \ldots$ arbitrary, $\ldots C_{2N}]$ means k=4. Thus by specifying k we can easily specify the size of A. For N = 2,3,4,5,6, the size of S, $|S|$, ranges from 10 to 3500. To measure the effectiveness of the A-segment method, we calculate the ratio of the variances of the LR estimates of the throughput for class 1 with respect to mean service rates without and with A-segmentation (averaged over all four mean service rates). Fig. 8 illustrates the results. These results are average over a minimum of 36 replications per experiment with at least 400 regenerative cycles per replication. It is clear that substantial gains in variance reduction are possible.

Exercise Provide an explanation to the above experimental evidence that there appears to be an optimum choice of the size of $|A|$ for a problem.
[Hint: Consider the two limits of A = singleton and A=the entire space S and the reason why the variance reduction from controlling the length of A-segment is significant]

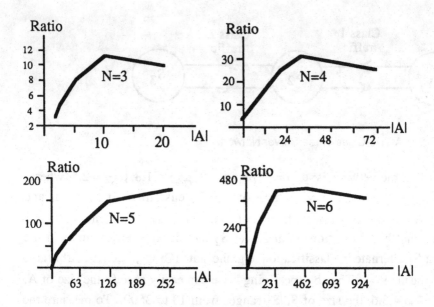

Fig. 8 Variance Ratios of LR Estimates   Without to With A-Segmentation of
Class 1 Average Throughput Sensitivity With Respect to Mean Service Times.

For the second set of experiments, we are in uncharted waters. There exists no alternative means to compute the performance of arbitrary Markov chains with state space in the range of 10^8 and 10^9. Consequently, our only means of checking the result is to use examples with known analytical answers. For this we take the same example as that of Fig.7 but change the mean service times for both classes at S_2 to $\mu_{2,1} = \mu_{2,2} = \mu_2 = 2$, which results in a product-form network. The following results are obtained.

N	$\partial TP_1/\partial \mu_i$	A-segment	95% Cf. Intl.	Theoretical
8	1	0.0274	0.0008	0.0284
	2	0.9806	0.0023	0.9783
	3	-0.0161	0.0007	-0.0157
10	1	0.0232	0.0010	0.0229

	2	0.9832	0.0021	0.9824
	3	-0.0133	0.0007	-0.0125
12	1	0.0196	—	0.0191
	2	0.9840	—	0.9852
	3	-0.0102	—	-0.0104
14	1	0.0144	—	0.0164
	2	0.9917	—	0.9872
	3	-0.0096	—	-0.0089

For these experiments, the lengths of the shortest regeneration cycles range from 33K to 105 million, the mean A-segment lengths range from 7.9 to 12,283 transitions, $|S| = 48K$ to 155 million, and the total number of transitions/ experiments = 18 million - 400 million[6]. The effectiveness of the A-segment methodology is clear.

Finally, we show one example of a Markov chain based on 5 servers with two classes of customers. The routing matrix for classes 1 and 2 are respectively

$$
\begin{array}{ccccc}
0.4717 & 0.3445 & 0.0405 & 0.0312 & 0.1121 \\
0.4837 & 0.1571 & 0.1027 & 0.1000 & 0.1565 \\
0.7030 & 0.1661 & 0.0201 & 0.0804 & 0.0304 \\
0.2291 & 0.3452 & 0.1237 & 0.1301 & 0.1719 \\
0.4814 & 0.2117 & 0.1070 & 0.1010 & 0.0989
\end{array}
$$

and

$$
\begin{array}{ccccc}
0.4102 & 0.3136 & 0.1501 & 0.0901 & 0.0360 \\
0.4586 & 0.3976 & 0.1415 & 0.0011 & 0.0012 \\
0.4330 & 0.2056 & 0.0707 & 0.2803 & 0.0104
\end{array}
$$

---

[6] At the high end of the experiments (N=12 or 14), even a single run of a hundred million transitions can be time consuming. Depending on the computing environment, a rate of 100K - 2 million transitions per minute is possible (translating into a couple of hours to a couple of days per run). Thus, no confidence intervals from multiple runs were calculated.

| 0.4737 | 0.2385 | 0.1551 | 0.1300 | 0.0027 |
| 0.4202 | 0.1701 | 0.1237 | 0.2001 | 0.0859 |

The service rates for both classes at each server are given by

| Server | S_1 | S_2 | S_3 | S_4 | S_5 |
| Rate | 9.7 | 7 | 3.3 | 13 | 2.5 |

The visiting ratio for each class is

| C1 | 0.476 | 0.266 | 0.069 | 0.068 | 0.121 |
| C2 | 0.435 | 0.316 | 0.137 | 0.094 | 0.019 |

The size of the state space is about 1.9 billion with $|A| = 184756$, and the mean length of A-segments is 136.2. The set A is taken to be the set of states when all customers are at server 1, the bottleneck server. We estimate the derivative of the throughput of class 2 customers with respect to service rate at S_1. The estimate is 0.7339 and the correct theoretical value is 0.7321. Four hundred million transitions were run[7].

---

[7] At 2 million state transitions/min on a dedicated SUN 4, this example took 20 hours to make one run.

Elements of Probability and Queueing Theory

The primary purpose of this appendix is to summarize for the reader's convenience those parts of the elements of probability theory, Markov process, and queueing theory that are most relevant to the subject matter of and in consistent notations with this book. Mathematical terminologies are used only for the purpose of introducing the language to the intended readers. No pretense at rigor is attempted. Except for some relatively new material and interpretations, the reader can find a more thorough introduction of this material in [Kleinrock 1975 Chapters 2 - 4] and [Cinlar 1975].

A.1 Elements of Probability Theory

Let (Ω, F, P) be a probability space, where F is a σ-field in Ω and P is a probability measure defined on F. A **random variable** X is a real (measurable) function defined on the probability space. For any $\omega \in \Omega$, $X(\omega) \in R$ is a real number. The distribution function of X is defined as

$$F(x) = P(\omega : X(\omega) \leq x).$$

Function f(x) is called the **probability density** of the random variable X if for any subset $A \in R$, we have

$$P(X \in A) = \int_A f(x)\, dx.$$

If F(x) is differentiable at x, then $f(x) = dF(x)/dx$. If X takes values only in a countable set of real numbers, X and F(x) are called *discrete*. A familiar discrete distribution is the *binomial* distribution:

$$p(X=r) = \binom{n}{r} p^r (1-p)^{n-r}, \qquad\qquad r=0,1,2,...,n.$$

If n=1, the distribution is called a Bernoulli distribution. For engineering purposes, we can think of F(x) or f(x) as a function defining or characterizing the random variable, X. To simplify the description, we often use gross characterizations, such as, the **mean** and **variance** which give respectively the center of gravity and the spread (along the x axis) of the shape of f(x). The mean (or the expected) value of a random variable X on Ω is the integral of X with respect to the distribution function F(x):

$$E[X] = \int_{-\infty}^{\infty} x dF(x) = \int_{-\infty}^{\infty} x f(x) dx.$$

The kth moment of X is

$$E[X^k] = \int_{-\infty}^{\infty} x^k dF(x) = \int_{-\infty}^{\infty} x^k f(x) dx.$$

In particular, the variance of X is the second central moment

$$var[X] = E[(x - E[X])^2 = E[X^2] - \{E[X]\}^2.$$

A **Random or Stochastic Sequence** is simply a sequence of random variables indexed by an independent variable, usually the time. A **random sequence** is defined by specifying the joint distribution function of all the random variables $X_1, X_2, X_3, \ldots, X_n, ..$ involved. If

$$f(X_{n+1}|X_n, X_{n-1}, X_{n-2}, \ldots) = f(X_{n+1}|X_n)$$

then the random sequence is said to possess the **Markov property** which colloquially can be described as "knowledge of the present separates the past and the future." Other modifying adjectives, such as **Gaussian**, **Stationary** (wide sense and strict sense), **Purely random** (the so-called "white" noise), **Discrete State, Semi-Markov, Renewal** etc are additional specifications and / or restrictions placed on the joint density function.

Exercise Make the above defining adjectives conceptually precise by spelling out these specifications on the joint density functions.

Ignoring mathematical intricacies, we can think of a **random or stochastic process** as the limit of a random sequence when the indexing variable becomes continuous. For a sequence of random variables, we are often interested in whether or not it converges, and if so, whether it converges to a given random variable or a deterministic constant. Let X_n be a sequence of random variables with distribution functions $F_n(x)$, $n = 1, 2, \ldots$ This sequence converges to a random variable X if $|X_n - X|$ becomes small in some sense as $n \to \infty$. The definitions for different convergences are given as follows.

(i) **Convergence in probability:** If for any $\varepsilon > 0$, $\lim_{n \to \infty} P[\,|X_n - X| \geq \varepsilon\,] = 0$, then X_n is said to converge to X in probability as $n \to \infty$.

(ii) **Convergence with probability one (almost sure (a.s.) convergence):** If for any $\varepsilon > 0$, $\lim_{n \to \infty} P[\max_{k \geq n} |X_k - X| \geq \varepsilon] = 0$, or If $P[\omega : \lim_{n \to \infty} X_n(\omega) = X(\omega)] = 1$, then X_n is said to converge to X with probability one (w.p.1), or almost surely, as $n \to \infty$.

(iii) **Convergence in mean or mean square:** If $\lim_{n \to \infty} E[\,|X_n - X|\,] = 0$ or $\lim_{n \to \infty} E[\,|X_n - X|^2\,] = 0$, then X_n is said to converge to X in mean or in mean square as $n \to \infty$. Convergence in mean square implies convergence in mean.

(iv) **Convergence in distribution and weak convergence:** If $\lim_{n \to \infty} F_n(x) = F(x)$ for every continuity point x of F, then X_n is said to converge to X in distribution as $n \to \infty$, and the distribution function $F_n(x)$ is said to converge weakly to $F(x)$.

Some examples will help to gain an intuitive understanding of the definitions.

Example 1 Let the sequence of X_i's be independently distributed according to

$$P(X_n = 1) = \frac{1}{n}$$

and

$$P(X_n = 0) = 1 - \frac{1}{n}.$$

(i)   This sequence clearly converges in probability to the degenerate random
      variable, zero, since

$$\lim_{n \to \infty} P(|X_n| \geq \varepsilon) = \begin{cases} \lim_{n \to \infty} \frac{1}{n} = 0 & \text{for } \varepsilon < 1, \\[2mm] \lim_{n \to \infty} 0 = 0 & \text{for } \varepsilon \geq 1. \end{cases}$$

(ii) However, this sequence does not converge to zero with probability one.
      Since

$$\lim_{n \to \infty} P\{\underset{k \geq n}{\text{Max}}|X_k| \geq \varepsilon\} = \lim_{n \to \infty} \{1 - \prod_{k=n}^{\infty}(1 - \frac{1}{k})\} = 1 - 0 = 1 \quad \text{for any n and } \varepsilon < 1.$$

(iii) Finally, this sequence does converge in mean or mean square because

$$\lim_{n \to \infty} E[X_n] = \lim_{n \to \infty} \{1 \times P(X_n=1) + 0 \times P(X_n=0)\} = \lim_{n \to \infty} \frac{1}{n} = 0,$$

and

$$\lim_{n \to \infty} E[X_n^2] = \lim_{n \to \infty} \{1^2 \times P(X_n=1) + 0^2 \times P(X_n=0)\} = \lim_{n \to \infty} \frac{1}{n} = 0.$$

Example 2 Consider the same setup as in Example 1 except

$$P(X_n = n) = \frac{1}{n^2}$$

and

$$P(X_n = 0) = 1 - \frac{1}{n^2}.$$

(i)   It is clear that convergence in probability still holds.

(ii) But now the sequence also converges with probability one since

$$\lim_{n \to \infty} P\{\underset{k \geq n}{\text{Max}}|X_k| \geq \varepsilon\} = \lim_{n \to \infty} \{1 - \prod_{k=n}^{\infty}(1 - \frac{1}{k^2})\} = 1 - 1 = 0, \quad \forall \text{ large n and } \varepsilon < 1.$$

(iii)   However, convergence in mean and mean square holds and no longer holds, respectively. This is because

$$\lim_{n \to \infty} E[X_n] = \lim_{n \to \infty} \{n \times P(X_n=1) + 0 \times P(X_n=0)\} = \lim_{n \to \infty} \frac{n}{n^2} = 0$$

and

$$\lim_{n \to \infty} E[X_n^2] = \lim_{n \to \infty} \{n^2 \times P(X_n=1) + 0^2 \times P(X_n=0)\} = \lim_{n \to \infty} \frac{n^2}{n^2} = 1.$$

Very roughly speaking, convergence with probability one is concerned with how fast deviations from the limit disappears as $n \to \infty$ while mean square convergence is concerned with how fast the amplitude of the deviation decreases as $n \to \infty$. Finally, convergence in distribution or weak convergence is concerned with whether or not the distribution of a random variable becomes statistically indistinguishable from that of another. In weak convergence, we do not make sample-path-wise comparisons. Both convergence with probability one and convergence in mean (mean square) imply convergence in probability which, in turn, implies convergence in distribution [Billingsley 1979]. Convergence with probability one and convergence in mean do not imply each other. However, if X_n is dominated by another random variable, Y, which has a finite mean, then X_n converges to X with probability one (w.p.1) implies that X_n converges to X in mean. This can be easily proved by using the Lebesgue dominated convergence theorem stated below.

Lebesgue dominated convergence theorem:
Let X_n and Y be random variables on Ω. If Y has a finite mean, $|X_n| \leq Y$ w.p.1, and $\lim_{n \to \infty} X_n(\omega) = X(\omega)$ w.p.1, then $\lim_{n \to \infty} E[X_n(\omega)] = E[\lim_{n \to \infty} X_n(\omega)] = E[X(\omega)]$.

Pictorially, the relationship among the various convergence definitions are as follows:

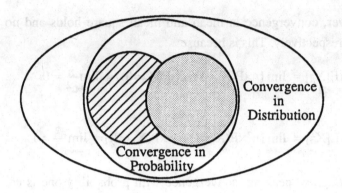

Another important theorem is:

The Law of Large Numbers:
Suppose that $\{X_n\}$ is an independent sequence and $E[X_n] = 0$. If Σ_n Var $[X_n] / n^2 < \infty$, then

$$\lim_{n\to\infty} \frac{1}{n} \sum_{k=1}^{n} X_k = 0, \qquad\qquad \text{w.p.1.}$$

A direct corollary of this result is: Suppose that $X_1, X_2, ...$ are independent and identically distributed (i.i.d) and $E[X_n] = E[X]$, then

$$\lim_{n\to\infty} \frac{1}{n} \sum_{k=1}^{n} X_k = E[X], \qquad\qquad \text{w.p.1.}$$

Finally, another concept used quite often in this book is the consistency of estimates. Let e be a quantity to be estimated and e_n, n=1,2,..., be estimates of e. The estimate e_n is said to be consistent for e if e_n converges to e in probability or with probability one. The former may be called weak consistency and the latter, strong consistency. For example, if $X_1, X_2, ...$ are i.i.d., then $\{X_1 + ... + X_n\}/n$ is a strong consistent estimate of $E[X]$.

A.2 Markov Processes

We review in this section the basic concepts and terminologies of the Markov processes basically following [Cinlar 1975].

A.2.1 Markov Chains

Let Ω be a sample probability space and P a probability measure on it. Consider a *stochastic sequence* $X = \{ X_n, n \in (0, 1,...) \}$ with a countable state space Ψ and indexed by time; that is, for each $n \in (0,1,...)$ and $\omega \in \Omega$, $X_n(\omega)$ is an element of Ψ. The elements of Ψ are customarily denoted as integers. Thus $X_n = j$ means the process is in state j at time n.

A stochastic process $X = \{X_n, n=0,1,...\}$ is called a Markov chain provided that $P(X_{n+1}=j \mid X_0, X_1, ..., X_n) = P(X_{n+1}=j \mid X_n)$ for all $j \in \Psi$ and $n \in \{0, 1,...\}$[1]. Let $P(i,j) \equiv P(X_{n+1}=j \mid X_n =i)$ be the transition probability of the Markov chain and $\mathbf{P} = [P(i,j)]$, $i,j \in \Psi$ be the transition matrix. Let $\pi(i)$, $i \in \Psi$, be the steady state probability of state i. Then $\pi(i)$ satisfies the following balance equation

$$\pi(i) = \sum_{j \in \Psi} \pi(j)P(j, i), \qquad\qquad i \in \Psi,$$

and the normalizing equation

$$\sum_{i \in \Psi} \pi(i) = 1.$$

Suppose the above set of equations has a unique solution. If at time n the state probability is $P(X_n = i) = \pi (i)$ for all $i \in \Psi$, then we have $P(X_{n+1} = i) = \pi (i)$ for all $i \in \Psi$. In particular, if $P(X_0 = i) = \pi (i)$, then $P(X_n = i) = \pi (i)$ for all n; such a Markov chain is called a *stationary* Markov chain.

A set of states in Ψ is said to be *closed* if no state outside the set can be reached from any state in the set. A closed set is *irreducible* if no proper subset of it is closed. A Markov chain is called an *irreducible* chain if the only closed

---

[1] In other words, knowledge of the present separates the past and the future.

set is the state space Ψ itself. In other words, in an irreducible Markov chain, the process in any state has a positive probability of reaching any other state in the state space. A state $i \in \Psi$ is called a *periodic* state, if X visits i periodically, i.e., if $X_k = i$, then $X_l = i$ only if $l = k + md$, $m = ...-2, -1, 0, 1, 2,...$, for some integer $d>1$. A Markov chain is said to be *aperiodic*, if no states are periodic.

An important property of a Markov chain is the *ergodicity*. Roughly speaking, if $X = \{X_n, n \in (0, 1,...)\}$ is an ergodic Markov chain, then for any function $f : \Psi \to R$, we have

$$\lim_{n\to\infty} \frac{1}{n} \sum_{k=0}^{n-1} f(X_k) = E[f(X)], \qquad\qquad \text{w.p.1.}$$

where $E[f(X)]$ is the steady state mean of $f(X)$:

$$E[f(X)] = \sum_{i \in \Psi} f(i)\pi(i).$$

Let $f = \chi_A$ be an indicator function of a subset A of Ψ, i.e.,

$$\chi_A(X_k) = \begin{cases} 1 & \text{if } X_k \in A, \\ 0 & \text{if } X_k \notin A. \end{cases}$$

Then we have

$$\lim_{n\to\infty} \frac{1}{n} \sum_{k=0}^{n-1} \chi_A(X_k) = \pi(A), \qquad\qquad \text{w.p.1.}$$

where $\pi(A)$ is the steady-state probability of the set A.

One very useful result regarding ergodicity is as follows. If $X = \{X_n, n \in (0, 1,...)\}$ is ergodic, then the Markov chain $Y = \{Y_n, n \in (0, 1,...)\}$, $Y_k = \phi(X_k, X_{k+1}, ...)$, is also ergodic where $\phi : R^\infty \to R$ is a measurable function [Breimen 1968]. Finally, if an irreducible Markov chain is aperiodic and has only finitely many states, then it is ergodic.

A.2.2 Markov Processes

The succinct discussion of Markov processes in this sub-section is extracted from [Cinlar 1975]. The approach taken there is based on the behavior of the sample paths, which is well-suited to the topics of this book.

A stochastic process $Y = \{Y_t, t \in [0, \infty)\}$ is called a Markov process with a countable state space Ψ, provided that for any $s > 0$ and $j \in \Psi$, $P(Y_{t+s} = j \mid Y_u, u \leq s) = P(Y_{t+s} = j \mid Y_s)$. For any $t \in [0, \infty)$, Y_t is a random variable on Ω. One difference between a Markov chain and a Markov process is that the indexing parameter of a Markov chain takes discrete values and that of a Markov process takes continuous values. A Markov process is said to be time homogeneous if $P(Y_{t+s} = j \mid Y_s)$ does not depend on s. We only deal with time homogeneous discrete state Markov processes in this book.

The transition function of a Markov process is defined as $P_t(i,j) = P(Y_{t+s} = j \mid Y_s = i)$. It satisfies the following properties:

$$P_t(i,j) \geq 0,$$

$$\sum_{k \in \Psi} P_t(i,k) = 1, \qquad\qquad i \in \Psi,$$

and the Chapman-Kolmogorov equation:

$$\sum_{k \in \Psi} P_t(i,k) P_s(k,j) = P_{t+s}(i,j).$$

In Matrix form, it can be written as

$$\mathbf{P_{t+s}} = \mathbf{P_t} \times \mathbf{P_s},$$

where $\mathbf{P_t}$ is the matrix whose (i, j)th entry is $P_t(i, j)$.

Let W_t be the length of the time that the process Y remains in state Y_t:

$$W_t = \inf \{ s > 0 : Y_{t+s} \neq Y_t \}.$$

Then

$$P\{W_{\triangleright u}| Y_t{=}i\} = e^{-\lambda(i)u} , \quad u{>}0$$

where $\lambda(i)$ is the transition rate of state i. Let T_n, n =0,1,2,..., be the nth transition time of the process and $X_n = Y_{T_n}$ be the state after the transition at T_n. Denoting $X_n = i$, we have

$$P\{X_{n+1}{=}j, T_{n+1}{-}T_n{>}u\, |X_0,..., X_n; T_0,..., T_n\} = Q(i,j)e^{-\lambda(i)u}$$

$Q(i, j)$ is the transition probability from state i to state j, which satisfies $Q(i, j) \geq 0$, $Q(i,i){=}0^2$ and $\Sigma_{k\in\Psi} Q(i, k) {=}1$.

The process $X = \{ X_n, n{=}0,1,2,...\}$ is a Markov chain with a transition probability $Q(i, j)$. This Markov chain is called an **Imbedded Markov Chain** of the Markov process Y. Of course, a Markov chain may also be imbedded in a stochastic process which is not a Markov process, e.g., while the transition probability among states may be Markov, the time duration between transition may be arbitrarily distributed. In such cases, we have a **semi-Markov process or generalized semi-Markov process (GSMP)** for even more general dependence on the past behavior (see Appendix B).

The transition function $P_t(i, j)$ of the Markov process is the solution to a set of differential equations. First, we define

$$A(i,j) = \begin{cases} -\lambda(i) & \text{if } i{=}j, \\ \lambda(i)Q(i,j) & \text{if } i{\neq}j, \end{cases}$$

where $Q(i,j)$ is as defined above (Note that $\Sigma_j A(i,j) = 0$). Then we have

$$\frac{d}{dt}P_t(i,j) = \sum_{k\in\Psi} A(i,k)P_t(k,j) = \sum_{k\in\Psi} P_t(i,k)A(k,j) \quad \text{for all } i,j\in\Psi.$$

---

[2] This is by convention. That is, when the embedded chain jumps from a state to itself, one observes no transition in the continuous time Markov process itself.

In Matrix form, we have

$$\frac{d}{dt} P_t = A P_t = P_t A$$

where A is a matrix whose components are $A(i, j)$. The solution to this equation is

$$P_t = e^{t A} \approx I + At + ..$$

For $t = \Delta t \to 0$ we have $A = \lim_{\Delta t \to 0} [P_{\Delta t} - I / \Delta t]$. Thus, the transition function can be generated by the matrix A. A is called the **infinitesimal generator** of the Markov process Y.

A.3 The Single Server Queue

A.3.1 Notations and Basic Ingredients

A single server queue is a basic element in a queueing network. A single server queue consists of a server, which provides service to customers, and a buffer or a queue, which accommodates customers while they are waiting for services. Customers arrive at the buffer in a deterministic or a stochastic manner. Each of them requires a certain amount of service from the server. In queueing theory, the arrival process of customers is modeled as a random process, and the service times required from the server are modeled as random variables. Each buffer has an infinite or finite size. A buffer with a size of m can accommodate only up to m customers. A buffer is said to be full if there are m customers in it, usually including the customer being served by the server. A customer arriving at a full buffer will be lost or will have to wait elsewhere. In this case we say the buffer is **blocking.** The first kind of blocking where the customers simply get lost is sometimes referred to as **communication blocking**, while the later kind which prevents the customer from departing the place from which it desires to leave is called **manufacturing blocking.**

It is conventional to adopt the four-part description A/B/1/m to specify a single server queue. The number "1" in the descriptor indicates that there is only one server in the system ; m indicates the size of the buffer; A and B describe the interarrival time distribution and the service time distribution, respectively. A and B take on values from a set of symbols, each letter representing a specific distribution. The most common single server queue is the M/M/1/∞ queue, where M/M indicates that both the interarrival time and the service time are exponentially distributed. (The state process of this system is a Markov process.) Other symbols that A and B may take and their meanings are the following: E_r for r-stage Erlangian distribution, H_R for R-stage hyper-exponential distribution, G for general distribution, D for deterministic distribution, and U for uniform distribution. If the buffer size is infinity, then m may be omitted in the descriptor. Thus, M/M/1 is the same as M/M/1/∞.

If it finds the queue empty, an arriving customer receives the service immediately. Otherwise, it is put in the queue, provided a space in the queue is available. The server provides services to the customers in the queue according to certain rules called *service disciplines*. Some commonly used service disciplines are as follows. (i) **First come first serve (FCFS)**: the customer who arrives at the queue first gets served first by the server. (ii) **Priority** scheme: customers are assigned different priorities and the customer with the highest priority in the queue gets served first. A priority scheme may be **preemptive or nonpreemptive**. If a customer in the process of being served is liable to be ejected from service and be returned to waiting status whenever a customer with a higher priority appears in the queue, then we say the queue is a preemptive priority queue. If such is not allowed, then we say the system is nonpreemptive. Furthermore, in a preemptive scheme a preempted customer may pick up his service where he left off. In this case we say the scheme is **preemptive resume.** The customer may lose credit for all service he has received so far. This is called **preemptive nonresume.** In a priority scheme, the service discipline for serving the customers with the same priority may be FCFS or LCFS (which is explained below). (iii) **Last come first serve (LCFS)**: the customer who arrives at the queue last receives service first. This scheme also has different versions: preemptive resume, preemptive nonresume, and nonpre-emptive. (iv) **Processor sharing (PS)**: the service power of the server is shared equally by all customers in the queue, i.e., if the service rate of the server is μ for one customer and there are n customers in the queue, then each customer will receive service at the rate μ/n.

An extension of a single server queue is a single station queue, where a station may consist of more than one server. All servers have the same service distribution and share a common buffer. A customer is served by one of the servers in the station. Such a system is denoted as A/B/k/m, where k is the number of servers in the station. Examples of such queueing systems are bank or airline service counters. All things being equal, one A/B/k system is superior to k A/B/1 systems in parallel from the viewpoint of efficiency, customer fairness, and satisfaction.

Exercise Explain the differences among the following arrangements which can be thought of as (A) the banking counter of yesterday, (B) that of today, (C) using the ATM (Automatic Teller Machines).

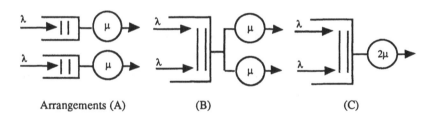

Arrangements (A)   (B)   (C)

A.3.2 Performance Measures and Formulas
The basic performance measures of single server queues are the average waiting time, response or system time (waiting plus service time), queue length, and utilization (percentage of the time during which the server is busy in total time). First, we shall state a simple but powerful formula in performance analysis of queueing systems, i.e., **Little's Law,** which is also applicable to any queueing network. This law can be illustrated by Fig. 1.

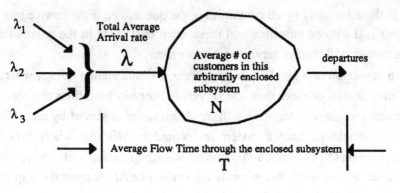

N = Average number of customers in the subsystem

λ = Average customer arrival rate

T = Average time a customer stays in the subsystem

Little's Law : $N=\lambda T$ (1)

Fig. 1 Little's Law

The validity of Little's Law can be graphically and intuitively demonstrated via Fig. 2. In the figure, $a(t)$ is the number of customers who arrive at the system in $[0,t]$ and $d(t)$ the number of customers who depart from the system in $[0,t]$. $N(t)$ = $a(t)-d(t)$ is the number of customer in the system at time t, and T_i is the time customer i spent in the system.

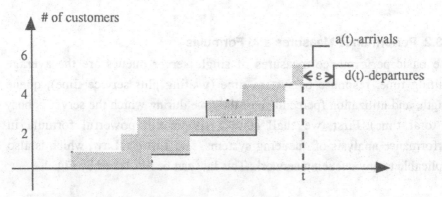

Fig. 2 Graphical Demonstration of the Little's Law

From Fig. 2, we have

$$\int_0^t N(\tau)d\tau = \sum_{i=1}^{a(t)} T_i - \varepsilon.$$

From definition,

$$T = \lim_{t \to \infty} \frac{\sum_{i=1}^{a(t)} T_i}{a(t)}; \quad \lambda = \lim_{t \to \infty} \frac{a(t)}{t}; \quad \text{and} \quad N = \lim_{t \to \infty} \frac{\int_0^t N(\tau)d\tau}{t}.$$

Little's law $N = \lambda T$ follows directly from

$$\lim_{t \to \infty} \frac{\int_0^t N(\tau)d\tau}{t} = \lim_{t \to \infty} \frac{\int_0^t N(\tau)d\tau}{a(t)} \times \lim_{t \to \infty} \frac{a(t)}{t} - \lim_{t \to \infty} \frac{\varepsilon}{t},$$

and $\lim_{t \to \infty} \frac{\varepsilon}{t} = 0.$

The other important fact in the single queue-server theory is the so-called **"memoryless triangle,"** which states that the following three facts are equivalent:

(i) The interarrival time distribution of the customers is exponential.

Let T be an exponentially distributed random variable with mean $s = 1/\lambda$. Then the conditional probability of T, given that T is greater than any finite value t_0, is the same as the original distribution, i.e.,

$$P(T \le t + t_0 \mid T > t_0) = 1 - \exp(-\lambda t) = P(T \le t). \tag{2}$$

This property, illustrated in Fig. 3, is called the memoryless property. The shaded portion of the unconditional density function is re-scaled to give the conditional distribution

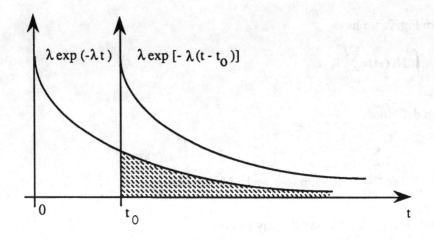

Fig. 3 The Memoryless Property of the Exponential Distribution

(ii) The probability of a single arrival in time Δt is equal to $\lambda \Delta t + o(\Delta t)$, where λ is the so-called constant arrival rate, and the arrivals are independent. Thus the probability of no arrival in $[0,t)$ (i.e., the arrival instant t_a is greater than t) can be obtained by dividing the interval into N equal increments of $\Delta t = t/N$, and

$$p(t_a > t) = \lim_{N \to \infty} (1 - \lambda \Delta t)^N = \lim_{N \to \infty} (1 - \lambda \frac{t}{N})^N = e^{-\lambda t},$$

which implies (i).

(iii) The probability of n arrivals in the interval $[0,t)$ obeys the Poisson distribution.

Let $P_n(t) \equiv$ Probability that there are n arrivals in time $[0,t)$, then (ii) implies that $P_n(t+\Delta t) = P_n(t)(1-\lambda \Delta t)+P_{n-1}(t)\lambda \Delta t$ which upon letting $\Delta t \to 0$ yields $dP_n/dt = -\lambda P_n+\lambda P_{n-1}$. An iterative solution to this differential equation for successive n gives the Poisson distribution $P_n(t) = [(\lambda t)^n e^{-\lambda t}]/n!$

Now, let us consider an M/M/1 queue. Because of the memoryless property, the residual service (interarrival) time has the same distribution as the service (interarrival) time. Thus, the system state can be simply denoted as n, the number of customers in the queue.

Let λ and μ ($\lambda<\mu$) be the mean interarrival rate and mean service rate, respectively, and p_n be the steady-state probability of state n. From the state transition diagram shown in Fig. 4, we can obtain the following flow balance equations for p_n:

$$(\lambda+\mu)p_n = \mu p_{n+1} + \lambda p_{n-1} \qquad n>0$$

and

$$\lambda p_0 = \mu p_1.$$

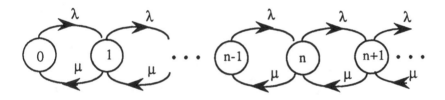

Fig. 4. State Transition Rate Diagram for the M/M/1 Queue

From this and $\Sigma p_n=1$, we have

$$p_n = \rho^n (1-\rho), \quad \rho \equiv \lambda/\mu, \quad n = 0,1,2,... \qquad (3)$$

An important measure of a queueing system is the average number of customers in the system, N. For an M/M/1 queue, this is clearly given by

$$N = \sum_{n=0}^{\infty} np_n = \frac{\rho}{1-\rho}. \qquad (4)$$

$N > 0$, since $\rho \equiv$ traffic intensity $= \lambda/\mu < 1$. Another measure is the average time (often denoted as the system time) that a customer spends in the system, T. This is obtained by applying Little's law

$$T = \frac{N}{\lambda} = \frac{\rho}{\lambda(1-\rho)} = \frac{1}{\mu-\lambda}. \tag{5}$$

The average waiting time a customer spends in the queue, W, equals T minus the average service time in the server, i.e.,

$$W = T - \frac{1}{\mu} = \frac{\rho}{\mu-\lambda}. \tag{6}$$

The average number of customers in the queue, N_Q, is then

$$N_Q = \lambda W = \frac{\rho^2}{1-\rho}. \tag{7}$$

Finally, we list below the formulas for an M/G/1 queue and an M/D/1 queue. In these formulas, m_2 is the second moment of the service time distribution. In an M/D/1 queue, $m_2 = (1/\mu)^2$, where μ is the (deterministic) service rate of the server. For an M/G/1 queue, we have

$$N = \rho + \frac{\lambda^2 m_2}{2(1-\rho)}, \qquad T = \frac{1}{\mu} + \frac{\lambda m_2}{2(1-\rho)}, \tag{8}$$

and

$$N_Q = \frac{\lambda^2 m_2}{2(1-\rho)}, \qquad W = \frac{\lambda m_2}{2(1-\rho)}. \tag{9}$$

For an M/D/1 queue, we have

$$N = \rho + \frac{\rho^2}{2(1-\rho)}, \qquad T = \frac{1}{\mu} + \frac{\rho}{2(\mu-\lambda)}, \tag{10}$$

and

$$N_Q = \frac{\rho^2}{2(1-\rho)}, \qquad\qquad W = \frac{\rho}{2(\mu-\lambda)}. \qquad (11)$$

Note that W and N_Q for an M/M/1 queue are just twice as big as those for an M/D/1 queue. This reflects the impact of randomness.

A.4 Network of Queues

A queueing network is a system consisting of a number of service stations. Each of them is associated with a buffer. A station may contain one or more servers that provide services to customers. Customers may belong to several different classes. Customers from different classes may require different services from the same server. After receiving service from one station, a customer may go to another station for service or may leave the system. The routing mechanisms for different classes of customers may be different. A queueing network may belong to one of three categories: an open network, if every customer arrives at the system from outside and eventually leaves the system after receiving services from stations in the network; a closed network, if every customer circulates among stations and never leaves the network and no customers may arrive at the system from outside; a mixed network, if for some classes of customers the network is open, and for other classes of customers it is closed. Examples of open networks are emergency wards of a hospital and the air/land side of an airport; those of closed networks are flexible manufacturing systems with a specific number of fixtures or transport mechanisms or a time-shared computer system where the number of time-sharing terminals is constant or the number of batch jobs allowed into the system is constant.

Queueing networks in general form are very complicated. It is extremely hard to develop analytical formulas for the steady-state probabilities. However, for a class of networks, the so-called **product-form networks**, analytical solution exists. First, we shall discuss the simplest network among this class, i.e., the Jackson network with single class customers and single-server stations.

A.4.1 The Jackson-Gordon-Newall Network

In an open Jackson network, there are M single-server stations numbered as server 1, 2, . . . , M, respectively. Each of them has a buffer with infinite capacity. The customers arriving at server i from external sources form a Poisson process with mean rate $\lambda_{0,i}$. After receiving the service from server i, a customer transfers to server j with a probability $q_{i,j}$ and leaves the system with a probability $q_{i,0}$. Of course, we have

$$q_{i,0} = 1 - \sum_{j=1}^{M} q_{i,j}.$$

The service time of server i is exponentially distributed with mean $s_i = 1/\mu_i$.

Because of the memoryless property of the exponential distributions, the system state can be denoted as $\mathbf{n} = (n_1, n_2, ..., n_M)$, where n_i is the number of customers in server i. Let λ_i be the mean arrival rate of the customers to server i, including the customers from outside of the network and the customers from the servers in the network. Then we have the following traffic flow equation from flow continuity considerations:

$$\lambda_i = \lambda_{0i} + \sum_{j=1}^{M} \lambda_j q_{j,i}, \qquad i = 1, 2, \ldots, M. \tag{12}$$

It has been proved that in an acyclic open Jackson network (i.e. a network, such as a treelike network, in which there is no feedback loop), the arrival process to any server is a Poisson process. However, if there are feedback loops in the network, the arrival process to a server is usually not a Poisson (even not renewal) process [Melamed 1979, Kleinrock 1975 pp.149-150]. Thus, in general, each individual server in the network is different from an M/M/1 queue. Nevertheless, if we consider each server as an M/M/1 queue, we can obtain its steady-state probability distribution, $P_i(n_i)$, as follows:

$$P_i(n_i) = (1-\rho_i)\rho_i^{n_i}, \qquad \rho_i = \lambda_i/\mu_i, \qquad i=1,2,\ldots,M \qquad (13)$$

An interesting results is that, despite the fact that the arrival process to each server in the open Jackson network may not be a Poisson process, the steady-state probability of the system state has the following form [Jackson 1963]:

$$P(\mathbf{n}) = P(n_1,n_2,\ldots,n_M) = P_1(n_1)P_2(n_2)\ldots P_M(n_M). \qquad (14)$$

This distribution shows that in an open Jackson network the steady-state distributions of the customer numbers in servers are independent of each other.

Gordon and Newell studied the closed form of the Jackson type networks [Gordon and Newell 1967]. A Gordon-Newell network is basically the same as a Jackson network, except that $\lambda_{0,,i} = 0$ and $q_{i,0} = 0$ for all i. Thus, in a Gordon-Newell network, no customers can arrive at or depart from the system. The number of customers in the network, N, is fixed. The state of the system is $\mathbf{n} = (n_1, n_2, \ldots, n_M)$ with n_i being the number of customers in server i. We have

$$\sum_{i=1}^{M} n_i = N.$$

In the following, we shall consider a general type of the Gordon-Newell network in which the service rate of a server may depend on the number of customers in the server. Let $a_i(n_i)\mu_i$ be the service rate of server i when there are n_i customers in it. "μ_i" represents the basic service rate of the server and "$a_i(n_i)$" the load factor. Such servers are called **load dependent servers** (see also Section A.4.3 and Chapter 3.5). The steady-state probabilities of the network satisfy the following flow balancing equation:

$$\{ \sum_{k=1}^{M} \varepsilon(n_k)a_k(n_k)\mu_k \} p(n_1,n_2,\ldots, n_M)$$

$$= \sum_{i=1}^{M} \sum_{k=1}^{M} \varepsilon(n_k)a_i(n_i+1)\mu_i q_{i,k} p(n_1,\ldots, n_k-1,\ldots, n_i+1,\ldots, n_M), \qquad (15)$$

where

$$\varepsilon(n_i) = \begin{cases} 0 & \text{if } n_i=0 \\ 1 & \text{if } n_i \neq 0 \end{cases}$$

Eq.(15) can be derived by balancing the rate at which probability flows out of a state $\mathbf{n} = (n_1, n_2, \ldots, n_M)$ with the rate at which probability flow into the state, assuming only one step transitions are allowed. Eq.(15) is known as the **Global Balance** equations. The solution to these equations is

$$P(\mathbf{n}) = \frac{1}{G(N)} \prod_{i=1}^{M} [\,(X_i)^{n_i} / A_i(n_i)\,], \tag{16}$$

where

$$G(N) = \sum_{n_1 + \ldots + n_M = N} \prod_{i=1}^{M} [\,(X_i)^{n_i} / A_i(n_i)\,] \tag{17}$$

is the normalizing constant,

$$A_i(n_i) = \begin{cases} 1 & \text{if } n_i = 0, \\ \\ \prod_{k=1}^{n_i} a_i(k) & \text{if } n_i > 0, \end{cases} \qquad i=1,2,\ldots,M, \tag{18}$$

and $(X_1, \ldots, X_M)$ is a solution to the equation:

$$\mu_j X_j = \sum_{i=1}^{M} \mu_i X_i q_{i,j}, \qquad i,j = 1,2,\ldots,M. \tag{19}$$

Note that $v_i = \mu_i X_i$ is the visit ratio of server i. If the matrix $Q = [q_{i,j}]$ is irreducible, then it can be shown that $v_i > 0$ for all i [Seneta 1981]. If the service rates are load independent, i.e., $a_i(n_i) = 1$ for all i and n_i, then the above equations can be simplified:

$$P(\mathbf{n}) = P(n_1, n_2, \ldots, n_M) = \frac{1}{G(N)} \prod_{i=1}^{M} X_i^{n_i}, \tag{20}$$

and

$$G(N) = \sum_{n_1 + \ldots + n_M = N} \prod_{i=1}^{M} X_i^{n_i}. \tag{21}$$

Despite the complexity of (20), it is conceptually analogous to (14) if we view (19) as the network version of the traffic intensity factor $\rho = \lambda/\mu$ (the analog of ρ is X). Similarly, (16) is a further generalization of (13) and (14) when service rates are load dependent. It is convenient to use Buzen's algorithm to calculate the constant G(N) [Buzen 1973]. We shall discuss only the load independent case. In the algorithm, we define

$$g(n, m) = \sum_{n_1 + \ldots + n_m = n} \prod_{i=1}^{m} X_i^{n_i}.$$

Then we have

$$g(n,m) = g(n,m-1) + X_m g(n-1,m) , \quad 1 \leq n \leq N , \ 1 \leq m \leq M.$$

The initial values are set to be

$$g(0,m) = 1 , \ G(n,1) = X_1^n , \quad 1 \leq n \leq N, \ 1 \leq m \leq M.$$

Note that

$$G(N) = g(N,M).$$

The algorithm also can be used to calculate marginal distributions:

$$p(n_i \geq n) = X_i^n \frac{G(N-n)}{G(N)} .$$

From this, we can easily obtain the throughput of server i, which is defined as $TP_i = \mu_i \, p(n_i \geq 1)$.

A.4.2 Other Product-Form Queueing Networks

The Jackson network discussed above is a special case of a class of general networks called BCMP queueing networks. [Baskett et al. 1975] derived the product-form solution for the steady-state probability distributions of the network. In a BCMP network, there are M service stations and R classes of customers. After receiving service, a class r customer at station i may change to class s and enter station j with probability $q_{i,r;\,j,s}$ or leave the system with probability $q_{i,r}$. A class r customer in station i is denoted as (i, r). In general, according to the probabilities $q_{i,r;j,s}$, all (i, r)'s can be decomposed into w indecomposable subsets, $E_1, E_2,..., E_w$. A customer in each subset has a positive probability of becoming a customer in the same subset, but he has zero probability of becoming a customer in a different subset (note w may equal to one). Let $n_{i,r}$ be the number of (i, r) customers, $n_{*,r}$ be the number of r class customers, $n_{i,*}$ be the number of customers in station i, k_u be the number of customers in subset E_u. Then

$$n_{*,r} = \sum_{i=1}^{M} n_{i,r}, \qquad n_{i,*} = \sum_{r=1}^{R} n_{i,r}, \qquad k_u = \sum_{(i,r)\,\in\,E_u} n_{i,r}.$$

Let N be the total number of customers in the network, then

$$N = \sum_{r=1}^{R} n_{*,r} = \sum_{i=1}^{M} n_{i,*} = \sum_{u=1}^{w} k_u.$$

The arrival process to the network may be modeled in two forms. In the first form, customers arrive at the network in a Poisson process with mean $\lambda(N)$, which depends on the number of customers in the network. Each arrival customer has a probability $p_{i,r}$ to enter station i as a class r customer. In the second form, there are w independent Poisson arrivals, one for each subset. The mean arrival rate for subset E_u is $\lambda_u(k_u)$. Similar to the first form, each arrival customer has a probability $p_{i,r}$ of entering station i as a class r customer.

The service stations in a BCMP network may have the following four forms:

1. FCFS single server station with exponential service time and with the same mean for all classes of customers.

2. Process sharing single-server station.

3. Service station with an infinite number of servers (i.e., no waiting queue in the station).

4. Single server station with LCFS preemptive-resume service discipline.

The service distributions of the stations above form 2 through 4 may be very general. An example of such a distribution is the Coxian distribution which has a rational Laplace transform. Cox [1955] proved that such a distribution can be achieved by a combination of service stages, each of them having an exponential distribution, in a configuration shown in Fig. 5. In the figure, a customer arrives at a station for service. The customer first enters stage 1 which provides a service time having an exponential distribution with mean s_1; then the customer enters stage 2 with a probability a_1 or leaves the station with a probability $b_1 = 1- a_1$. The customer at stage 2 receives an exponentially distributed service time with a mean s_2 and leaves the station with a probability b_2, and so on. The customer at the last stage, k, leave the station with a probability $a_k =1$. While a customer is present in such a coxian server (regardless of the stage), no other customer is allowed in. That is, the entire structure of Fig. 5 is to be considered as a single server. Since any continuous Laplace transform function can be approximate by a rational function, the Coxian distribution can be used to approximate other arbitrary distributions via this "Method of Stages" or its parallel-series generalizations (Kleinrock [1975] Chapter 4). The state of such a server-queue combination is described by a pair of numbers, (n,α), where n is the usual number of waiting customers, and α the current stage at which the customer being served is situated. Another way to look at the "method of stage" characterization of the state is to regard "α" as finite approximation to the continuous remaining service time of the customer at the station.

Basket et al. proved that the steady-state probability of the above described BCMP network has a product form. Each factor in the product depends only on the state of one station. We refer the readers to [Baskett et al. 1975] for the detailed formulas.

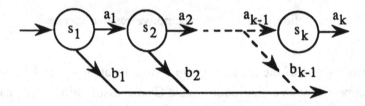

Fig. 5 The Structure of a Coxian Distribution

There are some other queueing networks for which the product-form solution exists (e.g. [Kelly 1979]). The result has been also extended to certain queueing networks with state-dependent routing [Towsley 1980] or a slightly more general service discipline called the last-batch-processor-sharing (LBPS) discipline [Noetzel 1979]. To characterize the product-form queueing networks, the properties of **local balance** and **station balance** in queueing networks were studied. These are additional equilibrium conditions (beyond the global balance Eq.(15) above) to be satisfied among subsets of the terms in each of the equilibrium state probability equations (15). Local balance is satisfied if the rate of probability out-flow of (or the rate of loss of the probability density of) a state due to the arrival of customers of class k from outside the system equals the rate of probability in-flow to (or the rate of gain of the probability density of) the state due to departure of customers from the same class to outside the system. Similarly, station balance is satisfied if the rate of gain of probability density of a state due to arrival at a station i is balanced by the rate of loss due to service completion at the station. It was proved in [Chandy et al. 1977] that if a state probability density satisfies station balance then it also satisfies local balance. If each queue of a closed queueing network satisfies local balance when isolated then the steady-state probability density of the states assumes the product form.

However, analytical results do not exist for many other queueing networks. Examples of such networks are those with general arrival and service distributions and FCFS service discipline, state dependent routing, those with finite capacity buffers (blocking), those with simultaneous resource possession requirements (e.g., a customer requires two or more servers simultaneously), and multiclass networks with different classes of customers requiring different service rates at an exponential server. Approximation and simulation seem to be the only approaches to analyzing these systems. On the approximation front, one of the best-known methodologies is a set of approximate equations known as the QNA (Queueing Network Analyzer) traffic equations [Whitt 1983] which relate the mean and the variance of the arrival rates to all nodes of the network, λ_i and c_i, i=1,2, . . ., to the parameters of the network. The mean traffic flow equation is simply Eq.(12) which is exact and generally applicable to networks with arbitrary arrival and service distributions. The relationship among the variances of arrival rates at each station is stated in terms of the coefficient of variation of the arrival rate,

$$c_i = \text{linear function of } (\lambda_j, c_j, \text{other system parameters}) \qquad (22)$$

In contrast to Eq.(12), (22) is an approximate formula from which c_i can be obtained by solving a system of linear equations. Knowing the arrival rate characterization at any node, the analysis of performance at that node is dependent only on the arrival and nodal parameters at that node, and hence can be decomposed from the rest of the network. (See also Chapter 5.4.3)

The other alternative to the study of general queueing networks and DEDS is direct simulation or experimentation which will be reviewed in Appendix B.

A.4.3 The Norton Theorem and Generalizations

The technique of aggregation is usually quite useful in the investigation of large and complex systems. The Norton theorem for queueing networks is an aggregation technique which provides exact results for product-form networks. The theorem shows that a subset of service stations in a queueing network can be replaced by a single server without affecting the marginal probabilities of the rest of the network [Chandy, Herzog, and Woo 1975]. We shall explain the theorem by using an example.

Consider the closed Gordon-Newell network shown in Fig. 6. In the figure, the network is decomposed into two sub-networks, A and B. Using the Norton theorem, we can replace subnetwork A by an equivalent server. The marginal probability distributions of subnetwork B in the resulting queueing network (shown in Fig. 7) will be the same as the marginal probability distributions of subnetwork B in the original network. The equivalent server is a load-dependent server. (This fact is indicated by an arrow on the circle representing the server.) To determine the service rates of the equivalent server for different loads, we "short" subnetwork B in the original network, i.e., set the service times of all servers in sub-network B to zero. The queueing network with a shorted subnetwork B is shown in Fig. 8. The service rate of the equivalent server, when there are k customers in it, $\lambda(k)$, equals the steady-state output rate on the shorted path in the network shown in Fig. 8 with k customers in it.

The most significant implication of the "Norton equivalent" for the subnetwork A is that it is "externally independent," namely, changing part B will not change the parameters of the equivalent load dependent server determined in Fig. 8 provided, of course, that the resultant network remains a Jackson network. The Norton theorem has been generalized to the multiclass BCMP networks [Kritzinger et al. 1982]. For more on the application of the concepts of aggregation and Norton equivalents, see Chapter 8 Sections 1 and 2.

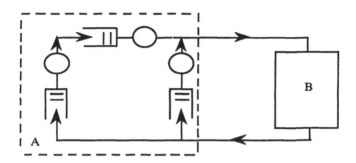

Fig. 6 A Gordon-Newell Network with Two Subsets

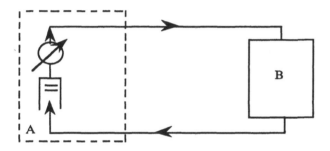

Fig. 7 The Equivalent Queueing Network

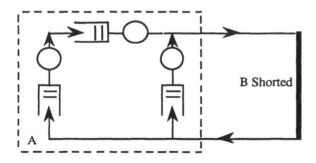

Fig. 8 The Network with One Subset Shorted

Elements of Discrete Event Simulation

There are many excellent texts for discrete event simulation to suit different purposes (e.g., [Brately, Fox, and Scharge, 1987] and [Fishman 1978]). It is not the purpose of this appendix to replace those sources of information. The primary objective of this appendix is the same as that stated at the beginning of Appendix A.

B.1 The Ingredients of a Discrete Event Simulation

In simplest terms, discrete event simulation is just writing a computer program to model a DEDS and is usually executing the program on a faster time scale. However, for reasons of efficiency one tries not to model the DEDS from scratch and not to write a different computer program for each different system. Discrete event simulation languages such as GPSS, SIMSCRIPT, SLAM, SIMAN, etc, take note of certain common constructions of all DEDS models and implement these constructions once for all to facilitate the coding of DEDS simulation programs. These commercial programs also make the collection of statistical data from the output easy again by implementing various commonly used formats and charts.

The most important quantity in a dynamic system, whether continuous variable or discrete event, is the **STATE**. For DEDS, we distinguish between the **physical state** x, which is typically discrete, such as the number of jobs waiting at each server or the status of certain logical variables, from the **mathematical state**, which is what is needed to uniquely determine the future evolution of the dynamic system. The evolution of a DEDS is driven by **events**, which can be both externally (e.g., the arrival of a job from outside the system) or internally generated (e.g., the completion of a task by a server). The occurrence of an event together with logical conditions (rules of operation) determines a state transition, which leads the system to a new state. In the new state, new future events are scheduled and existing events may be continued or terminated according to an event list. And the cycle repeats.

Generally speaking, the main ingredients of a discrete event simulation are:

(i)   mechanism for generating basic random numbers used to model all stochastic phenomena (used in Boxes 3 and 4 of Fig. 1 below).

(ii) mechanism for determining lifetimes of all events in a state (used in Box 3 below)

(iii) mechanism for scheduling future events based on their lifetimes (used in Boxes 2 and 3 below)

(iv) mechanism for keeping track of the physical and mathematical states of the DEDS and its transitions (used in Box 4 below)

A generic flow chart of discrete event simulation is illustrated in Fig. 1. The event oriented flow chart of Fig.1 obscures the fact that there are many objects (e.g., customers and servers) and concurrent processes (resources serving jobs) in a simulation. In terms of programming and implementation of the model, users are interested in interfaces that deal with objects and graphic displays that approximate the way the physical systems actually behave as well as convenient means for data collection and analysis. Thus, a simulation language design involves, in addition, considerations of data structure, object description, automatic statistics gathering, etc. Pursued to the limit, the trend in simulation language development is to have object-oriented graphic programming and animated outputs so that the simulation is not only numerically significant but also physically as life-like as possible within the limitation of computer displays[1].

---

[1] The power of animation should not be underestimated since it often prevents misunderstandings among users, programmers, and teams of workers.

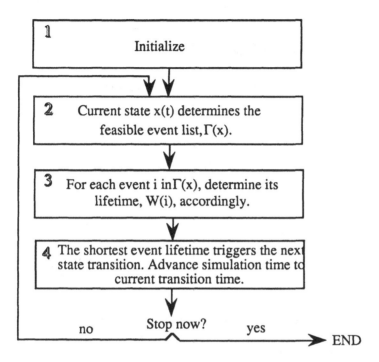

Fig. 1 Flow Chart of Event Oriented Discrete Event Simulation

B.2. Random Variable Generation

B.2.1 Random Number Generation and the Linear Congruential Method

The basic operation in discrete event simulation for generating arbitrarily distributed random variables is the generation of a sequence of uniformly, identically, and independently distributed (u.i.i.d.) random numbers over a finite interval, usually taken to be [0,1). Since there is no truly random device inherent in a computer, the actual mechanism used is a technique of "pseudo-random number generation." For example, starting with an n-digit integer, x_i, which is less than or equal to M, we recursively compute the next integer x_{i+1} by

$$x_{i+1} = \text{mod}_M(b + a\, x_i); \; i=0,1,2,....; \; x_0 \text{ is given} \qquad (1)$$

where a and b are given constant integers and x_0 is known as the initial seed. This is known as the **linear congruential method** for generating pseudo random numbers (see any of the simulation text mentioned at the beginning of this appendix). Let us illustrate the operation of this method by way of several simple examples[2].

Example 1 Let a=2, b=1, and M=16. Using Eq.(1) and various x_0's, we get

$x_0 =$	1	2	4	6	8	
$x_1 =$	3	5	9	13	1	
$x_2 =$	7	11	3	11	3	
$x_3 =$	15	7	7	7	7	
$x_4 =$	15	15	15	15	15	
$x_5 =$	15	15	15	15	15	*All sequences gets stuck*
$x_6 =$	•	•	•	•	•	*after the initial transients !*

Example 2 Let a = 3, b = 0, and M = 16. Similarly, we have

$$
\begin{array}{cccccc}
x_0 = & 1 \leftarrow & 2 \leftarrow & 4 \leftarrow & 5 \leftarrow & 8 \leftarrow & 10 \leftarrow \\
& 3 & 6 & 12 & 15 & & 14 \\
& 9 & & & 13 & & \\
& 11 & & & 7 & &
\end{array}
$$

Note that, depending on the initial seeds, the sequences get into cycles with different periods. But none of the sequences produce the maximal period of 0-15.

Example 3 Let a = 1, b = 3 , and M = 16. This time starting with *any* seed, we get the maximal period and the sequence [. . . , 1, 4, 7, 10, 13, 0, 3, 6, 9, 12, 15, 2, 5, 8, 11, 14, 1, ...]. This is nice. However, a plot of the sequence vs. time shows high correlation among successive numbers in the sequence as

---

[2] We are indebted to R. Suri for these examples.

illustrated in Fig. 2. Thus the numbers in the sequence are not at all independent.

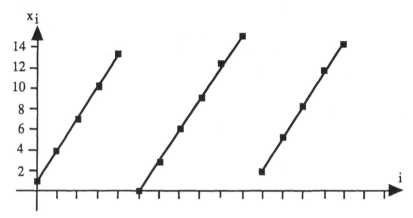

Fig. 2 Plot of Pseudo Random Sequence

Example 4 Let $a = 5$, $b = 3$, and $M = 16$. Once again we get a sequence of maximal period with any seed, [. . . . , 1, 8, 11, 10, 5, 12, 15, 14, 9, 0, 3, 2, 13, 4, 7, 6, 1, . . .]. A similar plot as in Fig.2 shows a reasonably random looking sequence.

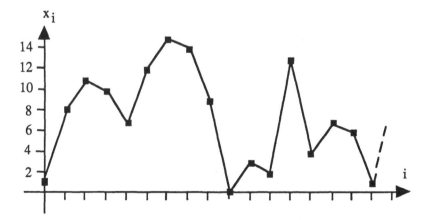

Fig. 3 Plot of Pseudo Random Sequence

The above examples show the considerations that go into the design of a pseudo - random number generator; they are: (i) periodicity - independence from the initial seed, (ii) uniformity - with maximal period we are assured of total uniformity, (iii) correlation - dependence of subsequences. Any practical scheme must have a very large M (= order of 10^{10}) and be subjected to rigorous statistical tests to qualify as a useful pseudo-random number generator. Given a satisfactory sequence of x_i's, we can simply re-scale it according to $u_i = x_i / M$ to obtain a u.i.i.d. sequence between [0, 1). Standard simulation texts, e.g., [Bratley et al. 1987], suggest good values for a, b, and M.

Exercise Does M have to be a prime number?

B.2.2 The Inverse Transform method
Given that we can generate a sequence of u.i.i.d. random numbers between [0,1), $u_1, u_2, \ldots$, etc., we can transform these into i.i.d. samples, $x_1, x_2, \ldots$ from an arbitrary distribution F(x) by using the inverse transform $x_i = F^{-1}(u_i)$, provided it exists. This is illustrated in Fig. 4.

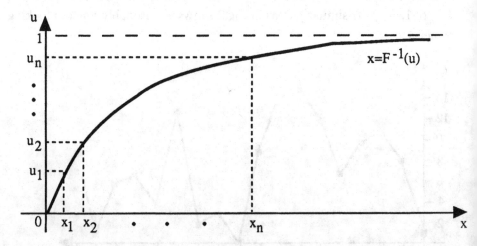

Fig. 4 Inverse Transform Method for Generating F(x)-Distributed Random Variables.

To see this, consider the probability

$$P(x \le a) = P(F^{-1}(u) \le a) = P(u \le F(a)) = F(a),$$

where the last equality is by virtue of the uniform distribution of the random variable u.

Exercise Repeat the above argument and show that $P(x \in [a, a+da]) = f(a)da$ where $f(x)$ is the density function of the random variable x.

B.2.3 The Rejection Method

The other well known method of generating random variables with arbitrary distributions is the so-called rejection method. Mathematically, consider a random variable x with probability density function $f(x)$. Assume that $f(x)$ is bounded and is nonzero only on some finite interval [a, b] and define c = max { $f(x) : a \le x \le b$ }. To generate samples of this random variable, the following procedure, called the rejection method, can be used

   1) generate u, a sample of a uniform random number U on [a, b);

   2) generate v, a sample of a uniform random number V on [0, c);

   3) if $v \le f(u)$ accept u as a sample of X; otherwise go to step 1.

Pictorially, steps (1 - 2) generate samples uniformly on the rectangular area defined by sides ac and ab. Step (3) simply throws away the samples in the shaded region. Those in the unshaded region are retained to yield samples from the density function $f(x)$. This is illustrated in Fig. 5. For more details see [Brately, Fox, and Scharge 1987].

B.3 The GSMP Formalism

An appropriate model for the description of a general discrete event dynamical system (DEDS) is the so-called Generalized Semi-Markov Process (GSMP).

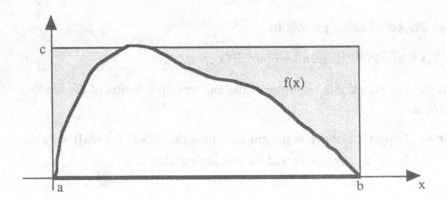

Fig. 5 Rejection Method of Random Variable Generation

The process was first proposed by [Matthes 1962] to study insensitivity properties of many applied probabilistic models. We can view this process as a formalization of the event scheduling approach of simulation models described in Section 1 and Fig. 1. Let (Ω, F, P) be the underlying probability space. Consider a continuous time stochastic process $\{x(t,\omega), t \geq 0\}$, $\omega \in \Omega$, defined on the probability space. The process takes values, called **states,** in a countable set X, called the **state space.** For each state $x \in X$, there is a uniquely defined finite set

$\Gamma(x) \subset N$ (the set of integers), each integer i in $\Gamma(x)$ represents a currently active (living) **event.** $\Gamma(x)$ is called the feasible **event list** associated with x. Each event $i \in \Gamma(x)$ has a **lifetime (clock reading),** $c(i)$. Once set, each of the clock readings runs down as time moves forward. Let $\Gamma = \cup_{x \in X} \Gamma(x)$ be the set of all events. A GSMP trajectory consists of piecewise constant segments, called the state sequence, and switches between two states at a transition instant, τ_i, $i=1,2, \ldots$, depending upon which of the event lifetimes (clock readings) associated with each of the $i \in \Gamma(x)$ reaches zero first. By convention, we say that an event occurs when the lifetime, or clock reading, of the event reaches zero. All the different events associated with a state compete for triggering (or scheduling) the next transition, and each event has its own particular jump distribution for determining the next state. The duration of the piecewise trajectory segment is called the **state holding time**. At each transition of a GSMP, new events are scheduled. For each of these new events, a lifetime (or clock reading) $c(i)$, indicating the future time at which this event will occur, is generated according to a probability distributions $\phi_i(x)$; $i \in \Gamma$. An event that is active but does not initiate a transition may be carried over into the next state with its remaining lifetime. Such GSMPs are called **noninterruptive**. In an interruptive GSMP, non-triggering events can be simply abandoned when state changes if they do not belong in the feasible event list of the new state. The sequence of states and their holding times determine the **trajectory or sample path** of the GSMP. The sequence of successive triggering events is called an **event trace** which sometimes is considered to be the more basic specification of the trajectory since it is often the only observed behavior of the DEDS.

More formally, given a set of probability distributions $\{\phi_i; i \in \Gamma\}$, and a set of transition probabilities $\{p(x';x,i), x \in X, i \in \Gamma\}$ ($p(x';x,i)$ denotes the probabi-lity of going from state x to state x' given event i triggers the transition), we specify a process $\{x(t,\omega), T(t,\omega), t \geq 0\}$ as follows. Starting at time $t=0$, a state $x_0 \in X$ is specified and a positive number $c_0(i)$ with the distribution ϕ_i is assign-ed to each element $i \in \Gamma(x_0)$ as its event lifetime. We define the first event occurrence as $i^* = \arg.\min_{i \in \Gamma(x_0)}\{c_0(i)\}$, the **triggering event** and the first event time $\tau_0 = c_0(i^*)$. In the period $0 \leq t < \tau_0$, let the state $x(t,\omega)=x_0$, and residual times $T(t,\xi) = (c_0(i_1)-t,...,c_0(i_{n_0})-t)$, where $i_1,...,i_{n_0}$ are the events in $\Gamma(x_0)$. At τ_0, the process jumps to a new state. The new state is defined as $x(t^*,\omega)=x_1 \in X$, where x_1 is selected according to the state transition rule, $p(x_1;x_0,i^*)$. Let the new set of events be $\Gamma(x_1)$. The new lifetimes for events $i \in \Gamma(x_1)$ at t^* are deter-mined as follows:

$$c_1(i) = \begin{cases} c_1(i) & \text{for } i \in \Gamma(x_1) \cap i \notin \Gamma(x_0), \\ c_0(i)-\tau_0 & \text{for } i \in \Gamma(x_1) \cap \Gamma(x_0), \end{cases}$$

where $c_1(i)$ is a newly generated random variable from ϕ_i. Now for $t > \tau_0$, the cycle repeats with the next event time determined by the shortest lifetime among $c_1(i)$, and $\tau_1=\tau_0+c_1(i^*)$. We define $x(t,\omega)=x_1$, and $T(t,\omega) = (\tau_1(i_1)-$

$t,...,\tau_1(i_{n_1})$-t) where i_j, $j=1,...,n_1$, are the events in $\Gamma(x_1)$ and $\tau_1(i\ j) = \tau_0+c_1(ij)$, $j=1,...,n_1$. The evolution of the process is then continued in a similar fashion starting from x_1. With the above formulation, we give the following definition (see Whitt [1980] and Schassberger [1977]):

Definition 1   The random process $x(t,\omega)$, $t\in R=[0,\ \infty)$, $\omega\in\Omega$ is called a *Generalized Semi-Markov Process* (GSMP) based upon $\Sigma =(X,\Gamma,p)$ by means of the family $\{\phi_i; i\in\ \Gamma\ \}$. The process $\{x(t,\omega),T(t,\omega\ )\}$ is called the *supplemented GSMP*[3]. A GSMP without the $\{\phi_i; i\in\ \Gamma\ \}$ specification is called a *Generalized Semi-Markov Scheme, (GSMS)*.

A GSMS can be thought of as specifying only the state/event sequence part of the trajectory, i.e., an untimed model of a DEDS (see Appendix D). It is also customary to make the following assumption:

Assumption 1   The distribution function ϕ_i concentrated on $[0,\infty)$ is absolutely continuous and has a finite mean $1/\mu_i$, for $i\in\Gamma$. The streams of the random sample $c(i)$'s and the streams of the states generated from ϕ_i and $p(\cdot;x,i)$, respectively, are i.i.d and are independent of each other.

Note that Assumption 1 implies that for any $n\geq0$, and events $i\neq j$, $i,j\in\Gamma(x_n)$, the event times are such that the probability of two events occurring at the same time is of measure zero. In what follows, we shall call the $c(i)$'s and the $\tau_i(j)$'s the event lifetimes and the event times, respectively. If all ϕ_i's are exponentially distributed, then the GSMP reduces to a discrete state Markov process. Finally, if we concentrate only on the sequence of discrete states of this GSMP, then we denote it as the imbedded Markov chain of the GSMP.

As in the case of Markov processes, we define the following concepts for GSMP.

---

[3] Glynn [1989] provides a more general definition for GSMP which permits more general dependence of the evolution of a sample path on the past history.

Definition 2 A GSMP is called *irreducible* if for every pair of states $x, x' \in X$, there exist two finite sequences $\{y_1, ..., y_n\}$, $y_k \in X$, and $\{j_0, j_1, ..., j_n\}$, $j_k \in \Gamma$, such that

$$p(x'; y_n, j_n) p(y_n; y_{n-1}, j_{n-1}) \cdots p(y_2; y_1, j_1) p(y_1; x, j_0) > 0.$$

A GSMP is *ergodic* if the supplemented GSMP is ergodic.

Let $\theta \in R$ be a parameter of the GSMP model. We also assume:

Assumption 2 The GSMP is irreducible and ergodic for all $\theta \in [a, b]$, for some $-\infty < a, b < \infty$.

In this book, we are interested in the behavior (sample path / trajectory) of a GSMP, denoted by $x_\theta(t, \xi)$, or sometimes simply (θ, ξ), with respect to a system parameter (or parameters) θ. We call the sample path, $x_\theta(t, \xi)$, the *nominal path* (NP). Likewise, the perturbed GSMP parametrized by $\theta' = \theta + \Delta\theta$, denoted by $x_{\theta'}(t, \xi)$, is called the *perturbed process*, and its sample path the *perturbed path* (PP). In Section 3.3, we shall define more formally the dynamical evolution of a GSMP trajectory using the ideas outlined here.

B.4 Elementary Output Analysis

Simulation is simply a computer experiment of the real thing. Like all experiments, their output must be analyzed and interpreted to yield meaningful results. In addition, since DEDSs are often stochastic in nature, statistical analysis must be employed to analyze the output. Generally speaking, we must perform many trials or exercise a DEDS over a long period of time to make sure that all conceivable behaviors occur with the appropriate frequency and the average performance thus derived have reasonable claim to accuracy. However, unlike ordinary statistical experiments, special care needs to be taken. In this section, we outline some elementary techniques for the analysis of DEDS experiments.

B.4.1 Sampling, the Central Limit Theorem, and Confidence Intervals

Consider independently and identically distributed (i.i.d.) random variables x_1, $x_2, \ldots, x_n$, with $E(x_i) = \mu$ and $Var(x_i) = \sigma^2$. Define $M_n = [(x_1+x_2+\ldots+x_n) - n\mu]/(n\sigma^2)^{1/2}$. As $n \to \infty$, the distribution of M_n converges to $N(0,1)$, i.e., the normal (Gaussian) distribution with mean zero and unit variance. This is known as the **Central Limit Theorem (CLT)**. The significance of CLT for experimental work lies in the fact that it enables us to predict the error of sampling. For example, suppose we take n samples of a random variable with mean μ. We may use $\bar{x} \equiv (x_1+x_2+\ldots+x_n)/n$ as an estimate for the unknown mean μ. Then M_n is the normalized error of the estimate. For large n, we can use standard tables for Gaussian random variables to calculate $P(-t < M_n < t)$ which is the probability that the error of the estimate for μ lies in $[-t, t]$. For example, if $t=1.96$, we get $P=0.95$; i.e., we are 95% confident that the interval $[\bar{x}-t (\sigma^2/n)^{1/2}, \bar{x}+t (\sigma^2/n)^{1/2}]$ contains the unknown mean μ. For a specified confidence and interval size, we can calculate how many trials of the experiment are needed.

The above confidence interval formula suffers from one drawback. It requires the knowledge of the variance of the random variable, σ^2. It hardly seems reasonable that we can know the value of σ when not even the mean μ is known. The common practice is to replace σ^2 by the sample variance,

$$\sigma_s^2 = \frac{1}{n} \sum_{i=1}^{n} (x_i - \bar{x})^2$$

in which case the formula is only approximate. However, if we do know that the random variable in question is Gaussian, an exact formula for the confidence interval can be stated in terms of the Student-t distribution using the sample mean, $\bar{x}$, and sample variance, σ_s^2 [Bratley, Fox, and Schrage 1987].

There is a multivariate version of the Central Limit Theorem which replaces μ and σ by their multidimensional version of μ and Σ and the denominator of M_n by $(\det \Sigma)^{n/2}$.

B.4.2 Nonparametric Analysis

In situations where we do not know the distribution of the output random variable, nonparametric methods are also used for confidence interval analysis. This is particularly appropriate in discrete event simulation of complex systems where often all one has is a set of independent samples, $x_1, x_2, \ldots, x_n$ from an unknown distribution. The basic fact that is utilized is based on a simple property of order statistics. Suppose the random variable x is continuous, and we order the samples, $x_1, x_2, \ldots, x_n$ as $x_{(1)} < x_{(2)} < \ldots < x_{(n)}$, then the distribution of the area under the density function between any two ordered samples is independent of the form of the density function. Define

$$u_i = \int_{-\infty}^{x_{(i)}} f(x)dx = F(x_{(i)})$$

The joint density function of the random variables $u_1, u_2, \ldots, u_n$ is then

$$f(u_1, u_2, \ldots, u_n) = n! \qquad 0 < u_1 < u_2 < \ldots < u_n \qquad (2)$$

which is independent of f(x) [Mood and Graybill 1963, Ch.16]. The expected value of u_i, $E(u_i) = i/(n+1)$. In other words, we expect on the average, the n ordered samples, $x_{(1)} < x_{(2)} < \ldots < x_{(n)}$, to divide the density function into n+1 equal areas which is entirely reasonable. Thus, using these order samples we can directly, or by simple interpolation, estimate the various population percentage points. In particular we can construct a confidence interval for the median (the 50% population point), υ, as follows. The probability that a sample falls to the left or right of υ is exactly 1/2 in either case. The probability that exactly i samples in a random sample of n fall to the left of υ is just

$$\binom{n}{i}\left(\frac{1}{2}\right)^n.$$

The probability that $x_{(r)}$, the rth ordered statistic, and $(x_{(n+r-1)}$, the (n-r+1)th ordered statistic), exceeds υ is then

$$P(X_{(r)} > \upsilon) = \sum_{i=0}^{r-1} \binom{n}{i} \left(\frac{1}{2}\right)^n,$$

and the probability that $x_{(n+r-1)}$, the $(n-r+1)$th ordered statistic, not exceeds υ is

$$P(X_{(n-r+1)} < \upsilon) = \sum_{i=n-r+1}^{n} \binom{n}{i} \left(\frac{1}{2}\right)^n.$$

From this we get the probability that $x_{(r)}$ and $x_{(n-r+1)}$ brackets the median υ as

$$\sum_{i=r}^{n-r} \binom{n}{i} \left(\frac{1}{2}\right)^n. \tag{3}$$

Other confidence intervals can be derived similarly [Mood and Graybill 1963, Chapter 16]. Since in any sampling scheme, one gets the ordered samples essentially free, Eq.(3) should always be considered along with the formula for the parametric case discussed in Section B.4.1.

B.4.3 Application to Discrete Event System Simulation

If we view simulation as a sampling process with each run of the experiment producing **one** sample value of the performance measure of interest, then a straightforward method of sampling and confidence interval analysis such as those of Sections B.4.1 - 4.2 can be directly employed. We only have to make sure that each run of the simulation experiment is i.i.d. by using different random seeds to start each run. In many cases, this is sufficient. However, often such a simple view is impractical. For example, if we are interested in the steady-state performance of a DEDS, then we will have to run many experiments, each of them has to reach steady state. This is very time consuming and in fact unnecessary. It seems reasonable that only one steady-state experiment is needed.

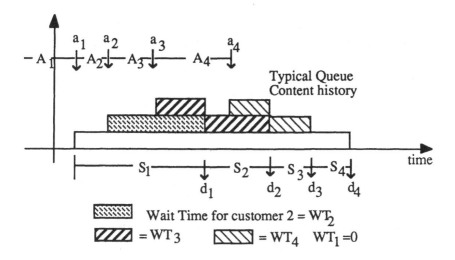

Fig. 6 A Typical Queue Content History

Consider the situation illustrated in Fig. 6, which shows a typical queue content history of a server. If we have to estimate the mean interarrival time λ, the mean service time μ, and the mean waiting time (WT) of the queue-server combination the following estimates may be considered:

$\lambda_{est} = (1/3)\ (A_2+A_3+A_4)$,

$\mu_{est} = (1/4)\ (S_1+S_2+S_3+S_4)$, and

$WT_{est} = (1/4)(WT_1+WT_2+WT_3+WT_4)$.

However, the situation is now more complicated. Because of the dynamic nature of DEDS, sample values of variables of interest (e.g., queue length at a particular server) are invariably correlated in time. In addition to making sure that enough sample values are taken, we now have to be concerned about the time-varying nature of the sample values due to transient and initial condition effects. Two "quick-and-dirty" approximations are:

(i) Use of a warm-up period – The DEDS simulation is run for a given amount of time before data collection begins. This, however, simply introduces another problem as to the length of the warm-up interval.

(ii) Use of batch data – The underlying idea here is the belief that correlation dies out in time and data collected at widely-spaced points in time can often be considered independent. Thus, for example, sample means collected at t_1 through t_{10}, t_{50} through t_{60}, t_{100} through t_{110}, etc., can in fact be regarded as i.i.d. Here again the question is how far apart is enough?

A more sophisticated method for statistical analysis of DEDS simulation is the method of **regenerative analysis** [Crane and Lemoine 1977]. The basic idea is that of a regenerative point in time which is analogous to the memoryless property in Markov process. This is best illustrated through an example. Consider a queue and server combination with exponential interarrival and service time distributions. Let us start the system at rest with no customer waiting or in service. We can let t_0 be the instant the first customer arrives. Subsequently, other customers arrive and queue for service until at some time t_{0f} the server finishes serving all waiting customers and the system is at rest again. The interval t_0 - t_{0f} is called a busy period. Now an idle period ensues until at t_1 when another customer arrives and starts a busy period once again. The interval t_0 - t_1 is a **re-generative cycle** in the sense that at t_1, and subsequent t_i, $i = 2, 3, \ldots$ similarly defined, everything starts anew due to the independent nature of the arrival process. One busy period behaves statistically the same as every other busy period. We can in fact consider every one of the intervals t_i - t_{i+1}, $i=0, 1, 2, \ldots$, as independent experiments, or simulation runs. The methods of Sections B.4.1-2 are now applicable with one proviso. Note the intervals t_i - t_{i+1} themselves are independent random variables. Thus, if we wish to compute sample averages, such as mean throughput or wait time, using each of these regenerative cycles, then these quantities must be normalized by the random interval length t_i - t_{i+1} before they can be combined. Regenerative analysis thus involves the ratio of sums of i.i.d. random variables. We shall not go into details here [Iglehart and Shedler 1980].

B.4.4 Importance Sampling and Change of Measure

From the simulation viewpoint, performance evaluation amounts to the computation of

$$E[L] \equiv \int L(\xi)P(\xi)d\xi = \lim_{n \to \infty} \sum_n L(\xi_n)P(\xi_n) \tag{4}$$

via the calculation of the sum

$$\{E[L]\}_{est} = \lim_{N \to \infty} \frac{1}{N} \sum_{n=1}^{N} L(\xi_n), \tag{5}$$

where $L(\xi)$ is the sample performance and $P(\xi)$ the probability of the sample value, ξ. The accuracy of (5) depends on the relative magnitudes of the term $L(\xi_n)P(\xi_n)$ for different n. Clearly if $P(\xi_i) > P(\xi_j)$, then simulation will automatically yield $L(\xi_i)$ more often than $L(\xi_j)$. This is good, particularly if $L(\xi_i)$ is also greater than $L(\xi_j)$. Since we cannot control L, we should attempt to change $P(\xi)$, i.e., "change the measure," or the probability, of the outcomes so that the more important outcomes occur more often. In other words, we should try to perform a simulation using $P'(\xi)$ with the ratio $L(\xi_i)/P'(\xi_i)$=constant; i.e., the vector $P' \equiv [P'(\xi_1), \dots, P'(\xi_N)]$ should be made parallel to the vector $L \equiv [L(\xi_1), \dots, L(\xi_N)]$. Fig. 7ab show this clearly in terms of the two-dimensional case (the simulation experiment consists of only two possible sample path outcomes). This notion of choosing a simulation distribution in order to compute more efficiently (e.g., less variance in the estimate of the performance) can be made rigorous. See [Glynn and Igelhart 1989].

Now the performance measure can be re-written as

$$\sum_n \left(L(\xi_n) \frac{P(\xi_n)}{P'(\xi_n)} \right) P'(\xi_n). \tag{6}$$

That is, we should simulate according to $P'(x_n)$ but re-scale each outcome of (5) as in (6) by $P(x_n)/P'(x_n)$. This is simply the "dual" of the likelihood ratio idea

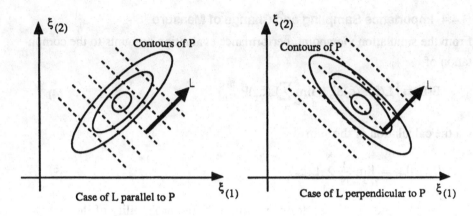

Fig. 7a Better Simulation   **Fig. 7b** Less Accurate Simulation

discussed in Chapter 7.   Instead of re-scaling the output of a simulation to
obtain the mean performance under a different distribution, we re-scale the
simulation output under a different distribution to compute efficiently the mean
performance of the original distribution. In importance sampling, we *choose*
the distribution under which the simulation is and re-scale the sample
values accordingly. In fact, we do so to minimize the variance in estimating the
performance.   The point being that the re-scaling ratio , $P(\xi_n)/P'(\xi_n)$, in importance sampling can be calculated accurately based on the given underlying
distribution. By choosing P' we can make rarely occurring but important
behavior more frequent. Performance can thus be efficiently simulated and
calculated. On the other hand, in likelihood ratio estimation we *choose* the re-
scaling of the sample value in order to get the mean performance value under a
different but *given*   distribution. We accept whatever variance property
associated with such a re-scaling. Oftentimes, not surprisingly, the variance
behavior of the likelihood ration method can be very poor. This discussion also
points out the difficulty of quantifying the accuracy of simulation – it depends
critically on the interactions of the performance measure and the underlying
randomness. For the same amount of computational effort, a simple change in
$L(\xi)$ or $P(\xi)$ can drastically change the accuracy. However, it is nevertheless
clear that estimating $E[L]$ is considerably easier than estimating $P(\xi)$ first and

then computing the estimate of E[L] directly through (4). This is, in fact, the intuitive reason that E[L] can often be accurately estimated in a fewer number of state transitions than the cardinality of the state space.

Elements of Optimization and Control Theory

It has been our thesis that much synergism can be derived from the time domain approach of optimal control theory, the probabilistic analysis of queueing net-works, and statistical simulation. Furthermore, the implied goal of performance analysis of DEDS is optimization. This section presents some basic concepts of optimization and control theory which we shall use at least implicitly in various parts of this book.

C.1 Optimization Basics

Consider a continuous and differentiable function $L(\theta)$ of n variables $\theta = [\theta_1, \ldots, \theta_n]$. A simple optimization problem can be stated as

$$\underset{\theta}{\text{Min}} \, L(\theta). \qquad (1)$$

The necessary and sufficient conditions for a local minimum are

$$\frac{\partial L}{\partial \theta} = 0 \qquad (2a)$$

and

$$\frac{\partial L}{\partial \theta} = 0 \, , \, \frac{\partial^2 L}{\partial \theta^2} > 0, \qquad (2b)$$

respectively, assuming these derivatives exist. Note that (2a) defines a set of n equations which, when solved, in general will yield one or more points in the θ-space; those points are candidates for the solution to the optimization problem of (1). Intuitively, we can think of $L(\theta)$ locally as a bowl-shaped surface in a three-dimensional (L, θ_1, θ_2) space. Eq.(2a) determines the flat bottom (zero slope of the tangent plane) of the bowl while Eq.(2b) the upward curvature of the bowl. A successive approximation scheme to determine the optimum is

$$\theta(i+1) = \theta(i) - \varepsilon \frac{\partial L}{\partial \theta}\Big|_{\theta=\theta(i)}; \quad i=1,2,\ldots, \varepsilon > 0, \tag{3}$$

which can be visualized as skiing downhill in a fog by following the fall line (the local direction of steepest descent).

In practice, optimization problems are complicated by the presence of equality and inequality constraints. In the case of equality constraints, Problem (1) is generalized to

$$\underset{\theta}{\mathrm{Min}}\, L(\theta) \quad \text{subject to } f(\theta)=0, \tag{4}$$

where $f(\theta)$ is a set of $m < n$ equations. Now, in general, there are only n-m truly free optimization variables since the other m variables are determined by the equality constraints $f(\theta) = 0$. It is convenient to define $\theta=[x,u]$, where x and u are of m and (n-m) dimensions, respectively, with u being the optimization variables and x determined by u through $f(\theta) = 0$. To derive the equivalent of Eq.(2), we introduce Lagrange multipliers λ to form the *Hamiltonian* $H(\theta,\lambda) = L + \lambda^T f$. Now, consider perturbations $d\theta=[dx, du]$. This results in a first order change of H

$$dH = \frac{\partial H}{\partial x}dx + \frac{\partial H}{\partial u}du = H_x dx + H_u du.$$

If we now choose for our convenience $\lambda^T = L_x [f_x]^{-1}$ then $H_x = 0$ and we are left with

$$dH = H_u du.$$

Direct argument leads to the intuitively pleasing necessary condition analogous to Eq.(2a)

$$\frac{\partial H}{\partial x} = 0, \quad \frac{\partial H}{\partial \lambda} = 0, \quad \frac{\partial H}{\partial A} = 0, \tag{5a}$$

which are a set of (m+n) equations of the (m+n) unknowns x, u, and λ. This accounts for the mnemonic property of the Lagrange multiplier, i.e., it permits us to solve the minimization problem by essentially ignoring the constraints. The particular choice of $\lambda^T = L_x[f_x]^{-1}$ also permits an easy analogy to Eq.(3). We have, to the first order, dH=$H_u$du with H_u interpretable as the gradient of L projected onto the m-dimensional space of f=0. Thus,

$$u(i+1) = u(i) - \varepsilon \frac{\partial H}{\partial u} \Big|_{x(i),u(i)}, \quad i=1,2,\ldots, \quad \varepsilon > 0, \tag{6}$$

is the equivalent of Eq.(3). A similar development to (5a) yields the second order equivalent of Eq.(2b) as

$$\frac{\partial^2 H}{\partial u^2} > 0 \quad \text{subject to } df = f_x dx + f_u du = 0,$$

or

$$\begin{bmatrix} f_u^T(f_x^{-1})T & I \end{bmatrix} \begin{bmatrix} H_{xx} & H_{xu} \\ H_{ux} & H_{uu} \end{bmatrix} \begin{bmatrix} I \\ f_x^{-1} f_u \end{bmatrix} > 0. \tag{5b}$$

When we further extend Problems (1) and (4) to inequality constraints, we enter the realm of mathematical programming. The problem now is

$$\underset{\theta}{\text{Min}} \, L(\theta) \quad \text{subject to } f(\theta) \leq 0. \tag{7}$$

Conceptually, we still have the analog of the necessary and sufficient conditions of Eqs.(2) and (5), and the steepest descent methods of Eqs.(3) and (6), as well as the intuitive interpretations behind these equations. Two major differences, however, now become important. Since not all inequality constraints are effective at every θ, and they are binding only in one direction even when they are effective, we have

(i) m need not be less than or equal to n; all λ must be of one sign.

(ii) any successive descent scheme must have efficient bookkeeping to keep track of all binding and nonbinding constraints and have a systematic method to switch from one set of binding constraints to another.

There is a huge literature on the subject with the simplex method of linear programming as the best known example of (i) and (ii).

Lastly, when the constraints become dynamic (i.e., differential or difference equations as constraints among x and u variables), we have optimal control problems. Conceptually, they still are no different from that of Eqs.(1-3). The optimal control function, $u(t)$, $t_0 \le t \le t_f$, can always be approximated for practical purposes by a staircase function

$$u(t) = u_i \text{ for } iT \le t \le (i+1)T, \, i = 0,1,2,...,K-1, \text{ and}$$

$$T = (t_f - t_0)/K, \, u = [u_0, u_1,...,u_{K-1}],$$

and Eqs.(1 - 7) are applicable again. However, there are both notational and analytical advantages of remaining in the continuous time. For this, we go to the next section.

C.2 Control Theory Fundamentals

The subject of control theory, as it is usually understood, applies to continuous variable dynamic systems (CVDS) governed by systems of ordinary differential equation as

$$\frac{dx_i}{dt} = f_i(x_1,\ldots, x_n; u_1,\ldots, u_m; t); \quad x_i(t_0) = \text{given}; \quad i = 1,\ldots,n, \, ,$$

or in vector matrix notation,

$$\frac{dx}{dt} = f(x,u,t); \quad x(t_o) = \text{given.} \tag{8}$$

The vector $x \in R^n$ is called the **state** vector, and u the **control** vector. Note that, in the context of control theory, x not only carries the usual physical meaning, such as the position and velocity, but also has the following mathematical implication: Given the value of x at any time t_o and the control input u(t) for $t \geq t_o$, the future values of x for all $t > t_o$ are completely determined by virtue of the uniqueness of the solutions to the differential equations. This is in contra-distinction to the convention in DEDS and particularly in GSMP models of DEDS, where the term "state" usually means the **physical state** of the systems, such as the number of customers waiting in each queue, and only sometimes means the **mathematical state** in the sense above.

The performance measure for CVDS is generally specified as

$$J = \phi(x(t_f) + \int_{t_o}^{t_f} L(x,u,t)dt, \tag{9}$$

where ϕ and L are given linear or nonlinear functions of x and u. The optimal control problem can then be stated as:

{Determine the control u(t), $t_o \leq t \leq t_f$,
to minimize (9) subject to (8)} $\qquad\qquad$ (10)

which is the equivalent of (1), (4), and (7) for dynamic optimization. As mentioned earlier, a conceptually simple way to develop the necessary condition for an optimum similar to Eq. (5a) is to discretize or approximate the continuous time u(t), $t_o \leq t \leq t_f$, into a series of piecewise constant steps $u = [u(t_o), u(t_1), \ldots, u(t_T)]$. Then Eq.(5a) is immediately applicable. However, a more elegant formu-lation exists. We consider the Lagrange multiplier function $\lambda(t)$, $t_o \leq t \leq t_f$, and define $H(x,u,\lambda,t) = L + \lambda^T f$. Now, augmenting Eq.(9) by (8) via $\lambda(t)$ and integrating by parts, we get

$$J = \phi(x(t_f)) + \int_{t_o}^{t_f} [L+\lambda^T(f - \frac{dx}{dt})dt = \phi - \lambda^T x\Big|_{t_o}^{t_f} + \int_{t_o}^{t_f} [H- \frac{d\lambda^T}{dt}x]dt \quad (11)$$

Now consider perturbations $\delta u(t)$, $t_o \le t \le t_f$, and the induced $\delta x(t)$, $t_o \le t \le t_f$, we get

$$\delta J = (\phi_x-\lambda^T)\delta x\Big|_{t_f}+\lambda^T \delta x\Big|_{t_o}+ \int_{t_o}^{t_f} \{[H_x-\dot{\lambda}^T]\delta x(t) +H_u\delta u(t)\}dt. \quad (12)$$

Choosing for convenience, $(\phi_x - d\lambda/dt) = 0$ at t_f and $d\lambda/dt = -H_x$, we are left with the analog to Eqs. (3) and (6)

$$\delta J = \int_{t_o}^{t_f} H_u\delta u(t)dt$$

$$u(t)|_{i+1} = u(t)|_i - \epsilon H_u; \ t_o \le t \le t_f; \ \epsilon > 0 \qquad (13)$$

and the necessary conditions

$$\frac{d\lambda}{dt} = -H_x^T \ ; \ \ \lambda^T(t_f) = \phi_{x(t_f)}$$
$$\frac{dx}{dt} = f(x,u,t) \ ; \ \ x(t_o) = \text{given}$$
$$H_u = 0 \qquad \text{for} \qquad t_o \le t \le t_f \qquad (14)$$

which is a two-point boundary value problem conceptually equivalent to the implicit equations of (5a). The sufficient condition (analog of Eq.(2b) and (5b)) for (10), which is known as the conjugate point condition and intuitively represents the properties of the second order functional space curvature $\partial^2 H/\partial u^2$, is more involved to state. Additional complications to Eqs.(13) and (14) arise if we impose further restrictions such as

$$u(t) \in U \text{ and } x(t) \in X, \ t_0 \le t \le t_f,$$

where U and X are subsets of R^m and R^n, respectively. We refer details to [Bryson and Ho 1969].

C.3 Linearization of General Dynamic System

The necessary condition of (14) can be understood from another constructive viewpoint. Consider the dynamic system defined by (8) and the small perturbations in the initial conditions $x(t_0)$, $\delta x(t_0)$, and in the control $u(t)$, $\delta u(t)$, $t_0 \le t \le t_f$. The resultant trajectory $x(t)$ will experience $\delta x(t)$, $t_0 \le t \le t_f$ These perturbations are governed by (to the first order) the linearized differential equations

$$\delta \dot{x} = f_x \delta x + f_u \delta u \equiv F \delta x + G \delta u, \quad \delta x(t_0) = \delta x_0, \tag{15}$$

where $F(t)$ and $G(t)$ are the $n \times n$ and $n \times m$ time varying matrices of partial derivatives $[\partial f_i / \partial x_j]$ and $[\partial f_i / \partial u_j]$, respectively, and are evaluated along the nominal trajectory $(x(t), u(t), t_0 \le t \le t_f)$, which is a solution of (8). Linear differential equation theory then gives the general solution of Eq.(15) as

$$\delta x(t) = \Phi(t, t_0) \delta x(t_0) + \int_{t_0}^{t} \Phi(t, \tau) G(\tau) \delta u d\tau, \tag{16a}$$

where

$$\frac{d\Phi(t, \tau)}{dt} = F(t) \Phi(t, \tau), \quad \Phi(\tau, \tau) = I; \tag{16b}$$

$$\frac{d\Phi(t, \tau)}{d\tau} = -\Phi(t, \tau) F(\tau), \quad \Phi(t, t) = I; \tag{16c}$$

and $\Phi(t, \tau)$ is known as the transition (or fundamental) matrix of the linear system $d\delta x/dt = F\delta x$ and Eq.(16c) the adjoint to (16b). If we take $\delta x(t_0) = 0$ as $x(t_0) =$ given and choose to write

$$\delta x(t) = \int_{t_0}^{t} \Phi(t, \tau) f_u \delta u d\tau, \quad \text{or} \ = \int_{t_0}^{t_f} \Phi(t, \tau) f_u \delta u d\tau \ \text{for } t = t_f, \tag{17}$$

then taking variations of Eq.(9) we get

$$\delta J = \frac{\partial \phi}{\partial x(t_f)} \delta x(t_f) + \int_{t_o}^{t_f} [\frac{\partial L}{\partial x} \delta x(t) + \frac{\partial L}{\partial u} \delta u(t)] dt$$

$$= \int_{t_o}^{t_f} \{[\phi_{x(t_f)} \Phi(t_f,t) - L_{x(t)} \int_{t}^{t_f} \Phi(t_f,t) dt] f_u + L_u\} \delta u(\tau) d\tau. \tag{18}$$

By virtue of Eqs.(16b) - (16c), the term in the square bracket of Eq.(18) can be shown by direct substitution as the solution of

$$\frac{d\lambda}{dt} = -f_x^T \lambda - L_x, \quad \lambda(t_f)^T = \phi_{x(t_f)}. \tag{19}$$

This finally gives

$$\delta J = \int_{t_o}^{t_f} (\lambda^T f_u + L_u) \delta u(t) dt \equiv \int_{t_o}^{t_f} H_u \delta u(t) dt \tag{20}$$

which is the same as Eq. (13). Thus, the necessary conditions for dynamic optimality can be viewed as the limit of a constructive (hill descending) approach to optimization via successive linearization. *Perturbation analysis -of DEDS can be thought of as the analog of (13) with the exception that we -are dealing with parameters rather than time functions. In other words, $\delta u(t)$=constant for all t.*

C.4 Stochastic Optimization and Control

Since noise and stochastic effects are ever present in real life, we have come to accept that in practice deterministic outcomes are seldom possible. Instead we should control or influence the mean, the variance, or the probabilities of the various possible outcomes so that on the average the system behavior is desirable or is the best. This naturally leads to stochastic optimization. The analog to problem (1) now becomes

$$\underset{\theta}{\text{Min}}\, E[L(\theta,\xi)] \equiv \underset{\theta}{\text{Min}}\, J(\theta), \tag{21}$$

where ξ represents all the random variables of the problem with given distribution, $F(\xi)$. Note that if it were possible to take expectation explicitly and to determine $J(\theta)$, then problem (21) is no different from (1). The difficulty comes from the fact that for most real problems it is impossible to determine $J(\theta)$ analytically. We must learn to descend the optimization hill using a noisy version of Eq.(3) which is known as stochastic approximation[1]. There are two versions of stochastic approximation algorithms known as Kiefer-Wolfowitz and Robbins-Monro algorithms depending on whether the noisy gradient is computed via the finite difference of $L(\theta+\Delta\theta,\xi) - L(\theta,\xi)$ or $\partial L(\theta,\xi)/\partial\theta$, respectively [Kiefer-Wolfowitz 1952], [Robbin-Monro 1951]. Conceptually, we can visualize the successive hill-descending process as follows. Initially, when we are far away from the optimum, the deterministic part of the gradient is likely to be large and to dominate the noisy part. Consequently, we can follow its direction without too much fear of "going the wrong way." As we gradually approach the optimum, the noisy part of the gradient will become overwhelming since the deterministic part approaches zero, and we should pay less and less attention to the value of the stochastic gradient. In the limit, if we wish to converge to the optimum, we must completely ignore it. The implication of this thinking is that we must choose with care the proportionality constant, ε, in (3). First of all, it must gradually decrease to zero in the limit if the algorithm is to converge at all. But at the same time it cannot decrease too rapidly for fear of stopping short. Theoretical analysis shows that $\varepsilon(t) = 1/t$ is an appropriate choice[2] if we can calculate $\partial L(\theta,\xi)/\partial\theta$ directly

---

[1] The analogy is "skiing downhill in a heavy fog".

[2] $\lim_{t\to\infty}(1/t)=0$ and $\int_{1}^{\infty}(1/t)\, dt = \infty$

without differencing. In the case of differencing, additional care must be taken regarding the size of the difference $\Delta\theta$ for striking a balance between the twin evils of the nonlinear effect (when $\Delta\theta$ is large) and the noise magnification effects (when $\Delta\theta$ is small) of $L(\theta+\Delta\theta,\xi) - L(\theta,\xi)$. Fig. 1 makes this clear.

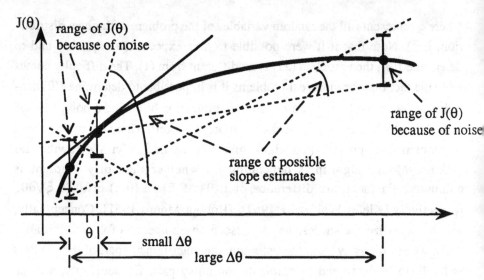

Fig. 1 The Twin Evils of Stochastic Estimation of Slope

The convergence rate of the Robbins-Monro (RM) algorithm is known to be superior to that of Kiefer-Wolfewitz (KW). In the context of this book, KW is akin to the brute force differencing method of determining the gradient of $J(\theta)$, while the PA approach directly provides the gradient estimate. Heuristically, this explains the observed superior variance properties of the PA approach.

In practice, a round-off error in computation also intrudes as a source of persistent disturbance. Empirically, it has been found that letting $\varepsilon(t) = 1/t$ often stops the hill descending process short of the optimum. A slower rate of decrease of $\varepsilon(t)$ is found to be necessary for good practical performance (this is also in accord with the estimation theory for dynamic system to be described below). A simple modification is to keep $\varepsilon(t)$ at a constant value and decrease it at the $1/t$ rate only when a descent step results in an increase in $L(\theta,\xi)$ [Wilde 1964, Chapter 6].

To describe the analog of (1) for dynamic systems in a stochastic environment, we must introduce one more major complication. Since the state $x(t)$ often cannot be measured directly with precision, we must contend with the fact that only a past history of indirect observations

$$z(t) = h(x(t), v(t), t) \quad ; \quad t_o \leq t \leq t_f \tag{22}$$

is available at time t, where h is some known linear or nonlinear function of x and v the measurement noise process. The measurement vector z is often of a dimension less than that of x, e.g., z is an angular measurement with respect to the local azimuth while x is the position and velocity of a particle in space. The dynamic system itself may also be subjected to disturbance. We modify Eq.(8) to

$$\frac{dx}{dt} = f(x,u,w,t); \quad x(t_0) \text{ with given distribution,} \tag{23}$$

where $w(t)$ is a given stochastic process describing the disturbance. The problem of determining

$$\hat{x}(t) \equiv E[x(t) / z(\tau), t_0 \leq \tau \leq t], \tag{24}$$

i.e., the mean of $x(t)$ conditioned on the past history subject to (22) and (23), is known as the optimal filtering problem in control theory. "Filtering" is important because the basis of stochastic control often requires knowledge of the best estimate of the current state $x(t)$. The simplest case of filtering corresponds to the situation where $x(t)$ is a scalar constant, i.e., $dx/dt=0$, and measurement $z(t) = x(t) + v(t)$, where $v(t)$ is Gaussian zero mean noise. In this case, common sense, as well as the filtering theory, dictates the optimal recursive estimate as

$$\hat{x}(t+1) = \hat{x}(t) + \frac{1}{t+1}[z(t+1) - \hat{x}(t)] \; ; \; \hat{x}(0) = 0 \tag{25}$$

$$\Longleftrightarrow \hat{x}(t) = \frac{1}{t}\sum z(i) \; \text{ or } \; \hat{x}(t) = \frac{1}{t}\int_0^t z(\tau)d\tau$$

which is also in agreement with the Robbin-Monro algorithm of stochastic approximation [Ho 1963]. If we further stipulate that $dx/dt = w$, where $w(t)$ is a purely random zero mean Gaussian process (white noise), then filtering theory requires that the weighting factor on $z(t)$ in Eq.(25) gradually converges to a constant rather than zero as in theoretical stochastic approximation. This again is in accordance with practice.

Finally, the problem of optimal stochastic control can be stated as determining

$$u(t) = k(z(\tau), \, t_o \le \tau \le t) \text{ to}$$

$$\underset{k \in \Gamma}{\text{Min}} E\left[\phi(x(t_f) + \int_{t_o}^{t_f} L(x, u, t)dt \Big/ z(\tau), \, t_o \le \tau \le t\right] \text{ subject to (22) and (23)} \tag{26}$$

where Γ is the admissible set of control laws (usually the set of measurable functionals on $z(\tau)$). This is a very difficult problem for which the only general solution known is for the case of (22) being Linear, ϕ and L being Quadratic, and w and v being Gaussian purely random processes, i.e., the so-called LQG problem in control theory. If we replace continuous time by discrete time, continuous state by discrete state, the dynamics of (23) by a probabilistic transition mechanism of

$$\text{Prob}[x(t+1)=j \, / \, x(t)=i, \, u(t)=k],$$

and (22) by direct and deterministic observation of $x(t)$, then problem (26) reduces to the Markov Decision Problem (MDP) of operations research [Hiller and Lieberman 1986, 4th edition Chapters .20 and 22]. In particular, [Alj and

Haurie 1983] has explicitly formulated an event driven version of the MDP. The dynamic programming approach to solve MDP is well known in operations research and has its counterpart in deterministic and stochastic control [Bryson and Ho 1969 Chapters 4, 5, 12, and 14].

C.5 On-Line Optimization of DEDS

Since PA offers the possibility of computing estimates of the performance gradient in real time as the trajectory evolves, on-line optimization of the performance of a DEDS becomes possible. Typical methodology involves what we just discussed in Section C.4 concerning stochastic approximation. However, when optimizing the performance of a DEDS, $J(\theta) = E[L(\theta,\xi)]$, where θ is the design parameter (or a set of parameters), using noisy estimates of $dJ(\theta)/d\theta$ in some iterative hill-climbing fashion, one encounters two fundamental problems. The first problem has to do with getting independent samples of $dL(\theta,\xi)/d\theta$ in order to compute an estimate of $dE[L(\theta,\xi)]/d\theta$ and proving its convergence. This usually means computing the estimate over regeneration cycles such as a busy period [Chong and Ramage 1990]. This, however, may require a long duration. Experimentally, [Suri and Leung 1989] and [Fu and Ho 1988] have found that much shorter durations can be used. But in such cases, convergence proof is difficult. The second problem is a trade-off decision. We often have a given computational budget, e.g., one is allowed a fixed number of noisy measurements of $dJ(\theta)/d\theta$ or $J(\theta)$ to be taken at various values of θ. One must decide on whether to spend more effort to get a better estimate of the gradient at fewer values of θ and hence a lesser number of hill-climbing steps or to do more iterative steps, i.e., to get estimates at more values of θ but content with less accurate gradient information for each value or each step. A related problem with stochastic hill climbing is the fact that old data are not remembered. At step n we have an estimate of $dJ(\theta)/d\theta$ based on data taken at $\theta=\theta_n$. The data, or the estimates of $dJ(\theta)/d\theta|_{\theta=\theta_i}$, $i=1,2, ..., n-1$, collected earlier, however, are simply discarded even though they contain information about the response curve $J(\theta)$ and can be useful for optimization purposes. The reason for this is probably due to hardware and memory limitations of earlier computers. In the following we suggest two alternative ideas that could be useful in DEDS optimization.

(1) Response Surface Methodology

A well-known approach to the optimization of $J(\theta)$ is the response surface method [Box and Draper 1987]. The method can be effectively combined with PA algorithms. This can be explained as follows. There is the basic response curve (surface) $J(\theta)$. Any performance data we take at any value of θ, say θ_i, represents a noisy (possibly very noisy) measurement of $J(\theta_i)$, call it $J(\theta_i)|_{est}$. As the optimization continues, we shall observe data $J(\theta_i)|_{est}$, $i=1,2,3,....$ Sooner or later we will accumulate enough data such that a least square fit of the entire response curve can be attempted. This is true regardless of how we optimize or hill climb. In fact, we suspect that under mild conditions, the least square fit will converge to the true $J(\theta)$ locally. A recursive way of doing the least square is also easy. What is needed is, of course, a parametrized model for $J(\theta)$ in order to perform the least square fit. For example, we may postulate that $J(\theta) = (a/2)\theta^2 + b\theta + c$ at least locally. And if there are noisy estimates of $dJ(\theta)/d\theta$ at $\theta = \theta_i$, $i=1,2,3, ...$, then a recursive least square fit of "a" and "b" can be easily carried out (see below). Knowing the estimates of "a" and "b," an estimate of the minimum of $J(\theta)$ can be attempted at $\theta = -b|_{est} / a|_{est}$. Thus, in this approach we attempt to take global steps instead of local gradient steps which may not be a bad idea in any case since it buys us quadratic convergence if it works.

Let z_i be the estimate of $dJ/d\theta$ at θ_i, $i=1,2,...,n$, then the recursive Bayesian (least square) estimates of "a" and "b" based on a quadratic model of $J(\theta)$ are given by [Bryson and Ho 1969, Chapter 12]

$$\begin{bmatrix} \hat{a}_{n+1} \\ \hat{b}_{n+1} \end{bmatrix} = \begin{bmatrix} \hat{a}_n \\ \hat{b}_n \end{bmatrix} + P_{n+1} \begin{bmatrix} \theta_{n+1} \\ 1 \end{bmatrix} [z_{n+1} - \hat{a}_n\theta_{n+1} - \hat{b}_n], \tag{27}$$

where

$$P_{n+1} = P_n - P_n \begin{bmatrix} \theta_{n+1} \\ 1 \end{bmatrix} \begin{bmatrix} \theta_{n+1} & 1 \end{bmatrix} P_n. \tag{28}$$

and P_0 is the covariance matrix of the prior estimate of "a" and "b". Adopting the logic of the second order iterative scheme, we set

$$\theta_{n+1} = - \frac{\widehat{b_{n+1}}}{\widehat{a_{n+1}}} \tag{29}$$

or if one is more conservative,

$$\theta_{n+1} = \alpha\, \theta_n + (1-\alpha) \frac{\widehat{b_{n+1}}}{\widehat{a_{n+1}}}, \quad 0 \le \alpha \le 1. \tag{30}$$

Furthermore, as we converge towards the optimum, gradual forgetting of old data can be implemented so that $J(\theta) = (a/2)\theta^2 + b\theta + c$ will be a better and better model of the response curve near the optimum. In summary, the following possible advantages of this approach are clear:

(i)   There is no waste of data. Everything is used to estimate $J(\theta)$ directly.

(ii)   Second order convergence and global iteration steps

(iii)   Flexibility in allocating computational effort through adaptive and/or weighted least squares.

(iv)   Reduced number of points required for surface fitting since PA give n additional pieces of information for each experiment at little additional cost where n is the dimension of the parameter vector θ.

Generalization of Eqs.(27 - 30) for multidimensional θ is straightforward.

(2) Simulated Annealing Method

In the response surface method, one has the problem of choosing optimally the initial set of points, θ_i, i=1,2, . . . , n to determine the initial fit. This is the well-known problem of experimental design. However, the simulated annealing method offers another rationale. Visualize the response surface $J(\theta)$ as a generalized conditional probability density function. This is always possible with appropriate translation of the origin and normalization. Then the problem of finding the maximum of $J(\theta)$ is equivalent to estimate the maximum likelihood of the density function. There are many ways to determine the estimate, almost all of which requires the generations of samples of θ from the density function $J(\theta)$. Simulated Annealing is a particular way of successively calculating the estimate which can be proved to converge w.p.1 to the maximum of the density function [Geman and Geman 1984]. The intuitive idea is to derive various values of θ by evaluating $J(\theta)$. The sample values of θ are chosen according to a parameterized version of the density function $J(\theta)$ itself. Due to the probabilistic interpretation of $J(\theta)$, the values of θ which will yield large $J(\theta)$ are chosen more often than not (which is, of course, what we want). The parameter in question is the variance of the density function. Initially the variance is chosen to be large so that a wide variety of θ can be tried. At any given time, we only retain the largest $J(\theta)$ obtained which can be regarded as a summary statistic of every point of $J(\theta)$ evaluated so far. As time goes on, the variance parameter is gradually shrunk so the sampling is more and more concentrated near the latest and the largest $J(\theta)$. It can be proved that if the variance parameter decreases in a particular way to zero, e.g. not too fast, then the successive θ will converge to the maximum of $J(\theta)$. This particular way of determining the values of θ at which to evaluate $J(\theta)$ can also be used as an alternative to the rationale of Eq.(29) to prevent being stuck at a local minimum.

A more complete survey of stochastic optimization ideas related to discrete event simulation can be found in [Jacobson and Schruben 1989]

A Simple Illustrative Example of the Different Models for DEDS

"Discrete-event dynamical systems exist in many technological applications, but there are no models of discrete-event systems that are mathematically as concise or computationally as feasible as are differential equations for continuous variable dynamical systems. There is thus no agreement as to which is the best model, particularly for the purpose of control." This quote from *Future Directions in Control Theory – A Mathematical Perspective*, Report of the Panel on Future Directions in Control Theory, Society for Industrial and Applied Mathematics [Fleming1988] describes the current situation in discrete event dynamic systems modeling. Many DEDS models have been proposed in recent years; examples include Markov chains and Markov jump processes, Petri nets, queueing networks, automata and finite-state machines, finitely recursive processes, models based on min-max algebra, and discrete event simulation and generalized semi-Markov processes. There has been considerable excitement and significant progress with many of these models. However, the rapid growth of literature has made it very difficult for a person who is not familiar with the field to gain a basic understanding of the fundamental concepts of these models within a relatively short period without an inordinate expenditure of effort.

In this appendix, we attempt to illustrate the main features of the above mentioned DEDS models by applying them to a very simple discrete event dynamic system. Our purpose is neither to do a detail analysis of any of these modeling tools, nor to give a complete survey of the existing literature. Readers who are interested in further study in the field can get a better picture from the special issue of IEEE Proceedings on DEDS [Ho 1989] and the papers referenced therein. The purpose here is primarily to give a flavor to rather than a definitive statement of the different modeling approaches. Furthermore, different models possess different features and are best suitable for different objectives. These features cannot always be shown to the maximal advantage by applying all these models to one standard example. Also, different models impose different assumptions on the system. We have to use slightly different versions of the standard example to prevent the illustration from becoming too contrived or trivial. We shall specify these assumptions when discussing each model.

The example system we adopted is a simple production line consisting of two machines in tandem with one buffer storage in between and an infinite supply of raw material (parts) at the input of the first machine (Fig. 1). The output of the first machine is the input to the intermediate buffer storage and ultimately becomes the input to the second machine. Each part (job) requires a certain amount of service from the two machines in succession. The service time characteristics will be specified for each model discussed below. For example, the service time for the part on each machine can be deterministic or stochastic. The service time can also incorporate breakdown and repair durations (viewed as an extra long service time).

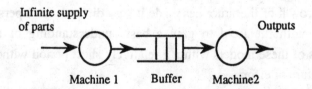

Fig. 1 The Two Machine System

The next seven sections attempt to model, in more or less detail, this simple and conceptual example using different approaches.

D.1 The Markov Process (Chain) Model

To apply the Markov process model, we assume that the service times of the two machines are exponentially distributed with service rates λ and μ, respectively. Under this assumption, the system can be modeled as a simple Markov process, i.e., the birth-death process, whose state transition diagram is illustrated in Fig. 2 below:

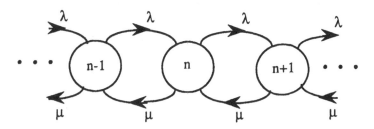

Fig. 2 A Birth-Death Process

The population variable, which is taken to be the number of parts in the buffer storage, is denoted by n. The process is a birth-death process with birth rate λ and death rate μ. Note that the death rate at state n=0 is zero; this assures that n $\geq$0. The stochastic service time can be visualized as to incorporate both the working and the breakdown/repair times for the machines.

A computer algorithm to implement this process can be carried out via the flow chart in Fig. 3. As an aside, we could easily have made this a finite-state version of the same problem by making the birth rate zero for n$\geq$N where N is the upper limit of the population size. In such a case, the birth-death process could be formulated as a finite-state Markov chain, where the elements of the transition probability matrix can be determined from the parameters λ and μ. This Markov chain would then become the discrete time version of the continuous time model above.

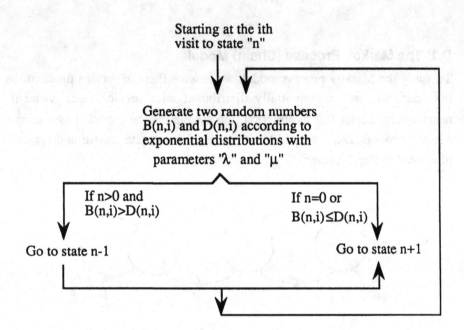

Fig. 3 Flow Chart for Fig. 1

D.2 The Equivalent M/M/1 Queue

There exists a well-known equivalent model for the simple birth-death process
in the form of an M/M/1 queue [Kleinrock 1975] as shown in Fig. 4.

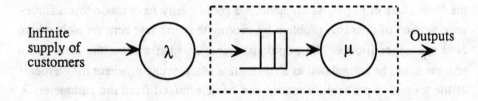

Fig. 4 An M/M/1 Queue

Here the first server, labeled λ, having an infinite supply of waiting customers, is used to simulate the arrival Markov process with parameter λ. The second server has an exponential service time distribution with parameter μ. The number of waiting customers in the queue is the state variable n, which has statistical behavior indistinguishable from that of the birth-death process in Fig. 2. However, the algorithmic implementation of Fig. 4 is, in general, different from that of Fig.3 as illustrated in Fig. 5. The algorithm maintains an event list which lists the future events and event times according to their order of occurrence.

Note that for the same number of birth (arrival) and death (service completion) events in Figs. 2 and 4, the algorithm of Fig. 5 generates half as many random variables as that of Fig. 3. In the birth-death model, the times to the next birth and next death are generated anew in each state. Only one of the times is used to determine the timing of the next event, which is either a birth or a death, while in the M/M/1 model, the remaining event time is not discarded but is kept to help determine the event time in the next state. Because of the memoryless property of the exponential distribution, these two methods of modeling are statistically indistinguishable. But the time behavior n(t) for Figs. 2 and 4 will be deterministically different. Even more dramatic will be the difference between the time behavior of the two models if we consider a small perturbation in the sample value of one of the random variables. Suppose we consider a particular pair of samples, B(n,i) and D(n,i), with B(n,i) = D(n,i) + ε, where ε is a small positive number. In the nominal case, the result of Fig. 3 indicates that the next state will be n-1. But suppose we introduce a perturbation into B(n,i), i.e., B'(n,i)=B(n,i)+Δ, where Δ is such that B'(n,i)<D(n,i). The perturbed next state will now be n+1. The subsequent nominal and perturbed paths will be deterministically different since they will start from different states. On the other hand, if in the model of Fig. 4 we perturb B(i) (see the flow chart shown in Fig. 5) by Δ such that the arrival now occurs after some service completion event (rather than before) in the future event list (i.e., a change of order of the occurrence of events), then the reader can verify for himself that only a small perturbation of the history of n(t) results. The perturbed path n(t)

quickly returns to the nominal path. Here we have an example showing that a different model of the same problem can give rise to very different time domain behaviors. In one case, the behavior is, in a sense, continuous with respect to small perturbations while in the other the behavior is not (see also Section 5.1).

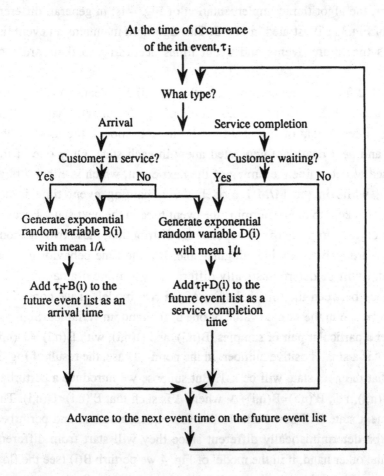

Fig. 5 Flow Chart for Fig. 4

Finally, if we assume that the service times of the two machines have general distributions, then we should apply the GI/G/1 queue as a model. Using general arrival and service time distributions may enable us to better model the real world of machine behaviors. Similarly, more complex machining or manufacturing systems can be modeled by more general Markov processes or networks of general queues.

D.3 The Generalized Semi-Markov Process (GSMP) and Discrete Event Simulation Model

The GSMP approach to DEDS modeling is an attempt to formalize discrete event simulation programs such as those of Figs. 3 and 5. The mathematical formalism is described in Appendix B Section 3. We shall briefly and informally review the main features here. The discrete part of the system, such as the number and location of customers, and the status of resources, are called "states." The continuous parts, such as the remaining service times, are called "clock readings," or "event lifetimes," which are determined as follows. For each state, we have an event list for all the feasible events for that state; for each event, we generate an event lifetime according to a given distribution. The smallest clock reading, or the smallest event lifetime, called the triggering event time, determines the next event and the state transition; at the time this event occurs, arbitrary logical conditions enabled by the current state, together with the appropriate distribution functions associated with the state and the triggering event, determine an instantaneous transition to the next state. A new event list and new lifetimes are determined and combined with the remaining lifetimes from the previous state to form the new event lifetime set. The process than repeats. In this way, the system evolves in time. Fig. 6 illustrates the GSMP model of the two-machine production line DEDS.

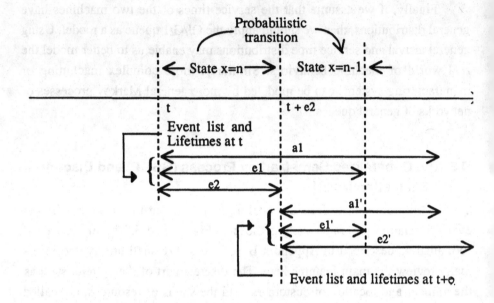

Interpretations:

i. e_1, e_2 are the residual service times of servers 1 - 2, respectively, and a_1 is the residual arrival time to server 1 at time t.

ii. At state x(t) two servers are busy.

iii. Server 2 finishes first (e_2 is shorter than e_1) and initiates a new service(e'_2) at server 2 for the next customer. At t+e_2 server 1 has a remaining lifetime e'_1, and the corresponding service completion becomes the next triggering event. a'_1 is the residual arrival time at t+e_2.

vi. State x(t+e_2) is n-1.

Fig. 6 Pictorial Illustration of a GSMP

Both the M/M/1 and GI/G/1 queues can be described by GSMP processes. The state space is $N = \{0,1,2,...,\}$. The event space is $\Gamma = \{\alpha, \beta\}$, where event α represents an arrival and event β is a departure. The event list for state n is $E(0) = \{\alpha\}$ and $E(n)=\{\alpha, \beta\}$ for $n>0$. The state transition functions are

$$p(m, n, \alpha) = \begin{cases} 1 & \text{if } m = n+1, \\ 0 & \text{otherwise,} \end{cases}$$

for $n \geq 0$, and

$$p(m, n, \beta) = \begin{cases} 1 & \text{if } m = n-1, \\ 0 & \text{otherwise,} \end{cases}$$

for $n>0$. For an M/M/1 queue, the clocking reading distributions for events α and β are exponential with means $1/\lambda$ and $1/\mu$, respectively. For a GI/G/1 queue they are, of course, general functions.

Exercise Consider the production line example in Fig.1 of the introduction with the exception that instead of an infinite supply at machine M_1, the jobs arrive randomly and queue for service at M_1.

(i)   Given the sequences of the interarrival times of jobs at M_1 and the service times of machines M_1 and M_2 as

$a_1, a_2, a_3, \ldots$	interarrival time sequence
$s_{1,1}, s_{1,2}, s_{1,3}, \ldots$	service time sequence of M_1
$s_{2,1}, s_{2,2}, s_{2,3}, \ldots$	service time sequence of M_2

write an algorithm for determining the trajectory of the two machine system. For uniformity of notations, let n_1 and n_2 be the number of jobs waiting at M_1 and M_2, t_0 the starting time, t_i the ith event time, and C_1, C_2, C_3 the customer arrival time, the departure time from M_1, and the departure time from M_2, respectively. There is no limitation on the buffer sizes.

(ii) Suppose each machine operates on a job for a cycle time of T units and then passes the job to the buffer storage of the next machine for the next cycle (i.e., everything is synchronized and evolves in T unit cycles). Machine M_i, i=1,2, may fail at the end of each operation cycle with a probability p_i. Once failed, M_i may be repaired during the next cycle with probability q_i. What is the throughput of the system?

(iii) What is the throughput of the system if there is no intermediate buffer storage at all? Let us assume that the failure of each machine will cause the shut down of the entire production line.

(iv) Discuss the connection of this problem with the materials in Sections D.1 - D.3.

D.4 The Automata and Finite-State Machine Approach

As the fourth approach to the same example, let us illustrate the workings of the finite-state machine model of DEDS taken from [Wonham and Ramage 1987]. We shall consider an extremely simple manufacturing system which is a slightly more restricted but more detailed version of the M/M/1 queue system considered in Fig. 4. Consider the first server with service rate λ as machine M_1, and the second server with service rate μ machine M_2. A part (customer) is first processed by M_1. Upon completion of the service at M_1, it is passed onto the buffer (queue), B, where it will be picked up by M_2 for further processing. The completed part leaves M_2 and the system. For simplicity of the model, let us assume that the size of the buffer is one. Thus, it can be in either one of the two states, empty or full (**E** or **F**). The machines can be in any one of three possible states, "**I**dle - and holding no part," "**W**orking - and holding one part," and "**D**own (broken) - and having discarded any part it was holding." The state transition diagrams for the machines and the buffer are shown in Fig. 7. Note that from the viewpoint of queueing theory, we can regard the machines as a black box server with service times which are a combination of working times and repair times. In this case, we can view the system as a G/G/1/1 system.

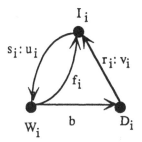

States:
I-Idle, W-Working, D-Down,
E-Empty, F-Full
Events:
s-start, f-finish, b-broken,
r-repaired
Control:
u,v = 1, or 0 (Enable/disable)

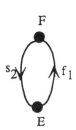

State diagram for machine i; i=1,2   State diagram for the buffer

Fig. 7 State Diagrams for Machines and Buffer

The transition between states is represented by an event which we assume to be observable. There are four types of events in this system: "start work," "finish work," "machine repair," and "broke while working." Note that the events "start work" and "machine repair" are enabled or disabled by the control variables "u" and "v," respectively. Furthermore, aside from being enabled or disabled by control variables, the mechanism for determining transition from one state to another and the actual transition are assumed to be specified. While working, a machine can legally go to the idle or the down state, but it is not of primary concern in which state it actually ends up. We are required only to observe the following set of operating rules:

(i) M_1 is allowed to start work only when B is empty. (This rule reflects the phenomenon of "blocking" in production lines.)

(ii) M_2 can start work only when B is full.

(iii) M_1 cannot start work when M_2 is down.

(iv) If both M_1 and M_2 are down, then M_2 gets repaired first.

Of the 18 ($2 \times 3 \times 3$) possible states, (i - iv) rule out the six states where M_1 is either working or down and B is full. The transition diagram for the remaining 12 states is shown in Fig. 8. The primary goal of the finite-state machine approach [Wonham and Ramadge 1987] is to design controllers in such a way that the control variables "u" and "v" are enabled and disabled based on the observed information, to insure that only the legally admissible state transitions in Fig. 8 will result. Now, if the complete state of the system is observable, it is easy to implement a feedback controller. Table 1 gives the function defining the feedback controller. However, if we make the further assumption that only the events are observable (a debatable assumption if we are modeling a manu-facturing system as opposed to a communication network), then we can raise the question of whether or not it is possible to implement the legal behavior of Fig. 8 based on observing the events of the system alone. One obvious solution is simply to create a copy of the system of Fig. 8 to run in parallel and in syn-chronism with the system. This way, we can certainly implement the appropriate control action by measuring the state of the copy. The function of Table 1 and the copy of the system together are then called the "supervisor." One last point, note that the outputs of the function in Table 1 are identical for certain states. Consequently, it is possible to reduce the state of the supervisor from 12 to 6, resulting in a state diagram for the supervisor as shown in Fig. 9. The reduced supervisor is obtained by projection from the complete supervisor and is called the *"quotient supervisor."*

For those readers familiar with the control theory approach, the example of the quotient supervisor design has an immediate parallel with the Linear-Quadratic-Gaussian optimal stochastic control problem. We show the conceptual similarity in the self-explanatory suggestive diagram below in Fig. 10 (the DEDS counterparts to that of the usual CVDS are indicated in the outline font).

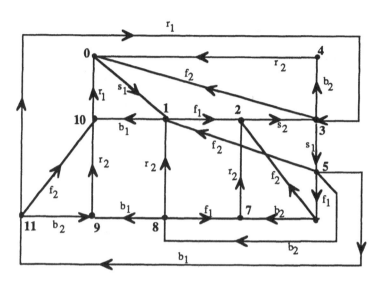

State	0	1	2	3	4	5	6	7	8	9	10	11
M1	I	W	I	I	I	W	I	I	W	D	D	D
B	E	E	F	E	E	E	F	F	E	E	E	E
M2	I	I	I	W	D	W	W	D	D	D	I	W

Fig. 8 Legally Admissible State Transitions

State	0	1	2	3	4	5	6	7	8	9	10	11
u_1	1	1	0	1	0	1	0	0	0	0	-	-
v_1	-	-	-	-	-	-	-	-	-	0	1	1
u_2	0	0	1	0	0	0	1	-	0	0	0	0
v_2	-	-	-	-	1	-	-	1	1	1	-	-
reduced state	0	0	1	0	3	0	1	2	3	4	5	5

Legend: 1 = enable,   0 = disable,   - = immaterial.

Table 1 State Feedback Controller

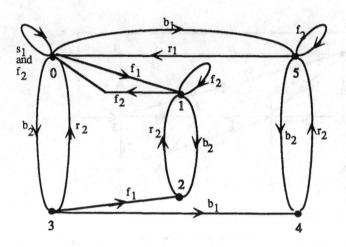

Fig. 9 The Quotient Supervisor

Given the above description of our simple problem, we can now informally state the approach of the finite state machine model. Consider a set of states Q and a set of events Σ. Transitions between states are triggered by events and enabled/disabled by control variables. The sequence of events describes the output behavior of a system. A collection of event sequences is a *language*. The so-called *Supervisor Control Problem* is to determine whether or not a given legally admissible language can be implemented by proper manipulation of the control variables based on the observed event sequence of the system, and if so, how? The major results in this area concern the characterization of such languages and structural insight into the required controllers. A more complete description of the formalism in connection with some ideas of perturbation analysis can be found in Section 7.2. Further results dealing with analogous concepts of controllability, observability, stability, etc. from CVDS control theory in the finite state machine model setting can be found in [Wonham 1988,1989] and [Ozveren and Willsky 1991].

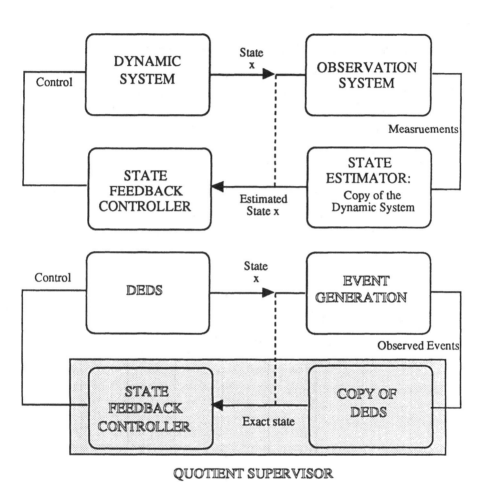

Fig. 10 The LQG Problem Analogy to the DEDS Supervisor Problem Obtained by Collapsing the Copy of the DEDS and the State of Feedback Controller

D.5 THE Min-Max Algebra Approach

The min-max algebra model [Cohen et al.1985] mainly deals with systems with deterministic events. (A recent work [Olsder et al. 1990] contains certain stochastic extensions of this model.) To illustrate this approach, we assume that the service times of machines M_1 and M_2 are exactly a and b, respectively. Thus, the M/M/1 queue in Fig.3 becomes a D/D/1 queue. To ensure that the queue is stable, we assume a>b. The min-max algebra approach studies the transient as well as the steady-state behavior of a system. Let x_n, and y_n, n>0, be the n th service completion times of machines M_1 and M_2. Let x_0 and y_0 be the earliest available times of these two machines. For convenience we assume that $x_0=0$ and that the queue length at time 0 is 0. It is easy to see that

$$x_1 = x_0+a; \quad x_2 = x_1+a \tag{1}$$

$$y_1 = \max\{ y_0+b, x_1+b \} \tag{2}$$

Using the min-max algebra, the above two equations can be written in a neat way. In min-max algebra, the two operations are defined as follows:

Product: $a*b \equiv a+b$

Addition: $a\#b \equiv \max\{ a, b \}$

Therefore, Equations (1) and (2) can be written as

$$x_2 = x_1*a$$

$$y_1 = x_1*b \# y_0*b$$

It is easy to prove that the operation # (= max) is commutative, associative over $\Re = \{ -\infty, +\infty \}$, and the operation * is distributive over #. Thus they can be used to define an algebra. In order to write these two equations in a matrix form using the product "*" and addition "#" convention, we arbitrarily choose a negative number $-H < x_1 - y_0$. Then we have

$$\begin{bmatrix} x_2 \\ y_1 \end{bmatrix} = \begin{bmatrix} a & -H \\ b & b \end{bmatrix} \begin{bmatrix} x_1 \\ y_0 \end{bmatrix}.$$

To avoid confusion, it is important to remember that all the formulas under this algebra must be interpreted by the rule: $a*b=a+b$ and $a\#b=\max(a,b)$. If we let

$$z_n = \begin{bmatrix} x_{n+1} \\ y_n \end{bmatrix} \quad \text{and} \quad M = \begin{bmatrix} a & -H \\ b & b \end{bmatrix}$$

Then we have $z_1 = M z_0$. Similarly, we have $z_2 = Mz_1 = M^2 z_0$ and $z_n = M^n z_0$.

Now let us calculate the value of M^n. Let $c=a-b$, and $J=H+a$. Then $b = a-c=a*(-c)$ and $-H=a*(-J)$. Therefore,

$$M = a \begin{bmatrix} 0 & -J \\ -c & -c \end{bmatrix} = aK \quad \text{where} \quad K = \begin{bmatrix} 0 & -J \\ -c & -c \end{bmatrix}.$$

Thus, $z_n = M^n z_0 = a^n K^n z_0 = (na)K^n z_0$. We also have

$$K^n = \begin{bmatrix} 0 & -J \\ -c & \max(-nc,-J-c) \end{bmatrix}$$

which is equivalent to

$$K^n = \begin{cases} \begin{bmatrix} 0 & -J \\ -c & -nc \end{bmatrix} & \text{if } (n-1)c < J \\ \begin{bmatrix} 0 & -J \\ -c & -J-c \end{bmatrix} & \text{if } (n-1)c > J \end{cases}$$

In min-max algebra, a matrix M is said to be order d periodical if and only if there is an integer n_0 such that for all $n > n_0$, $M^{n+d} = M^n$. Thus, matrix K is order 1 periodical. Let

$$Z = \begin{bmatrix} 0 & -J \\ -c & -J-c \end{bmatrix}, \quad z_0 = \begin{bmatrix} x_1 \\ x_1-c \end{bmatrix} = \begin{bmatrix} a \\ a-c \end{bmatrix}.$$

Then for a sufficiently large n, we have $z_n=(na)Zz_0$ or $x_{n+1}= (n+1)a$ and $y_n= (n+1)a +(-c) = na +b$. The explanation of this result is trivial: after the initial period, the service completion time of machine M_2 always equals that of machine M_1 plus the service time b of machine M_2. The periodicity of matrix K is just a mathematical description of the above statement.

In more general deterministic and periodic systems, the min-max algebra theory provides a powerful tool for analyzing the system behavior. To illustrate further the theory, we examine matrix M in the above illustrative system. Associated with M is the graph shown in Fig. 11. Each arc is weighted by a number which equals the corresponding entry in matrix M. In the graph there are 3 circuits whose lengths (the number of arcs in the circuit) are 1, 1, and 2, respectively. The weights of these circuits (the sum of the weights of all arcs in the circuit) are a, b, and b-H, respectively. The average weights (weight/length) are a, b, and (b-H)/2, respectively. The circuit with the maximum average weight is called a *critical circuit*. In Fig.11, the critical circuit has an average weight w= a and a length d=1. The following theorem was given in [Cohen et al.1985]:

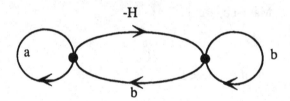

Fig. 11 Critical Circuits of the Simple Example

Theorem If the unique critical path has a length d and an average weight w, then the matrix M/w is of order d periodic. (The division is defined in the sense of min-max algebra.)

It is also proved in [Cohen et al.1985] that the maximum average weight is the eigenvalue of M under the min-max algebra. As another example, we consider the matrix

$$M = \begin{bmatrix} 1 & 5 \\ 3 & 2 \end{bmatrix} = 4 \begin{bmatrix} -3 & 1 \\ -1 & -2 \end{bmatrix} = 4K \quad \text{where} \quad K = \begin{bmatrix} -3 & 1 \\ -1 & -2 \end{bmatrix},$$

and we have

$$K^2 = \begin{bmatrix} 0 & -1 \\ -3 & 0 \end{bmatrix}, \quad K^3 = \begin{bmatrix} -2 & 1 \\ -1 & -2 \end{bmatrix}, \quad K^4 = \begin{bmatrix} 0 & -1 \\ -3 & 0 \end{bmatrix}.$$

Thus, K is of order 2 periodic. The eigenvalue of M is 4. In fact, the graph associated with matrix M is the same as that of Fig.11 with a, -H, the b on the right circuit, and the b on the middle circuit replaced by 1, 5, 2, and 3, respectively. M has a unique critical path with length 2 (the middle circuit), average weight $(3+5)/2=4$, and 4 is an eigenvalue of M. For any $e>0$, $[e+1, e]$ is an eigenvector which can be shown as follows:

$$\begin{bmatrix} 1 & 5 \\ 3 & 2 \end{bmatrix} \begin{bmatrix} e+1 \\ e \end{bmatrix} = 4 \begin{bmatrix} e+1 \\ e \end{bmatrix}.$$

D.6 The Petri Net Approach

In this sub-section we shall discuss the Petri net model and its application to the study of discrete event systems. A Petri net is a mathematical representation of a system [Peterson 1986]. Thus, it is another approach to the modeling of DEDS. Petri nets are designed specifically to model systems with interacting concurrent components. Many discrete event systems consist of components which exhibit concurrency. For example, in a computer system, peripheral devices, such as line printers, tape drives, disks, and so on, may all operate concurrently under the control of the computer. In a manufacturing system, machines, pallets, tools and control units usually operate in parallel. Thus, Petri nets are one of the important tools for modeling these systems.

In graphical representation, the structure of a Petri net is defined by three sets: a set of *places* P, a set of *transitions* T, and a set of directed *arcs* A. An arc connects a transition to a place, or a place to a transition. A place is an *input* to a transition if an arc exists from the place to the transition. A place is an *output* of a transition if an arc exists from the transition to the place. A formal definition of a Petri net is

$$PN = (P, T, A)$$

$$P = \{ p_1, p_2, ..., p_n \}$$

$$T = \{ t_1, t_2, ..., t_m \}$$

$$A = A_i \cup A_o$$

$$A_i \subseteq P \times T, A_0 \subseteq T \times P.$$

In the definition, p_1, p_2,...,p_n are places and t_1,t_2,...,t_m are transitions. An arc can also be considered as a mapping from P to T or from T to P. Graphically, places are drawn as circles, and transitions as bars, arcs are arrows from a place to a transition or from a transition to a place.

The dynamic feature of a Petri net is represented by *tokens*. Tokens are assigned to the places of a Petri net. The number and position of tokens may change during the execution of a Petri net. Tokens are drawn as small dots inside the circles representing the places of a Petri net. A place containing one or more tokens is called a *marked place*. A Petri net containing tokens is called a *marked Petri net*. A *Petri net marking* is an assignment of tokens to the places of a Petri net.

The execution of a Petri net is controlled by the number and distribution of tokens in the Petri net. A Petri net executes by firing transitions. A transition may fire if it is enabled. A transition is enabled if each of its input places contains at least one token. When a transition fires, one token is removed from each input place and one token is added to each output place. Firing a transition will in general change the marking of the Petri net to a new marking. Note that the number of input places of a transition may not be equal to the number of the output places of the transition. Thus, the number of tokens may also change during the execution.

Markings of a Petri net correspond to a state of the system being modeled. The firing of a Petri net transition corresponds to the occurrence of an event in the system. The occurrence of an event causes the system to move to a different state. Therefore, the successive firing of transitions (and the resulting changes in markings) in a Petri net represents the evolution of the system through different states.

The standard Petri net model described so far is effective in representing the logical relationships between the components of a system and the sequence of events in the system. It is, however, unable to represent the timing of activities which is important to all performance studies. The notion of time is introduced in Petri nets by associating a *firing delay* with each transition. A firing delay is the delay between the enabling and the firing of a transition. Petri nets which include the timing behavior are called *timed Petri nets*. The firing delay of a timed transition can be either deterministic, or random with a specified distribution. Petri nets which use random variables to model firing delays are called *stochastic Petri nets*. Pictorially, an immediate transition is drawn as a thin bar; a timed transition is drawn as a hollow box with the firing delay distribution written next to it.

Now, let us apply the Petri net model to the example discussed in the previous subsections. First, the behavior of the second machine, machine M_2, can be modeled by a Petri net shown in Fig. 12.

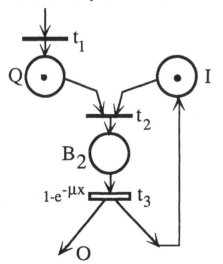

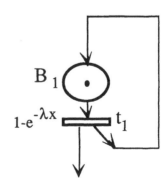

Fig. 12 A Petri Net Model of Machine M_2 Fig.13 A Petri Net Model of Machine M_1

In the figure, transition t_1 is connected to an input arc; each firing of t_1 adds a token to place Q and represents an arrival of a part to the machine. Q is a place representing the queue (or the buffer) of the machine. The number of tokens in Q equals the number of parts waiting in the buffer. t_2 is a transition corresponding to the event of "service starts" and I is a place representing that the machine is idle. The transition t_2 is an immediate transition which fires once there is a token in place I and at least one token in place Q. This models the fact that the machine starts to work on a new part once there is a part in the buffer and the machine is not working on other parts. Place B represents the busy status of the machine. A token in it indicates that the machine is busy. t_3 is a stochastic transition. The time delay associated with it represents the service time of the machine. In this example, it is exponentially distributed with mean $1/\mu$. After t_3 is enabled, i.e., after a token arrives at place B_2, the transition waits for a random time and then starts to fire. After the firing of t_3, a token moves to place I, indicating that the machine is idle, and another token enters place O, which represents an output.

Machine M_1 can be considered as a special case of machine M_2 with infinity many parts waiting for services. In this case, M_1 is always busy. In the Petri net representation, places Q and I can be omitted. The corresponding Petri net model is shown in Fig. 13. A complete representation of the two machine system can be obtained by combining the two Petri-nets in Figs. 12 and 13 together and is shown in Fig. 14.

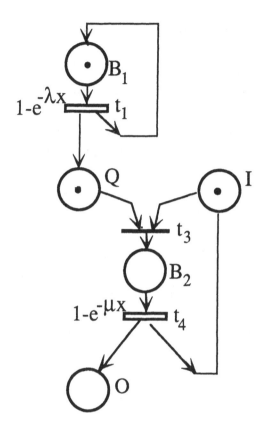

Fig.14 A Petri Net Model of the Two Machine System

D.7 The Finitely Recursive Process Formalism

One of the basic forms of output of any model of DEDS is a string of events, $\{e_1, e_2, e_3, \dots\}$. The different models of DEDS devise different internal structures to regenerate such event strings. This is accomplished by capturing the various "rules of operation" of the DEDS in terms of a particular internal structure, such as a Markov Chain, a queueing network, a finite state machine, a min-max algebra, a Petri-net, etc. The algebraic approach to DEDS models, for which the Finitely Recursive Process (FRP) Formalism is one representative example, attempts to describe DEDS in terms of a set of functions which operate on event strings called processes in FRP. In particular, the set of functions includes those which combine, concatenate, and modify event strings to reflect various parallel, series, enabling or disabling operations of one or more interacting DEDS. In essence, the FRP formalism makes precise a colloquial description of a DEDS and codifies various ad hoc "rules of operation" as formal operators or functions on event strings. The descriptive power of FRP is measured in some sense by what kind of interesting event strings can be generated by how simple an internal structure. Simplicity in turn is measured by how few "constructs," "definitions," and "functions" one needs to specify the formalism. FRP as defined in [Inan and Varaiya 1988] can be shown to be more powerful than Petri nets which in turn are more powerful than Finite state Machines in the sense of the set of event strings it can generate.

To describe the two-station example used in this appendix via the FRP algebra, it is essential to introduce a number of definitions, which translate words into symbols to capture the colloquial description of the example. We shall not give these definitions. Instead, a short description of FRP which informally capture the main idea of the approach is outlined below.

Again consider the two-machine system with only one buffer space. The event sequence of machine M_2 may be represented by a string of symbols $s=\alpha\beta\alpha\beta\alpha\beta\ldots$, where α represents an arrival and β a departure. The length of the string s may be any integer. A process P is a set of strings together with an *event function*, which specifies the next event that the process may execute after executing a string, and a *termination function*, which determines the termination of a string. For example, all the event sequences of M_2 form a process containing strings of the type $\alpha\beta\alpha\beta\alpha\beta\ldots$ Let Π be the space of all processes. We can define operators on processes. These operators are functions that map Π into Π. Five fundamental operators are defined in [Inan and

Varaiya 1988]. These operators are a mathematical formulation of the colloquial description of their manipulation of the strings. For example, the *Deterministic Choice Operator* works as follows: Applying this operator to an event α and a process P generates a process denoted as $Q=(\alpha\rightarrow P)$; any string in Q starts with event α followed by a string in the process P. Next, note that almost all DEDS must perform some repetitive tasks, e.g., machine M_2 starts to serve the next part after it completes the service to its first one. This feature makes it possible to determine the event string recursively. That is, the process associated with a DEDS can be represented by a recursive equation using the fundamental operators. For example, the process P for M_2 (consisting of all the strings of the type $\alpha\beta\alpha\beta\alpha...$) can be represented by the following equation

$$X = (\alpha\rightarrow(\beta\rightarrow X)).$$

This equation can be viewed as a **finite** representation of the process P which consists of **infinitely** many string symbols.

Generally, a process Y is called a **finitely recursive process** (FRP) if it can be represented as

$$X = f(X)$$

$$Y = g(X)$$

where f and g are functions on Π and f satisfies certain conditions. The obvious advantage of this representation is that these equations are analogous to the difference equations

$$x(t+1) = f(x(t)), \; y(t) = g(x(t)), \qquad t = 0, 1, 2, \ldots$$

for control systems. A more detailed and similar two-machine example illustrating the FRP approach can be found in [Inan and Varaiya 1988 Ex. 3].

This concludes our attempt to introduce the various proposed DEDS models. The modeling of DEDS is still at its early stage. There exist many theoretical as well as practical problems in the field. It is perhaps useful to repeat the criteria set forth by [Inan and Varaiya 1988] in evaluating any model formalism along the following dimensions:

(1)   *Intrinsic properties of formalism:*

(i)   descriptive power

(ii)   language complexity – e.g., L(FSM) < L(P-Nets) i.e., language of Petri nets is larger than that of finite state machines

(iii) algebraic complexity – how many interesting systems can be made up from how few building blocks

(2) *Connections to the Real World:*

(i)   implementation

(ii)   performance evaluation.

These criteria should be considered in connection with the desiderata discussed in Chapter 1.

Finally, It is also important to distinguish between models of DEDS and techniques that can be used to analyze such models. For example mean value analysis is an algorithm for calculating certain performance measures of DEDS modelled as a network of queues while formal language theory is the tool for describing finite state machines. In a similar vein, perturbation analysis and likelihood ratio methods are most suited for GSMP models.

A Sample Program of Infinitesimal Perturbation Analysis

As an example, in this appendix we show a program which incorporates IPA in a simulation program of a closed Jackson network. The intention is to show the efficiency of the IPA approach. The network consists of M servers and N customers with $q_{i,j}$ being the routing probabilities and s_i the mean service time. The programs contains about 400 lines, of which only three are for IPA. All the other lines are for the simulation of the network. These three lines are shown below in capital letters. To save space as well as to illustrate the main idea, only the basic functions of each procedure in the simulation part of the Jackson network are stated.

A Pascal-like program for simulating a closed Jackson network and implementing IPA.

```
Program PAofJacknet
Var         M:        (number of servers);
            N:        (number of customers);
         q[i,j]:      array [1..M, 1..M] of real numbers
                      (routing probabilities);
           s[i]:      array [1..M] of real numbers (service times);
           t[i]:      array [1..M] of real numbers (service completion times);
           n[i]:      array [1..M] of integers (number of customers at each
                      server);
         nsd[i]:      array [1..M] of integers (the number of customers served
                      by each server);
```

nd:	integer (the number of customers served by all the servers);
TP[i]:	array [1..M] of real numbers (throughput of each server);
ELS[i]:	array [1..M] of real numbers (the elasticity of throughput with respect to s_i);
PERT[i,j]:	array [1..M, 1..M] of real numbers;
	(perturbation of server j if server i's mean service time is perturbed)
T:	current event time;
nextsvr :	integer (the next event is a service completion of this server);

procedure input;
    < input the values for all the parameters >

procedure initialize;
    < sets initial values for all the variables, including the initial service times>

procedure randomNo;
    < A random number generator; when the procedure is called, it returns a
    [0,1) uniformly distributed random number, named xi >

procedure service (i);
    < determines the service time and service completion time of server i>

    begin
        call procedure randomNo;
        s[i] := - s_i * ln(1- xi);   < service time of server i >
        t[i] := T + s[i];   <service completion time of server i >
    end;

procedure destination(i);
    < determines the destination of the customer in server i>

procedure eventsearch;

 < determines the server which completes its service earlier than other
  servers; Note that a perturbation is generated for the server .>

Var s: real; i,j : integer;

 <recall the variable "nextsvr" is defined by : the next event is a customer
 transition from this server>

 begin

  s := infinity;

  for i := 1 to M do

  if (n[i] > 0) and (t [i] <s) then s:= t[i], nextsvr := i;

   < s is the smallest service completion time, which will be the next
   event time. >

  T := s; < advance the current time to the next event time >

  n[i] := n[i] +1;

  nd := nd +1;

  nsd[i] := nsd[i] +1;

(1) PERT [NEXTSVR, NEXTSVR]:= PERT [NEXTSVR, NEXTSVR]
  + S[NEXTSVR];

     < perturbation generation>

 end;

procedure update(i);

 <update queues after a customer transition from server i>

 begin

  n[i] := n[i] -1;

  called procedure destination(i);

   <which returns a destination of server i, named destsvr>

  n[destsvr] := n[destsvr] +1;

  if n[i] >0, then call procedure service(i);

  if n[destsvr] = 1, then

```
          begin
               call procedure service(destsvr);
(2)            FOR J := 1 TO M DO PERT [J, DESTSVR ] := PERT [J, I];
                    < perturbation propagation from server i to server destsvr >
               end;
     end;

procedure results;
     < calculates the results. >
     begin
          for i := 1 to M do
               begin
                    TP[i] := nsd[i] / T;
(3)                 ELS[I] := PERT [I,J] / T;
                    < For a large T, the differences among PERT[I,J] for
                    different J's are negligible. This will be explained below. >
                    end;
     end;

                         THE MAIN PROGRAM
begin
     input;
     initialization;
     repeat
          eventsearch;
          update(nextsvr);
     until (nsd= a prespecified number);
     results;
end.
```

In the program, the only memories added to the simulation for perturbation analysis is a [1..M] array ELS[i] for storing the values of all the estimated elasticities and a [1..M, 1..M] array PERT[i,j] for storing the perturbations of every server during the simulation. Thus, the memories increase for implementing IPA is only $M \times (M+1)$ real numbers. PERT[i,j] is proportional to the perturbation of server j when server i's mean service time is perturbed. Line (1) in the program implements the perturbation generation: If the next event is a service completion of server i, then a perturbation is generated at the server. According to the perturbation generation rule of exponential distributions, the perturbation added to server i should be $s[i] \times (\Delta s_i / s_i)$. But the factor $(\Delta s_i / s_i)$ will be cancelled when calculating the estimate of the elasticity by using $(s_i / \Delta s_i) \times (\Delta T / T) = (s_i / \Delta s_i) \times (\Delta t[i, j] / T)$, where $\Delta t[i,j]$ is the perturbation of the server j's service completion time due to the change of s_i. Thus, if we add only $s[i]$ to PERT[i,i], then PERT[i,j] $= (s_i / \Delta s_i) \times (\Delta t[i, j])$ and the elasticity becomes simply PERT[i, j]/T. Note that for a sufficiently large T, most perturbations generated during the simulation have already been realized or lost at the end of the simulation. Thus, the differences of PERT[i,j] among different j's are relatively small compared with T. This explains line (3), which calculates the estimates of the elasticities. Line (2) carries out perturbation propagation. The condition n[destsvr] = 1 implies that an idle period of the destination server is just terminated by the customer transition. Therefore, we propagated the perturbations (corresponding to the changes of the mean service times of different servers) of server i to the destination server by copying PERT [J, I] to PERT [J, DESTSVR].

The program clearly shows that the IPA algorithm is very efficient. The computing time for these three extra lines is even not noticeable. Also, as discussed throughout this book, the main PA methodology is basically distribution independent. Thus, adaptation of this program to general service time distribution and finite buffer involve minimal changes (e.g. see Chapter 3 and the exercises in therein).

The flow chart of the program is shown in Fig. 1. In the diagram, The block "Search for the next event" corresponds to the procedure "Eventsearch," where "Perturbation generation" is implemented by the sentence labeled (1). The block "State update" corresponds to the procedure "Update," where "Determine destination" is done by calling the subprocedure "Destination," "Assign new service times" is done by calling the subprocedure "Service," and perturbation propagation is implemented by the sentence labeled (2). The elasticity is simply calculated by the sentence labeled (3).

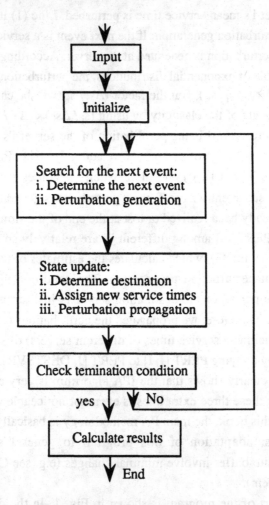

Fig. 1 Flow Chart of the Simulation Program

REFERENCES

Aleksandrov, V.M., Sysoyev, V.J., and Shemeneva, V.V., "Stochastic Optimi-zation," *Engineering Cybernetics*, (5), 11-16, (1968).

Alj, A. and Haurie, A., "Dynamic Equilibrium in Multi-generation Stochastic Games," *IEEE Transactions on Automatic Control* 28, (2), 193-203 (1983).

Balbo, G. and Denning, P. J., "Homogeneous Approximations of General Queueing Networks," *Performance of Computer Systems*, ed Arato, M., Butrimenko, A., and Gelenbe, E., North-Holland, Amsterdam (1979).

Balsamo, S. and Iazeolla, G., "Aggregation and Disaggregation in Queueing Networks - The Principle of Product Form Synthesis," *Proceedings of the Mathematics of Computer Performance and Reliability,* ed G. Iazeolla, P.J. Courtois, and A. Hordijk, Elsevier Publishing Co. (1984).

Basket, F., Chandy, K.M., Muntz, R.R., and Palacios, F.G., "Open, Closed, and Mixed Network with Different Classes of Customers," *Journal of ACM* 22, (2), 248-260, April (1975).

Bello, M., "The Estimation of Delay Gradient for Purpose of Routing in Data Communication Networks," *S.M. Thesis, Electrical Engineering Department, M.I.T.* (1977).

Billingsley, P., *Probability and Measurement,* Wiley, New York, (1979).

Box, G.E.P. and Draper, N.R., *Empirical Model-Building and Response Surfaces,* Wiley, New York, (1987).

Brandwajn, A., "Equivalence and Decomposition in Queueing Systems - A Unified Approach," *Performance Evaluation* 5, 175-186 (1985).

Bratley, P., Fox, B., and Schrage, L., *A Guide to Simulation (2nd Edition)*, Springer-Verlag, New York, (1987).

Breiman, L., *Probability*, Addison-Wesley, Reading, Massachusetts, (1968).

Bremaud, P, *Point Processes and Queues - A Martingale Approach*, Springer-Verlag Series in Statistics, New York, (1981).

Bryson, A.E. and Ho, Y.C., *Applied Optimal Control*, Blaisedell,, (1969).

Buzen, J., "Computational Algorithms for Closed Queueing Networks," *Communication of ACM* 16, (9), 527-531, (1973).

Cao, X. R., "Convergence of Parameter Sensitivity Estimates in a Stochastic Environment," *IEEE Transactions on Automatic Control* AC-30, 834-843, (1985).

Cao, X. R., "Realization Probability in Closed Queueing Networks and its Application," *Advances in Applied Probability* 19, 708-738, (1987a).

Cao, X. R., "Sensitivity Estimators Based on One Realization of Stochastic System," *Journal of Statistical Computation and Simulation* 27, 211-232, (1987b).

Cao, X. R., "First Order Perturbation Analysis of a Simple Multiclass Finite Source Queue," *Performance Evaluation* 7, 31-41, (1987c).

Cao, X. R., "A Sample Perforcmance Function of Jackson Queueing Newtorks", *Operations Research* 36, (1), 128-136, (1988a).

Cao, X. R., "Realization Probability in Multi-Class Closed Queueing Networks", *European Journal of Operations Research* 36, 393-401 (1988b).

Cao, X. R., "The Convergence Property of Sample Derivative in Closed Jackson Queueing Networks," *Stochastic Processes and Their Applications* 33, 105-122 (1989a).

Cao, X. R., "System Representations and Performance Sensitivity Estimates of Discrete Event Systems," *Mathematics and Computers in Simulation* 31, 113-122, Elsevier, North-Holland (1989b).

Cao, X. R., "Calculations of Sensitivities of Throughputs and Realization Probabilities in Closed Queueing Networks with Finite Buffers," *Advances in Applied Probability* 21, 181-206 (1989c).

Cao, X. R., "A Comparison of the Dynamics of Continuous and Discrete Event Systems," *Proceedings of the IEEE (Special Issue)* 77, (1), 7-13 (1989d).

Cao, X. R., "The Predictability of Discrete Event Systems", *IEEE Transactions on Automatic Control* 34, 1168-1171 (1989e).

Cao, X. R., "Perturbation Analysis of Closed Queueing Networks with General Service Time Distributions", *IEEE Transactions on Automatic Control* 36, to appear (1991).

Cao, X. R. and Ho, Y. C., "Sensitivity Estimator and Optimization of Produc-tion Line with Blocking," *IEEE Transactions on Automatic Control* 32, (11), 959-967 (1987a).

Cao, X. R. and Ho, Y. C., "Estimating Sojourn Time Sensitivity in Queueing Networks Using Perturbation Analysis," *Journal of Optimization Theory and Applications* 53, 353-375, (1987b).

Cao, X. R. and Ma, D. J., "Perturbation Analysis of General Performance Measures of Closed Queueing Networks with State Dependent Rates," *Applied Mathematics Letters* submitted (1991).

Cao, X. R., Gong, W. B., and Wardi, Y., "Nondifferentiability of Perfor-mance Functions of Queueing Networks," submitted, (1991).

Cassandras, C. G., "On-Line Optimization for a Flow Control Strategy," *IEEE Transactions on Automatic Control* 32, (11), 1014-1017 (1987).

Cassandras, C. G. and Strickland, S. G., "An 'Augmented Chain' Approach for On-Line Sensitivity Analysis of Markov Processes," *Proceedings of the 26th IEEE Conference Decision and Control* 1873-1880 (1987).

Cassandras, C. G. and Strickland, S.G., "Sample Path Properties of Timed Discrete Event Systems," *Proceedings of the IEEE* 77, 59-71, (1989a).

Cassandras, C. G. and Strickland, S. G., "On-Line Sensitivity Analysis of Markov Chains," *IEEE Transactions on Automatic Control* AC-34, (1), 76-86, Jan (1989b).

Cassandras, C.G. and Strickland, S.G., "Observable Augmented Systems for Sensitivity Analysis of Markov and Semi-Markov Processes," *IEEE Transactions on Automatic Control* AC-34, (10), (1989c).

Cassandras, C., Gong, W.B., and Wardi, Y., "Gradient Estimation of Queueing Systems with Real Time Constraints," *Proceedings of the IEEE Conference on Decision and Control* (1990).

Cassandras, C.G., Gong, W.B., and Lee, J.I., "Robustness Properties of Perturbation Analysis Estimators for Queueing System," *Journal of Optimization Theory and Application* submitted (1990).

Cassandras, C.G., Lee, J.I., and Ho, Y.C., "Efficient Parametric Analysis of Transient Performance Measures in Networks," *IEEE Journal of Selected Areas in Communications* 8, #9, 1709-1722 (1990).

Cassandras,C.G., Strickland, S.G., and Lee, J.I., "Discrete Event Systems with Real-Time Constraints: Modeling and Sensitivity Analysis Control," *Proceedings of the 27th IEEE Conference on Decision and Control* 220-225, Dec. (1988).

Chandy, K. M., Howard, J.H. and Towsley, D.F., "Product Form Solution and Local Balance in Queueing Networks," *Journal of ACM* 24, (2), 250-263 (1977).

Chandy, K.M., Herzog, U. and Woo, L., "Parameteric Analysis of Queueing Networks," *IBM Journal of Research and Development* 19, 36-42 (1975).

Chong, E.K.P. and Ramadge, P.J., "Convergence of Recursive Optimization Algorithms using IPA Derivative Estimates," *Proceedings of American Control Conference* (1990).

Cinlar, E., *Introduction to Stochastic Processes,* Prentice-Hall, (1975).

Cohen, G., Dubois, D., Quadrat, J.P., and Voit, M., "A Linear-System-Theoretic View of Discrete Event Process and its Use for Performance Evaluation in Manufacturing," *IEEE Transactions on Automatic Control*, 30, 210-220 (1985).

Courtois, P. J., *Decomposability:Queueing and Computer System Applications*, Academic Press, (1977).

Courtois, P. J., "On Time and Space Decomposition of Complex Structures," *Communications of ACM* 28, (6), 590-603 (1985).

Cox, D. R., "A Use of Complex Probabilities in the Theory of Stochastic Processes," *Proceedings Cambridge Philosophical Society* 51, 313-319 (1955).

Crane, M.A. and Lemoine, A.J., *An Introduction to the Regenerative Method for Simulation Analysis*, Springer, New York, (1977).

Davenport, W. and Root, W., *Introduction to the Theory of Random Signals and Noise*, McGraw Hill, (1958).

Denning, P. and Buzen, J., "The Operational Analysis of Queueing Network Models," *Computing Survey* 10, (3), 225-261 (1978).

Dille, J., "The Experimental Investigation of Perturbation Analysis Tech-niques," *Ph.D. Thesis, Harvard University* (1987).

Fishman, G.S., *Principles of Discrete Event Simulation*, Wiley (1978).

Fishman, G.S., "Accelerated Accuracy in the Simulation of Markov Chains," *Operations Research* 31, (3), 466-487, (1983).

Fishman, G.S., "Sensitivity Analysis for the System Reliability Function," *Technical Report UNC/OR/TR-87/6, Department of Operations Research, University of North Carolina* (1987).

Fishman, G.S., "Sensitivity Analysis Using the Monte Carlo Acceptance-Rejection Method", *Technical Report UNC/OR/TR-88/3, Department of Operations Research*, University of North Carolina (1988).

Fleming, W. H. (chair), *Future Directions in Control Theory - A Mathematical Perspective*, SIAM Panel report (1989).

Fu, M.C., "On the Consistency of Second Derivative Perturbation Analysis Estimators for the M/G/1 Queue," *Applied Mathematics Letters* 2, (2), 193-197 (1989).

Fu, M.C., "Convergence of a Stochastic Approximation Algorithm for the GI/G/1 Queue Using Infinitesimal Perturbation Analysis," *Journal of Optimi-zation Theory and Application* April (1990).

Fu, M.C., and Ho, Y.C., "Using Perturbation Analysis for Gradient Estima-tion, Averaging, and Updating in a Stochastic Approximation Algorithm," *Proceedings of the Winter Simulation Conference* 509-517 (1988).

Fu, M.C. and Hu, J.Q., "Consistency of Infinitesimal Perturbation Analysis for the GI/G/m Queue," *European J. of Operational Research* to appear (1991a).

Fu, M.C. and Hu, J.Q., "On Choosing the Characterization for Smoothed PA," *IEEE Transactions on Automatic Control* 36, to appear (1991b).

Geman, S. and Geman, D., "Stochastic Relaxation, Gibbs Distributions, and the Bayesian Restoration of Images," *IEEE Transaction on Pattern Analysis and Machine Intelligence* PAMI-6, (6), 721-741, November (1984).

Glasserman, P., "Equivalence Methods in the Perturbation Analysis of Queueing Networks," *Ph.D. Thesis, Division of Applied Science, Harvard University*, (1988a).

Glasserman, P., "IPA of a Birth and Death Process," *Operations Research Letters*, 7, (1), 43-49 (1988b).

Glasserman, P., "Performance Continuity and Differentiation in Monte Carlo Optimization," *Proceedings of the Winter Simulation Conference*, ed M. Abrams, P. Haigh, J. Comfort, 518-524 (1988c).

Glasserman, P., "Derivative Estimates from Simulation of Continuous Time Markov Chains," *Technical Report, AT&T Bell Laboratories*, Holmdel, New Jersey, (1989).

Glasserman, P., "On the Limiting Value of Perturbation Analysis Derivitive Estimates," *Stochastic Models* 6, (2), 229-257 (1990a).

Glasserman, P., "Structural Condition for Perturbation Analysis Derivative Estimation : Finite Time Performance Indices," *Journal of the ACM,* to appear (1990b).

Glasserman, P., *Gradient Estimation via Perturbation Analysis,* Kluwer Academic Publisher, (1990c)

Glasserman, P. and Gong, W.B., "Derivative Estimates from Discontinuous Realizations: Smoothing Techniques," *Proceedings of the Winter Simulation Conference,* ed E. A. MacNair, K. J. Musselman, and P. Heidelberger, 381-389 (1989a).

Glasserman, P. and Gong, W.B., "Smoothed Perturbation Analysis for a Class of Discrete Event Systems," *IEEE Transactions on Automatic Control* 35, (11), 1218-1230, November (1989b).

Glasserman, P. and Ho, Y.C., "Aggregation Approximation for Sensitivity Analysis of Multi-Class Queueing Networks," *Performance Analysis* 10, 295-308 (1989).

Glasserman, P. and Yao, D., "Monotonicity in Generalized Semi-Markov Processes," Manuscript, October (1989).

Glasserman, P. and Yao, D., "Algebraic Structure of Some Stochastic Discrete Event Systems with Applications," *Journal of Discrete Event Dynamic Systems* 1, (1), to appear (1991).

Glasserman, P., Hu, J.Q., and Strickland, S.G., "Strongly Consistent Steady State Derivative Estimates," *Technical Report, Division of Applied Sciences, Harvard University* submitted for publication (1990).

Glynn, P., "Likelihood Ratio Gradient Estimation: an Overview," *Proceedings of the Winter Simulation Conference,* 366-375 (1987).

Glynn, P., "A GSMP Formalism for Discrete Event Systems," *Proceedings of the IEEE (Special Issue)* 77, (1), 14-23, (1989).

Glynn, P. and Inglehart, D., "Importance Sampling for Stochastic Simula-tions", *Management Science* 35, (11), 1367-1392 (1989).

Glynn, P. and Sanders, J.L., "Monte Carlo Optimization in Manufacturing Systems: two new approaches," *Proceedings of ASME Computer In Engineering Conference,* Chicago (1986).

Gong, W.B. and Ho, Y.C., "Smoothed Perturbation Analysis of Discrete Event Dynamic Systems," *IEEE Transactions on Automatic Control* AC-32, (10), 858-866, (1987).

Gong, W.B. and Hu, J.Q., "The Light Traffic Derivatives for the GI/G/1 Queue," *Journal of Applied Probability,* to appear (1991).

Gong, W.B., Cassandras, C., and Pan, J., "The RIPA Algorithm for M/G/1(∞,K) Queue," *1990 IFAC Congress Proceedings,* Pergamon Press (1990).

Gordon, W. J. and Newell, G. F., "Closed Queueing Systems with Exponen-tial Servers," *Operations Research* 15, 254-265 (1967).

Hajek, B., "Optimal Control of Two Interfacing Service Stations", *IEEE Transactions Automatic Control* 29, 491-499 (1984).

Heidelberger, P., Cao, X.R., Zazanis, M.R., and Suri, R, "Convergence Properties of Infinitesimal Perturbation Analysis Estimates," *Management Science,* 34, (11), 1281-1302 (1988).

Heymann, M, "Concurrency and Discrete Event Control," *IEEE Control Systems Magazine* 10, (4), 103112, (1990).

Hiller, F.S. and Lieberman, G. J., *Introduction to Operations Research,* 4th Edition, Hoplden-Day, (1986)

Ho, Y.C., "On Stochastic Approximation and Optimal Filtering," *Journal of Mathematical Analysis and Applications,* 6, (1), 152-154 (1963).

Ho, Y.C., "Parameter Sensitivity of a Statistical Experiment," *IEEE Transactions on Automatic Control,* 24, (6) (1979).

Ho, Y.C., "Is It Applications or Is It Experimental Science? - Editorial", *IEEE Transactions on Automatic Control,* 27 (1982).

Ho, Y.C., "Performance Evaluation and Perturbation Analysis of Discrete Event Dynamic Systems," *IEEE Transactions on Automatic Control,* 32, (7), 563-572 (1987).

Ho, Y.C., "Perturbation Analysis Explained," *IEEE Transactions on Automatic Control* 33, (8), 761-763 (1988).

Ho, Y.C. (Editor), "Dynamics of Discrete Event Systems," *Proceedings of the IEEE* 77, (1), 1-232, January (1989).

Ho, Y.C. (Editor), *Discrete Event Dynamic Systems*, IEEE Press, (1991).

Ho, Y.C. and Cao, X. R., "Perturbation Analysis and Optimization of Queueing Networks," *Journal of Optimization Theory and Application* 40, (4), 559-582 (1983).

Ho, Y.C. and Cao, X. R., "Performance Sensitivity to Routing Changes in Queueing Networks and Flexible Manufacturing Systems Using Perturbation Analysis," *IEEE Journal on Robotics and Automation* 1, 165-172 (1985).

Ho, Y.C. and Cassandras, C. G., "A New Approach to the Analysis of Discrete Event Dynamic Systems," *Automatica* 19, (2), 149-167 (1983).

Ho, Y.C. and Gong, W.B., "A Note on Filtering in Queueing Networks and Discrete Event Dynamic Systems," *IEEE Transactions on Automatic Control* 31, (6), 898-790, (1986).

Ho, Y.C. and Hu, J.Q., "An Infinitesimal Perturbation Analysis Algorithm for a Multiclass G/G/1 Queue," *Operations Research Letters* 9, (1), 35-44, to appear (1990).

Ho, Y.C. and Li, S., "Extensions of the Perturbation Analysis Techniques for Discrete Event Dynamic Systems," *IEEE Transactions on Automatic Control* 33, (5), 427-438 (1988).

Ho, Y.C. and Yang, P.Q., "Equivalent Networks, Load Dependent Servers, and Perturbation Analysis - An Experimental Study", *Teletraffic Analysis and Computer Performance Evaluation*, (editor) O.J. Boxma, J.W.Cohenm, H.C. Tijms, North-Holland (1986).

Ho, Y.C., Cao, X. R., and Cassandras, C. G., "Inifinitesimal and Finite Per-turbation Analysis for Queueing Networks," *Automatica* 19, 439-445 (1983).

Ho, Y.C., Eyler, A., and Chien, T. T., "A Gradient Technique for General Buffer Storage Design in a Serial Production Line," *International Journal on Production Research,* 17, (6), 557-580 (1979).

Ho, Y.C., Eyler, A. and Chien, T. T., "A New Approach to Determine Parameter Sensitivities on Transfer Lines," *Management Science* 29, (6), 700-714 (1983).

Ho, Y.C., Li, S., and Vakili, P., "On the Efficient Generation of Discrete Event Sample Paths under Different Parameter Values," *Mathematic and Computation In Simulation* 30, 347-370 (1988).

Hogarth, R.M., "Generalization in Decision Research: the Role of Formal Models" *IEEE Transactions on Systems, Man, and Cybernetics* 16, (3), 439-449, (1986).

Hoitomt, Debra J., Luh, P. B., Pattipati, K.R., "A LaGrangain Relaxation Approach to Jobshop Scheduling Problems," *Proceedings of the IEEE Robotic and Automation Conference* May (1990).

Holtzman, J.M., "On Using Perturbation Analysis to do Sensitivity Analysis," *Proceedings of the 28th IEEE Conference on Decision and Control* 2018-2023, (1989).

Hordijk, A., Iglehart, D.L., and Schassberger, R., "Discrete Time Method for Simulating Continuous Time Markov Chains," *Advance in Applied Probability* 8, 772-788 (1976).

Hsiao, M. T. and Lazar, A., "Optimal Flow Control of Multiclass Queueing Network with Partial Information," *IEEE Transactions on Automatic Control* 35, (7), 855-860, (1990).

Hu, J.Q., "On Steady-State Sample Path Derivative Estimators," *Applied Mathematics Letters* 3, (3), 57-60 (1990a).

Hu, J.Q., "Convexity of Sample Path Performance and Strong Consistency of Infinitesimal Perturbation Analysis," *IEEE Transactions on Automatic Control* to appear (1990b).

Hu, J. Q., "Consistency of Infinitesimal Perturbation Analysis Estimators with Load-DependentRates," *Queueing Systems*, to appear (1991).

Hu, J.Q. and Fu, M.C., "Second Derivatives Sample Path Estimators for the GI/G/M Queue," *Management Science*, submitted (1989).

Hu, J.Q. and Fu, M.C., "On Smoothed Perturbation Analysis for Second Derivative Estimation of the GI/G/1 Queue," *IEEE Transactions on Automatic Control*, to appear (1991).

Hu, J.Q. and Strickland, S., "Strong Consistency of Sample Path Derivative Estimates," *Applied Mathematic Letters* (1990).

Iglehart, D. and Shedler, G., *Regenerative Simulation of Response Time of Queueing Networks*, Springer-Verlag Lecture Note in Control and Information Sciences No.26 (1980).

Inan, K. and Varaiya, P., "Finitely Recursive Process Models for Discrete Event Systems," *IEEE Transactions on Automatic Control* 33, (7), 626-639, (1988).

Jackson, J.R. "Netwrok of Waiting Lines," *Operations Research*, 5, 518-521 (1957).

Jacobnon, S.H., and Schruben L.W., "Techniques for Simulation Response optimization" *Operations Research Letters*, 1-9, (1989).

Kalman, R.E., "Mathematical Description of Linear Dynamicl Systems," *SIAM Journal on Control and Optimization* 1, (2), 152-192 (1963).

Keilson, J., *Markov Chain Models - Rarity and Exponentiality*, Springer-Verlag, (1979).

Kelly, F.P., *Reversibility and Stochastic Networks*, Wiley, New York, (1979).

Kemeny, J. and Snell, J. L., *Finite Markov Chains*, Van Nostrand, Princeton, New Jersey, (1960).

Kiefer, J. and Wolfowitz, J., "Stochastic Estimation of the Maximum of a Regression Function," *Annals of Mathematical Statistics* 23, 462-466 (1952).

Kleinrock, L., *Queueing Systems, Vol I: Theory*, Wiley, (1975).

Kreitzinger, P. S., van Wyk, S., and Krzesinski, A. E., "A Generalization of Norton's Theorem for Multiclass Queueing Networks," *Performance Evaluation* 2, 98-107 (1982).

Krogh, B.H., Holloway, L.E., "Synthesis of Feedback Control Logic for a class of Controlled Petri Nets" *IEEE Transactions on Automatic Control* 35, (5), 514-523, May (1990).

Kushner, H. J. and Clark, D. S., *Stochastic Approximation Method for Constrained and Unconstrained Systems*, Springer-Verlag, New York, (1976)

Lazar, A., "Optimal Flow Control of a Class of Queueing Networks in Equilibruim," *IEEE Transactions on Automatic Control,* AC-28, (11), 1001-1007, (1983).

Lee, J.I., Cassandras, C.C., and Ho, Y.C., "Real time Parametric Analysis of Discrete System Trajectories," *Proceeding of the 28th IEEE Conference on Decision and Control* (1989).

Lewis, P.A.W. and Shedler, G.S., "Simulation of Nonhomogeneous Poisson Processes by Thinning," *Naval Research Logistics Quarterly* 26, (3), 403-413, Sept. (1979).

Li, S., "Extended Perturbation Analysis of Discrete Event Dynamic Systems," *Ph.D. Thesis, Harvard University*, (1988)

Li, S. and Ho, Y.C., "Sample Path and Performance Homogeneity of Discrete Event Dynamic Systems," *Automatica* 25, (6), 907-915, November (1989).

Lin, F., and Yao, D.D., "Generalized Semi-Markov Processes: A View Through Supervisory Control," *Proceedings of the 28th IEEE Conferecne on Decision and Control* 1075-1076 (1989).

Lindley, D. V., "The Theory of Queues with a Single Server," *Proceedings Cambridge Philosophical Society* 48, 277-289 (1952).

Matthes, K., "Zur Theorie der Bedienungsprozesse," *Proceedings of Third Prague Conference on Information Theory* Prague (1962).

Melamed, B., "Characterization of Possion Traffic Streams in Jackson Queue-ing Networks," *Advances in Applied Probability* 11, 422-439 (1979).

Mood, A.M. and Graybill, F.A., *Introduction to the Theory of Statistics, 2nd Edition*, McGraw Hill (1963).

Morse, P.M. and Kimball, G., *Methods of Operations Research*, MIT Press and Wiley, (1951).

Neuts, M., "A Celebration in Applied Probability," *Journal of Applied Proba-bility* (special issue), ed J. Gani 25A, 31-43 (1988).

Neuts, M. F., "Computer Experimentation in Applied Probability," *Journal of Applied Probability (special issue)* , ed J. Gani 25A, 31-43 (1988).

Noetzel, A.S., "A Generalized Queue discipline for Product Form Network Solutions" *Journal of ACM* 27, (4), 779-793 (1979).

Olsder, G. J., Resing, J. A.C., deVries, R. E., and Hooghiemstra, G., "Discrete Event Systems with Stochastic Processing Times," *IEEE Transactions on Automatic Control* 35, 299-302 (1990).

Ostroff, J.S. and Wonham, W.M., "A Temporal Logic Approach to Real Time Control of Discrete Event Systems," *Proceedings of the 24th IEEE Conference on Decision and Control* (1985).

Ozveren, M. C. and Willsky, A.S., "Aggregation and Multitlevel Control in Discrete Event Dynamic Systems," *Automatica*, to appear (1991).

Peterson, J.L., *Petri-Net Theory and the Modeling of Systems*, Prentice-Hall, (1986).

Pflug, G. Ch., "Sampling Derivatives of Probabilities", *Computing* 42, 315-328 (1989).

Ramadge, P. J., "Control and Supervision of Discrete Event Processes," *Ph.D. Thesis, University of Toronto*, (1983).

Ramage, P.J. and Wonham, W.M., "Supervision of Discrete Event Processes," *Proceedings of the 21st IEEE Conference on Decision and Control* 1228-1229 (1982).

Ramage, P.J. and Wonham, W.M., "Modular Feedback Logic for Discrete Event Systems," *SIAM Journal on Control and Optimization* 25, (5), (1987).

Reiman, M. I. and Simon, B., "Light Traffic Limits of Sojourn Time Distribu-tions in Markovian Queueing Networks," *Comm. Statist.-Stochastic Models* 4, (2), 191-233 (1988a).

Reiman, M.I. and Simon, B., "An Interpolation Approximation for Queueing Systems with Poisson Input," *Operations Research* 36, (3), 454-469, (1988b).

Reiman, M. I. and Simon, B., "Open Queueing Systems in Light Traffic," *Mathematics of Operations Research* 14, (1), 26-59, (1989).

Reiman, M.I. and Weiss, A., "Sensitivity Analysis via Likelihood Ratios," *Proceedings of the 1986 Winter Simulation Conference,* 285-289 (1986).

Reiman, M.I. and Weiss, A., "Light Traffic Derivatives via Likelihood Ratios," *IEEE Transactions on Information Theory* 35, (3), 648-654 (1989a).

Reiman, M.I. and Weiss, A. "Sensitivity Analysis for Simulatioins via Likelihood Ratios," *Operations Research* 37, (5), 830-843 (1989b).

Reiser, M. and Lavenberg, S.S., "Mean Value Analysis of Closed Multichain Queueing Networks," *J. of ACM* 27, (2), 313-322 (1980).

Robbins, H. and Monro, S., "A Stochastic Approximation Method," *Annals of Mathematical Statistics* 22, 400-407 (1951).

Rockafellar, R. T., *Convex Analysis*, Princeton University Press, (1970).

Ross, S.M., *Introduction to Stochastic Dynamic Programming*, Academic Press, (1983a).

Ross, S.M., *Stochastic Processes* , Wiley, (1983b).

Rubinstein, R., *Monte Carlo Optimization, Simulation, and Sensitivity Analysis of Queueing Networks*, Wiley, (1986a).

Rubinstein, R., "The Score Function Approach of Sensitivity Analysis of Computer Simulation Modelsputer Simulation ModesSimulation," *Math ematics and Computation in Simulation*, 28, 1-29, (1986b).

Rubinstein, R. and Samorodnitsky, "Variance reduction by the Use of Common and Antithetic Random Variables," *Journal of Statistical Computation and Simulation* 22, 161-180 (1985).

Rubinstein, R. and Szidarovszky, F., "Convergence of Perturabtion Analysis Estimates for Discontinuous Sample Functions: A General Approach," *Advances in Applied Probability*, 20, 59-78 (1988).

Schassberger, R., "On the Equilibrium Distribution of a Class of Finite State Generalized Semi-Markov Processes," *Mathematics of Operation Research* 1, (4), 395-40 (1976).

Schassberger, R., "Insensitivity of Steady-State Distributions of Generalized Semi-Markov Processes, I," *Annals of Probability* 5, 87-99 (1977).

Seneta, E, *Non-negative Matrices and Markov Chains*, Springer-Verlag, New York, (1981)

Shanthikumar, J.G. and Yao, D.D., "Second Order Stochastic Properties in Queueing Systems," *Proceedings of the IEEE* 77, 162-170, (1989).

Singh, M. (Editor), *Systems and Control Encyclopedia*, Pergamon Press, (1987)

Strickland, S. G. and C. G. Cassandras, "Sensitivity Analysis of Discrete Event Systems with Non-Markovian Event Processes," *Proceedings of the 28th IEEE Conference on Decision and Control* 111-116 (1989).

Suri, R., "Robustness of Queueing Network Formulas," *Journal of ACM* 30, (5), 564-594 (1983a).

Suri, R., "Implementation Sensitivity Calculation on a Monte-Carlo Experiment", *Journal of Optimization theory and Applications* 40, (4), 625-630, September (1983b).

Suri, R., "Infinitesimal Perturbation Analysis of Discrete Event Dynamic Systems - A General Theory," *Journal of ACM* 34, (3), 686-717 (1987).

Suri, R., "Perturbation Analysis: The State of the Art and Research Issues Explained via the G1/G/1 Queue," *Proceedings of the IEEE* 77, 114-137 (1989).

Suri, R. and Leung, Y.T., "Single Run Optimization of Discrete Event Simulations - An Empirical Study using the M/M/1 Queue," *IIE Transactions* 21, (1), 35-49 (1989).

Suri, R. and Zazanis, M.A., "Infinitesimal Perturbation Analysis and the Regenerative Structure of the GI/G/1 Queue," *Proceedings of the 26th IEEE Conference on Decision and Control* 677-680 (1987).

Suri, R. and Zazanis, M.A., "Perturbation Analysis Gives Strongly Consistent Sensitivity Estimates for the M/G/1 Queue," *Management Science* 34, (1), 39-64 (1988).

Towsley, D., "Queueing Network Models with State-Dependent Routing," *Journal of ACM* 27, 323-337 (1980).

Vakili, P., "Three topics on Perturbation Analysis of Discrete-Event Dynamic Systems," *Ph. D. Thesis, Harvard University* (1989).

Vakili, P., "Using Uniformization for Derivative Estimation in Simulation,"
Proceedings of the American Control Conference, 1034-1039, (1990a).

Vakili, P., "A Standard Clock Technique for Efficient Simulation," *Operations Research Letters*, submitted (1990b).

Vakili, P. and Ho,Y.C., "Alternative Representation and Perturbation Analysis in a Routing Problem," *Proceedings of the 28th IEEE Conference on Decision and Control* (see also Proceedings of the twenty-fifth Allerton Conference, 279-287, (1987).), 1082-1083, (1989).

Van Schuppen, J., "Filtering, Prediction and Smoothing for Counting Process Observations, a Martingale Approach," *SIAM Journal on Applied Mathematics* 32, (3), 552-570 (1977).

Vantilborgh, H., "Exact Aggregation in Exponential Networks," *Journal of ACM* 25, 620-629 (1978).

Walker, A.J., "An Efficient Method for Generating Duiscrete Random Variables with General Distribution," *ACM Transactions on Mathematical Software* 3, (3), 253-256, September (1977).

Wardi, Y., "Simulation Based Stochasric Algorithm for Optimizing GI/G/1 Queues," *Manuscript, Georgia Institute of Technology* (1988).

Wardi, Y., "Discontinuous Sample Derivative and PA," *manuscript* (1989).

Weld, D.S., *Theories of Comparative Analysis*, MIT Press, (1990).

Whitt, W., "Bivariate Distributions with Given Marginals," *Annal of Statistics* 4, 1280-1289 (1976).

Whitt, W., "Continuity of Generalized Semi-Markov Processes," *Mathematics of Operations Research* 5, (4), 494-501 (1980).

Whitt, W., "The Qyueueing Network Analyzer," *The Bell System Technical Journal,* 62, (9), 2779-2815, November (1983).

Wilde, D., *Optimum Seeking Methods,* Prentice Hall, (1964).

Williams, A. C. and Bhandiwad, R. A., "A Generating Function Approach to Queueing Network Analysis of Multiprogrammed Computers," *Networks,* 6, (1), 1-22 (1976).

Wonham, W. M., "A Control Theory for Discrete Event Systems," *NATO ASI Series on Advanced Computing Concepts and Techniques in Control Engineering,* Spring-Verlag, (1988).

Wonham, W. M., "On the Control of Discrete Event Systems:Three Decades of Mathematical System Theory," *Lecture Notes in Computer and Information Science,* ed H. Nijmeier and Schumacher, J.M. 135, 542-562, Spring-Verlag (1989).

Wonham, W. M. and Ramadge, P. J., "On the Supremal Controllable sublanguage of a given language," *SIAM Journal on Control and Optimization* 25, 637-659 (1987).

Woodside, C.M., "Response Time Sensitivity Measurement for Computer Systems and General Closed Queueng Networks," *J. of Performance Evaluation,* 4, 199-210 (1984).

Zazanis, M. A., "IPA Estimate for the Moments of the System Time for an M/M/1 Queue," *Technical Report 87-07 Northwestern University,* August (1987a).

Zazanis, M. A., "Weak Convergence of Sample Path Derivatives for the Waiting Time in a Single Server Queue," *Proceedings of Allerton Conference on Communication, Computing, and Control,* 297-304, (1987b).

Zazanis, M. A., "An Expression for the Derivative of the Mean Response Time of a GI/G/1 Queue," *Technical report 87-08 Northwestern University* (1987c).

Zazanis, M. A., "Compensators and Statistical Estimation for Discrete Event Systems," *Proceedings of Allerton Conference on Communication, Computing, and Control,* 549-555, (1988).

Zazanis, M. A., "Infinitesimal Perturbation Analysis for Moments of the System Time on an M/M/1 Queue", *Operations Research,* 37, (6) (1989).

Zazanis, M. A. and Suri, R., "Perturbation Analysis of the GI/G/1 Queue," *Queueing Systems: Theory and Applications,* submitted .

Zazanis, M. A. and Suri, R., "Comparison of Perturbation Analysis with Conventional Sensitivity Estimates for Regenerative Stochastic Systems," *Operations Research,* submitted (1988).

Zeigler, B., *Theory of Modelling and Simulation,* Wiley, (1976).

Zhang, B., Performance Gradient Estimation for Very Large Markov Chains, *Ph.D. Thesis, Harvard University,* (1990).

Zhang, B. and Ho, Y.C., "Variance Reduction for Likelihood Ratio Methods," *Proceeding of the 28th IEEE Conference on Decision and Control,* 145-150, (1989).

Zhang, B. and Ho, Y.C., "Improvements in the Likelihood Ratio Methods for Steady-State Sensitivity Analysis and Simulation," *Performance Evaluation,* to appear (1991a).

Zhang, B. and Ho, Y.C., "Performance Gradient Estimation for Very Large Markov Chains," *IEEE Transactions on Automatic Control,* to appear (1991b).

INDEX